Telephone (+44) 1243 779777

Email (for orders and customer service enquiries): cs-books@wiley.co.uk
Visit our Home Page on www.wiley.com

Other Wiley Editorial Offices

John Wiley & Sons Inc., 111 River Street, Hoboken, NJ 07030, USA

Jossey-Bass, 989 Market Street, San Francisco, CA 94103-1741, USA

Wiley-VCH Verlag GmbH, Boschstr. 12, D-69469 Weinheim, Germany

John Wiley & Sons Australia Ltd, 42 McDougall Street, Milton, Queensland 4064, Australia

John Wiley & Sons (Asia) Pte Ltd, 2 Clementi Loop #02-01, Jin Xing Distripark, Singapore 129809

John Wiley & Sons Canada Ltd, 6045 Freemont Blvd, Mississauga, ONT, Canada L5R 4J3

Wiley also publishes its books in a variety of electronic formats. Some content that appears in print may not be available in electronic books.

Anniversary Logo Design: Richard J. Pacifico

Library of Congress Cataloging in Publication Data

Arrillaga, J.
Flexible power transmission—the HVDC options / J. Arrillaga, Y.H. Liu, N.R. Watson.
p. cm.
ISBN 978-0-470-05688-2 (cloth)
1. Electric power transmission. 2. Electric power distribution. I. Liu, Y. H.
II. Watson, N. R. III. Title.
TK3111.A76 2007
621.319'12—dc22

2007019888

British Library Cataloguing in Publication Data

A catalogue record for this book is available from the British Library

ISBN 978-0-470-05688-2 (HB)

Typeset in 10/12pt Times by Integra Software Services Pvt. Ltd, Pondicherry, India
This book is printed on acid-free paper responsibly manufactured from sustainable forestry in which at least two trees are planted for each one used for paper production.

Flexible Power Transmission

The HVDC Options

J. Arrillaga
University of Canterbury, New Zealand

Y.H. Liu
Inner Mongolia University, China

N.R. Watson
University of Canterbury, New Zealand

BICENTENNIAL 1807 WILEY 2007 BICENTENNIAL

John Wiley & Sons, Ltd

Flexible Power Transmission

The HVDC Options

Contents

Preface

The structure and characteristics of HVDC (High-Voltage Direct Current) converters have remained practically unaltered for the first 40 years of commercial operation. Restricted by the switching characteristics, first of the mercury arc valve and later of the silicon-controlled rectifier, this technology requires substantial extra support at the link terminals to ensure stable operation.

More recently the development of power semiconductors with improved characteristics has provided the basis for a flexible AC transmission system (FACTS) technology. This technology covers a variety of power electronics controllers created to enhance the performance of the traditional grid. The individual members of the FACTS family are designed to solve a specific problem, e.g. active or reactive power flow control, short-circuit current limitation, etc. So it is the complete family that provides transmission flexibility, rather than the individual controllers.

The new power semiconductors have also, in the past decade, changed the attitude towards HVDC transmission, and a variety of converter configurations have been developed to take advantage of the higher controllability and switching frequencies of the new devices.

Although the main market for HVDC is still thyristor based, a transistor-based technology has recently been developed, and is already being used throughout the world. The new HVDC technology can provide most of the enhancements of the individual FACTS controllers, i.e. permit large stable power transfers, deliver or absorb the required reactive power to maintain the specified voltages at the interconnected buses, contain fast emergency controls to avoid large fault current levels, be designed (if required) to control sub-synchronous resonances, etc. Moreover, the DC link is the only practical way of connecting asynchronous systems and systems of different frequencies. For a given HVDC configuration, all these tasks can be achieved purely by control action.

Therefore, a modern HVDC interconnection is potentially the most flexible power transmission system. However, the provision of greater HVDC transmission flexibility comes at a price, in terms of either reduced efficiency or increased structural complexity. Thus, when considering a new scheme, it is important to decide on the degree of flexibility required for the particular application (i.e. taking into account power ratings, transmission distances, extent of ancillary services expected, etc.).

A critical review of the HVDC options already available and under consideration constitutes the purpose of this book, which therefore complements recent titles describing the FACTS technology to help power system engineers to make informed decisions on the planning, design and operation of future power transmission systems. It is also a useful reference text for students taking advanced courses in power transmission.

The first five chapters describe the principles and components of existing converter technology. Chapters 6 and 7 discuss alternative proposals for self-commutating conversion and Chapters 8, 9, 10 and 11 the application of the various converter configurations to HVDC transmission.

The authors would like to acknowledge the main sources of information that have made the book possible and in particular the material reproduced, with permission, from CIGRE Study Committee B4 and ABB industry documents.

At the personal level they wish to acknowledge the services of Greta Arrillaga (who half a century ago also typed the first book on the subject of HVDC by Adamson and Hingorani!), the advice received from Alan Wood and Nick Murray of Canterbury University, Dennis Woodford of Electranix and Gunnar Asplund of ABB.

They also wish to thank The University of Canterbury and The University of Inner Mongolia for providing the facilities for their work.

1

Introduction

1.1 The Conventional Power Grid

The power sources in conventional power systems must operate at exactly the same frequency and in perfect synchronism. Each generator controls the magnitude of its terminal voltage by the excitation current and the phase angle of this voltage by means of the mechanical torque developed by the turbine. The generators are designed to produce relatively low voltages, and thus the generated power undergoes a number of voltage transformations, from low to high voltage (for efficient power transmission) and from high to medium and low voltage (for economic and safe power distribution). These changes are implemented by power transformers.

Within a national grid, the use of a fully interconnected primary transmission system, to which the new power stations are connected, has traditionally been the generally accepted philosophy behind the development of an efficient power system.

The expansion of the primary transmission system was normally continued until the rated switchgear fault level was exceeded. Beyond that point a new primary transmission system, of higher voltage and fault levels, was created, while the previous one continued expanding into several separate (secondary) systems. Each of these secondary transmission systems in turn supplied a number of distribution (normally radial) feeders. So the conventional power grid has traditionally been grouped into three separate parts, i.e. generation, transmission and distribution, all of them inflexibly tied by the synchronous constraints.

1.1.1 Power Transfer Mechanism

Transformers, generators and transmission lines are predominantly inductive, and most loads have an inductive component as well. The presence of inductance delays the current response of these components to the voltage variation across them, and this effect causes phase shifts between the voltage and current waveforms which affect the efficiency of the power transmission process.

Flexible Power Transmission: The HVDC Options J. Arrillaga, Y.H. Liu and N.R. Watson

The instantaneous power (p) associated with a power system component is the product of the instantaneous values of the voltage (v) and current (i) at its terminals ($p = vi$). The integration of the instantaneous power variation over a complete cycle divided by the period of repetition, i.e.

$$(1/T)\int_{t}^{t+T} pdt$$

provides the average or ***active power***. If both the voltage and current vary sinusoidally at the same frequency, in terms of rms (root mean square) voltage (V) and current (I) quantities, the active power is expressed as

$$P = VI\cos(\phi) \tag{1.1}$$

where ϕ is the phase angle between the voltage and the current fundamental frequency waves.

As the rms values are always positive, the product VI (referred to as volt-ampere or apparent power), gives no indication of the active power sign. It is the sign of cos (ϕ) (the ***power factor***) that determines whether the circuit component is generating or absorbing power.

In Figure 1.1, using the voltage as the phase angle reference and resolving the current into in-phase (I_p) and quadrature (I_q) components, the product of V and I_p is clearly the active power, while the product of V and the quadrature component I_q, i.e.

$$Q = VI\sin(\phi) \tag{1.2}$$

is referred to as ***reactive power***.

Reactive power is needed to establish the magnetic and electrostatic fields; it is temporarily stored and then released (i.e. it consists of positive and negative regions within the cycle). In fact the energy associated with the reactive power oscillates between the element and the rest of the circuit (at the rate of two reversals per period). Although the reactive power has a zero average value, it still represents real reciprocating energy that must be present by virtue of the inductance or capacitance of the network.

When ϕ, the phase angle difference in Equation (1.2), is between 0 and π, $\sin(\phi)$ is positive and the circuit element is said to be a consumer of Q; similarly, when ϕ is between π and 2π, $\sin(\phi)$ is negative and the element is said to be a generator of Q. The convention

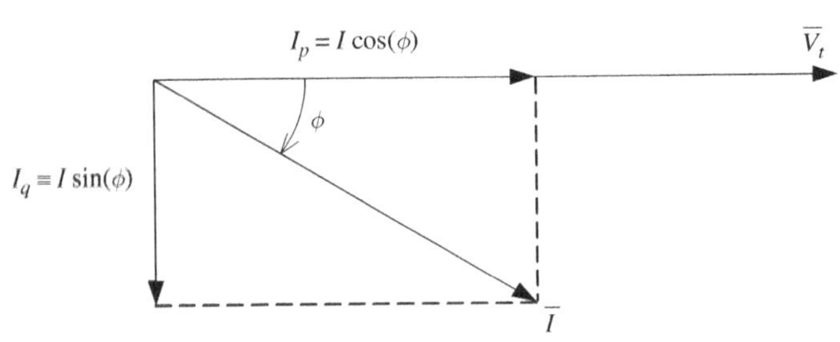

Figure 1.1 In-phase and quadrature current components

used is that when Q is positive the current lags the voltage and when Q is negative the current leads the voltage

Squaring the expressions of P and Q in Equations (1.1) and (1.2) and adding them gives

$$(VI\cos(\phi))^2 + (VI\sin(\phi))^2 = (VI)^2 \tag{1.3}$$

and

$$VI = \sqrt{P^2 + Q^2} \tag{1.4}$$

Equations (1.3) and (1.4) can be represented in a four-quadrant complex diagram, as shown in Figure 1.2, with the axes labelled $\pm P$ and $\pm jQ$.

Power transfer between active sources

Figure 1.3 shows a purely inductive line interconnecting two ideal voltage sources V_1 and V_2 (which can be either generators or nodes of a synchronous system). The phasor diagram in Figure 1.4 represents the operating condition when the voltage at terminal 1 leads that of terminal 2 by an angle δ (referred to as the power angle) and the current at terminal 2 lags its voltage by an angle ϕ (referred to as the power factor angle). Using the voltage of terminal 2 as a phase reference, the following expressions are derived from this diagram:

$$I_2 X\cos(\phi) = V_1\sin(\delta) \tag{1.5}$$

$$I_2 X\sin(\phi) = V_1\cos(\delta) - V_2 \tag{1.6}$$

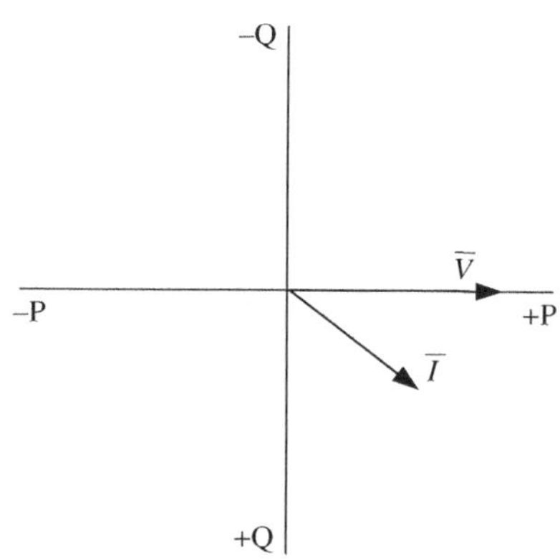

Figure 1.2 Four-quadrant diagram with the voltage as reference

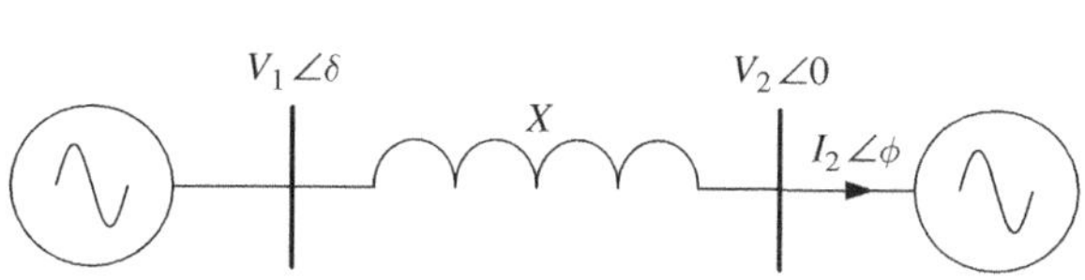

Figure 1.3 Interconnection between two synchronous systems

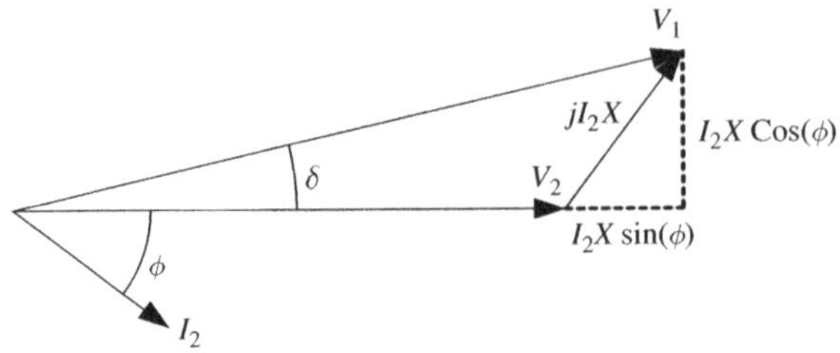

Figure 1.4 Phasor diagram for the interconnection of Figure 1.3

From Equations (1.5) and (1.6) the active and reactive power transfers become

$$P = V_2 I_2 \cos(\phi) = \frac{V_1 V_2 \sin(\delta)}{X} \tag{1.7}$$

$$Q = V_2 I_2 \sin(\phi) = \frac{V_2 (V_1 \sin(\delta) - V_2)}{X} \tag{1.8}$$

Thus to control the P and/or Q transfers it is necessary to vary one or more of the four variables V_1, V_2, δ and X in Equations (1.7) and (1.8). As indicated earlier, the generated voltage phase and magnitude values can be controlled by the turbine governor and generator excitation respectively. However, from the power transmission viewpoint, the generator controls are slow and inefficient: the slow control imposes a power transmission restriction on the steady-state operating point, as the power angle δ in Equation (1.7) has to be kept low in order to preserve transmission stability following large disturbances; also the relatively large requirement of reactive power (Equation (1.8)) will overload unnecessarily the generation and transmission systems.

Power transfer to a consumer load

Consumer loads are connected to radial feeders, normally at the end of the power distribution network. Low power factor loads have a detrimental effect on the load voltage and, therefore, on the power transfer capability. This effect is illustrated with reference to Figure 1.5 on the assumption that the feeder and the primary system behind it are represented by a voltage source (V_s) in series with a total system reactance (X_s). To maintain the active power constant when the power factor reduces (i.e. angle ϕ increases) requires an increase in the load current, i.e. $I'_L > I_L$; this increase causes a higher voltage drop in the system reactance which, in turn, reduces the load voltage (V_L).

Thus to maintain the required power level, either the source voltage must increase or some means of voltage support must be provided locally. For instance, the latter can be achieved by connecting a capacitance in parallel with the load. This will add a quadrature component to the load current and will reduce the overall current in the feeder; this solution is referred to as ***power factor correction***. However, the use of local compensation by means of passive components, although efficient, is neither fast nor continuous and increases the likelihood of low-order harmonic resonance with the system impedance.

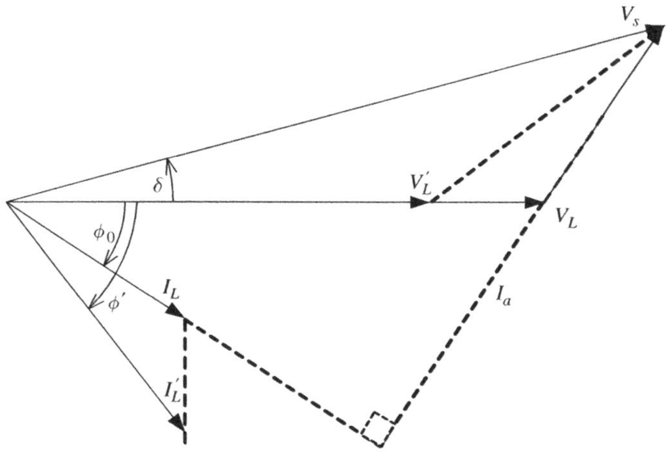

Figure 1.5 Effect of the load power factor on the load voltage

1.2 Towards a More Flexible Power Grid

A variety of technical, economical and environmental reasons affecting the generation, transmission and utilisation of power are forcing a rethink on the conventional power system development philosophy. The dilemma is that, on the one hand, there is growing opposition to the acceptance of new transmission lines and ever increasing primary transmission voltages. On the other hand, there is the realisation that power system interconnections bring undisputable benefits, such as economies of scale, wider choices of generating plant, reductions in reserve capacity, diversity in demand, supply reliability, pooling opportunities, etc.

Clearly an important factor in the solution is the possibility of increasing the power carrying capability of the transmission lines. In this respect conventional AC transmission is severely restricted by the need to keep the two systems interconnected by the line in synchronism following disturbances (i.e. when the phase difference between the terminal voltages increases rapidly), a condition referred to as transient stability. Therefore increases in the steady-state power carrying capability are linked to improvements in the transient stability levels, which in turn require faster controllability. Controllability and flexibility are used in power transmission as synonymous terms; in other words, greater flexibility implies greater and faster controllability. The latter has been made possible by the development of power semiconductors (discussed in Chapter 2) and their application to the control of power apparatus and systems, commonly referred to as power electronics.

1.2.1 Power Electronics Control

The advent of power electronics technology has been the catalyst for the provision of greater grid flexibility. A power electronics controller can be broadly described as a matrix of static switches connecting a number of input nodes to a number (not necessarily the same) of output nodes and the power flow may be in either direction. The circuits behind these nodes may be either DC or AC and predominantly inductive or capacitive.

The application of power electronics in power transmission systems has led to the development of two complementary technologies, referred to as HVDC (High-Voltage Direct Current) [1–5] and FACTS (Flexible AC Transmission System) [6–7]. However, only the latter has so far used the term flexibile, which is often interpreted as an exclusive characteristic of this technology. This book widens the flexibility concept by discussing HVDC technology under the umbrella of flexible power transmission.

Both HVDC and FACTS make extensive use of AC–DC static power conversion; the basic characteristics of this process are discussed in the next section.

Static power conversion

Within the power electronics discipline, the designation AC–DC static power conversion is used for the processes of rectification and inversion, which provide the basis for fast power controllability. As well as improving controllability, these two processes, when applied to DC transmission, also remove the synchronous constraints. The first consideration in the process of static power conversion is how to achieve instantaneous matching of the AC and DC voltage levels, given the limited number of phases and switching devices that are economically viable. The following circuit restrictions are imposed on a static power converter by the characteristics of the external circuit and of the switching components:

1. If one set of nodes (input or output) of the matrix of switches is inductive, the other set must be capacitive so as not to create a loop consisting of voltage sources (or capacitors and voltage sources) when the switches are closed or a cut set consisting of current sources when the switches are opened.

2. The combination of open and closed switches should not open-circuit an inductor (except at zero current) or short-circuit a capacitor (except at zero voltage).

For stable conversion some impedance must, therefore, be added to the switching circuit of Figure 1.6(a) to absorb the continuous voltage mismatch that inevitably exists between the two sides. If the impedance is exclusively located on the AC side (as shown in Figure 1.6(b)), the switching devices transfer the instantaneous direct voltage level to the AC side and, thus, the circuit configuration is basically a voltage source converter (referred to as VSC), with the possibility of altering the DC current by controlling the turn-on and turn-off instants of the switching devices.

If, instead, a large smoothing reactor is placed on the DC side (as shown in Figure 1.6(c)), pulses of constant DC flow through the switching devices into the AC side. Then, a basically current source converter results (referred to as CSC), again with the possibility of adjusting the direct voltage by appropriate switching control. CSC is the more practical alternative when using thyristor switches without turn-off controllability.

Another classification relates to the source of the commutation process between the converter valves. When the source is the AC system voltage the converter is said to be line-commutated (LCC). LCC relies on the natural current zeros created by the external circuit for the transfer of current from switch to switch. CSC-LCC is the only

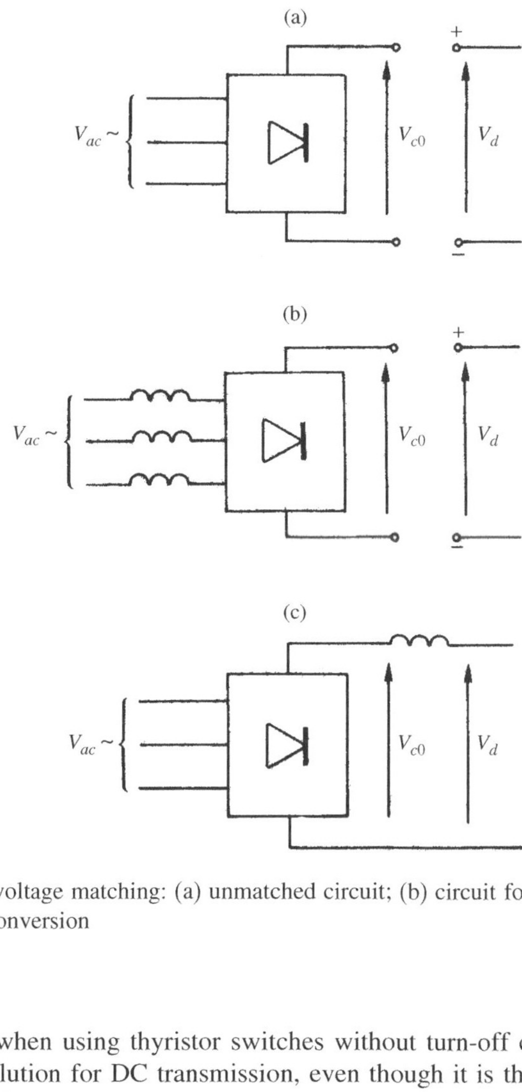

Figure 1.6 AC–DC voltage matching: (a) unmatched circuit; (b) circuit for voltage conversion; (c) circuit for current conversion

practical alternative when using thyristor switches without turn-off capability. This is still the most common solution for DC transmission, even though it is the least flexible, and is discussed in Chapter 3.

The alternative is self-commutation (a process described in Chapter 4), which is achieved independently from the external circuit components by the use of advanced switching devices with turn-off capability and, thus, provides greater flexibility.

Static power converters are also classified by the principle, and related configurations, used to produce the output waveforms. The most flexible alternative, in this respect, is the use of a simple (normally two-level) converter configuration and a complex sub-cycle waveform control. This subject, discussed in Chapter 5, uses a high-frequency carrier signal to modulate the required waveform and is, thus, referred to as pulse width modulation (PWM).

The second principle in the self-commutating category is the derivation of a stepped (multi-pulse or multi-level) waveform using only fundamental frequency switching. This concept, discussed in Chapters 6 and 7, is more efficient but less flexible than PWM, and it involves a greater number of components.

The main technical requirements of static converters for high-power applications are:

1. The provision of high-voltage valves with balanced voltages across the series-connected individual switches during the off-state and in the dynamic regions.
2. The ability to achieve high-quality output waveforms.
3. Limitation of the *dv/dt* rate across the switches and other converter components to simplify insulation coordination and reduce RF interference.
4. The ability to achieve high efficiency by reducing on-state and switching losses.
5. Simplification of the topology to reduce component costs.
6. Flexibility in terms of active and reactive power controllability.

1.3 HVDC Transmission

The original motivation for the development of DC technology was transmission efficiency, as the power loss of a DC line is lower than that of a corresponding AC line of the same power rating. However, this required the use of HVDC and, therefore, the development of conversion switches capable of withstanding high voltages

The invention of the high-voltage mercury valve half a century ago paved the way for the development of HVDC transmission. By 1954, the first commercial DC link came successfully into operation and was soon followed by several other schemes orders of magnitude larger. The success of the new technology immediately triggered research and development into an alternative solid-state valve, which by the mid-1960s had already displaced the use of mercury arc valves in new schemes. The early history and technical development of the HVDC technology are described in [8].

Substantial progress made in the ratings and reliability of thyristor valves has increased the competitiveness of HVDC schemes. DC transmission has lower transmission losses and cost than equivalent AC lines, but requires terminal equipment which adds to the cost and power losses. Thus traditionally, the DC option has been found economically viable only when the distance involved is long and the amount of energy to be transferred large. However, there are other factors that must be taken into consideration in the selection of an HVDC interconnection. An important factor in the economic comparison between AC and DC interconnections is to determine whether synchronisation of the previously separate systems is feasible and economical.

Issues affecting the feasibility of the interconnection include:

- whether the cable (in the case of a submarine interconnection) exceeds its capacity to carry its own charging current (for sea cable interconnections with distances over 50 km, DC is the only practical solution);
- whether the link is capable of maintaining synchronism of the two systems under all but extreme operating conditions;

- whether it is practical to arrange generation and frequency control in the joint system on a common basis;
- whether the synchronous interconnection exceeds the fault levels of the interconnected systems.

All the above issues can be avoided when using the DC alternative, which offers the following advantages:

- lack of technical limitations on the length of a submarine cable;
- the interconnected systems do not need to operate in synchronism;
- no increase in the short-circuit capacity is imposed on the AC systems switchgear
- any power transfer can be set independently of impedance, phase angle, frequency and voltage;
- the receiving end of the link operates like a generator, i.e. it can supply power according to any prespecified criteria (load flow, frequency control, voltage regulation, etc.);
- the interconnection can be used as a fast system's generation reserve to be able to provide power immediately;
- the DC link can be operated to improve the stability of one or both AC systems by modulating the power in response to the power swing.

HVDC links are built for a variety of purposes and the specific factors to be taken into account in their economic evaluation will depend on the purpose of the link. As well as identifying the main economic risks and uncertainties, the assessment must take into account the environmental and institutional changes bringing competition to the electricity supply.

Very often the projected links are intended to deliver firm energy from remote sources to a loading centre, in which cases the economic evaluation will be mainly concerned with the cost of generation and transmission.

The main purpose of recently emerging merchant plants (or IPPs, for Independent Power Producers) will be to calculate the cost of energy delivered to the proposed market.

In schemes developed for a one-way flow of energy an important extra consideration will be the variability of the price of energy purchased and sold. Some schemes are justified in terms of the seasonal variations in the interconnected systems. In such cases the economic analysis will be based on the impact of the interconnection on total costs in both systems.

HVDC interconnections are also considered for mutual support, such as reserve sharing, generation capacity savings, etc. In these cases the economic evaluation must also take into account the impact of the interconnection on the total capital and operating costs of the two systems to be connected.

The place of conventional and modern HVDC in power systems can be found in [9] and [10] respectively. The reliability of HVDC schemes, expressed in the permissible number of forced outages (and therefore the unavailability of the system), is published by CIGRE every two years [11]. The economic assessment of HVDC transmission has been the subject of a recent CIGRE Working Group document [12] and the procedure to calculate HVDC losses has been established by the IEC [13].

1.3.1 Thyristor-Based CSC Transmission

The thyristor, or silicon-controlled rectifier (SCR), has been the only solid-state switch used in the process of HVDC conversion, prior to the availability of high-power turn-off switching devices. The thyristor converter is still the most cost-effective solution for large-power and long-distance power transmission.

With reference to the complex diagram of Figure 1.2, converter operation in the two lower quadrants permits bidirectional transfer of active power but only absorption of reactive power. The DC current flow is unidirectional, but the DC voltage can be reversed. This mode of operation is achieved purely by thyristor turn-on control.

Since the static conversion process provides rectification and inversion, its use in power transmission permits the interconnection of synchronous and asynchronous systems. The elimination of the synchronous restrictions permits a better utilisation of the transmission systems. In half a century of existence the static converter terminals used in DC technology have gone from tens to thousands of megawatts, with present individual SCR ratings of over 8 kV and 4 kA. In fact the modular design of series SCR structures permits the use of valves of any power and voltage rating; this is an important factor because bipoles with only one converter group per pole appear to be more reliable than bipoles with two converter groups per pole (probably because the more items of equipment, the more there is to go wrong). However, the converter size may, of course, be limited by the interface transformer.

SCR-based conversion provides active power controllability (i.e. rectification and inversion) at the expense of large and varying demand of reactive power. Both the rectification and inversion processes consume reactive power, because the commutation is performed by the voltage source and the leakage reactance of the converter transformer generally dominates the commutation circuit. Therefore, the fundamental component of the current always lags that of the voltage. Moreover, for inverter operation, as the commutation overlap is not known at the instant of firing, an extinction angle of the order of 15° must be allowed to prevent commutation failure and this increases further the consumption of reactive power, which will be of the order of between 50 and 60 % of the active power in normal operation. As the reactive power consumption varies with load, the filters and extra capacitors must be switched by means of circuit breakers to match the converter reactive power requirement. If the commutation circuit inductance is compensated by means of series capacitance, unity power factor is possible; this is the purpose of the so-called CCC (Capacitor-Commutated Converter or Conversion), a configuration described in Chapters 3 and 9.

A minimum short-circuit level is needed for stable operation of LCC, because an active power change requires a corresponding change in reactive power, which causes voltage fluctuations. Moreover, AC system voltage drops cause further reactive power consumption which may lead to a voltage instability. To avoid this instability the short-circuit ratio (or SCR), defined as the ratio between the AC system short-circuit power and the rating of the converter, is rarely below two.

The reversal of power flow requires a polarity change of the DC system voltage. This is not a problem in two-terminal DC transmission systems, but complicates the operation of multi-terminal DC schemes, which would need the use of mechanical switches to achieve individual terminal power reversals.

Despite their limited controllability, static LCCs of high reliability are now used extensively in the power transmission field, either as long-distance interconnections or back-to-back asynchronous links. The conventional thyristor-based LCC technology is at present superior to self-commutating VSC transmission in terms of capital cost, power losses and reliability for large-scale HVDC transmission.

Over 75 GW of installed capacity is already in existence and more capacity is added every year, despite its inherent reactive power restriction. Present schemes include both overhead and cable point-to-point schemes as well as zero-distance (back-to-back) interconnections.

Chapters 8 and 9 describe the present state and future developments of LCC-CSC transmission and Chapter 11 discusses the potential application of thyristor-type self-commutating multi-level alternatives.

1.3.2 VSC Transmission Based on the Integrated Gate Bipolar Transistor (IGBT)

The flexibility of the transmission system can be greatly improved by the use of VSC-based four-quadrant static power conversion, the subject of Chapter 10. VSC transmission permits the flow of active power, as well as the provision of reactive power, in either direction at each end of the link. It is particularly effective when combined with cable transmission, because in this case the lack of polarity reversal greatly simplifies the cable design. However, their power rating is restricted to about 300 MW, although larger ratings are under development, due to the voltage limitations in the presently used PWM technology.

Self-commutated VSC does not need an AC system voltage source for the commutations and therefore can operate stably with any (even zero) SCRs. VSC can be controlled to generate or absorb reactive power independently from the active power flow. The maximum active power that can be exchanged with the AC system is limited only by the reactance of the AC system viewed from the VSC terminals. VSC, whether the PWM or multi-level control concept is used, reduces substantially the generation of harmonics; therefore the need for filters is eliminated or their size reduced to absorb only the higher harmonics. The low rating of passive filtering required by VSC also eliminates the problem of overvoltages following converter disconnection. As it is possible to change the current flow direction, fast power reversal can be achieved at each terminal of a multi-terminal DC scheme without the need for switching operations.

The following applications have been identified for VSC transmission [14]:

1. The supply of power to isolated areas without generating sources, as it avoids the need to install expensive synchronous compensators. In this application the inverter end controls the frequency and the voltage of the receiving system.

2. The interconnection of two or more synchronous or asynchronous AC systems, where each converter end controls its own AC voltage and all, except one of them, their DC power contribution, while the remaining converter controls the DC voltage.

3. To bring power from an offshore wind farm to an onshore substation. At the sending end, the control of frequency, voltage and power can be coordinated with the generators, which can be induction machines, as well as with the turbine pitch controller and the wind velocity.

   The variable speed of the wind turbine using static converters allows a reduction in mechanical load and much greater control of the output. This can also be achieved by a doubly fed induction generator (DFIG). Variable speed turbines can then be used by injecting a variable voltage into the rotor of the induction generator via slip rings at the slip frequency. In this case a low-voltage DC link using two AC–DC IGBT-based VSCs injects the rotor voltage. The converter rating required for the variable speed range is of the order of 25 % of that of the generator. When running at super-synchronous speed the DFIG wind turbine will deliver power from the rotor through the converters to the network, while at sub-synchronous speed the DFIG rotor will absorb power.

   However, distributed generation systems based on the direct connection of induction generators may often be impractical because of the fault-level increase, and in such cases a DC link interface can be a more effective solution.

4. The direct connection of generators to DC links [15] avoids the need for generator transformers and AC filters and reduces considerably the switchgear requirements. Voltage control can be exercised entirely by the generator excitation and, thus, converter transformer tap-changers are not needed. While no use has been made to date in conventional power stations, the idea is now being used to transmit DC power from synchronous generators in oil rigs and is equally applicable to wind generation. Also, with the use of flexible static power conversion, which eliminates the need for the supply of reactive power, the unit connection concept can be extended to non-synchronous generators.

1.3.3 Multi-terminal HVDC

The extension of CSC technology to multi-terminal HVDC has been discussed for over four decades [16, 17], the interest peaking in the 1980s, by both academic researchers and the power industry. However, only one fully multi-terminal scheme was constructed for commercial operation [18]. Its object was to convert the Hydro-Quebec–New England link (commissioned in 1986) into a five-terminal scheme with the addition of three further terminals. However, the original two-terminal link (between Des Cantons and Comerford) was never integrated into the multi-terminal DC network because of anticipated performance problems. No other (higher than three) multi-terminal scheme has been considered since then based on CSC technology. When at the planning stage a third terminal extension is to be made in the future, this is equivalent to reducing the transmission distance, and makes the DC solution less competitive.

Normally the power rating of a proposed third-terminal tap is relatively small compared with the main transmission link and, therefore, the high transmission voltages of the bipolar interconnection will require expensive converter equipment for a parallel-connected third terminal. To try and reduce the cost, consideration has been given to the use of series, rather than parallel, tapping of the additional terminal. There have been many contributions on the series tapping concept to show that it is technically feasible, but none has been built so far.

The transistor-based VSC technology is better suited to multi-terminal HVDC transmission. However, due to present limitations on the transmission voltage, the multi-terminal applications being discussed relate to medium-voltage DC grids interconnecting different alternative energy sources (such as wind farms, solar panels, etc.) with local loads and the distribution system. This topic will be discussed in Chapter 10.

1.3.4 The Flexibility Concept Applied to HVDC

The term flexibility has been defined with reference to AC transmission and it implies the use of power electronics control. In this respect the conventional thyristor-based option described in Section 1.3.1 is already a flexible system. However, this option is severely restricted by the switching characteristics of the SCR, or thyristor, and requires considerable support at the terminals for stable operation. The recent incorporation of advanced switching devices permitting self-commutation (described in Section 1.3.2) has greatly increased the degrees of freedom of HVDC transmission and, therefore, its flexibility. The new HVDC converters are capable of controlling their terminal voltage and power conditions independently from the AC system parameters. In a fully flexible DC system, the fundamental component of the converter output voltage can be controlled in magnitude and phase, and its harmonic content practically eliminated.

HVDC is also being introduced in distribution systems, especially in the presence of distributed generation, to decouple the voltage and frequency of the various energy renewable and non-renewable generating sources from the nominally fixed values of the conventional grids. In this respect it is an interesting matter to discuss the appropriateness of the term HVDC for use in distribution systems. A recent CIGRE document [19] has considered the use of the acronym FDS (Flexible Distribution System) for a distribution system that employs power electronics controllers to make the distribution of electricity more reliable, controllable and efficient. The advantage of the term FDS is that it includes FACTS and HVDC. However, the Working Group finally adopted the more conservative approach and decided to continue using the presently accepted terminology, i.e. FACTS-D and HVDC.

The question of whether the acronym HVDC needs to be further extended to differentiate between transmission and distribution has not been discussed. In this book we have decided to consider the whole subject under the umbrella of HVDC transmission. The reasons for the decision are the lack of clear boundaries (in terms of voltage or power levels) between conventional transmission and distribution systems and the changing nature of distribution systems, including the use of dispersed generation, which is often installed to supply power to the network. HVDC transmission is sufficiently general to cover all present and future HVDC applications, even multi-terminal ones, and should provide a more permanent term.

1.4 Relative Power Carrying Capability of AC and DC Transmission Lines

(1) For a given insulation length, the ratio of continuous-working withstand voltages is

$$k = \frac{\text{DC withstand voltage}}{\text{(rms) AC withstand voltage}} \tag{1.9}$$

If an overhead line is passing through a reasonably clean area, k may be as high as $\sqrt{2}$, corresponding to the peak value of rms alternating voltage. The DC advantage is even greater for cable transmission, where k is at least two.

(2) A transmission line has to be insulated for overvoltages expected during faults, switching operations, etc. To meet such requirement the AC lines need levels of insulation corresponding to an AC voltage of 2.5 to 3 times the normal rated voltage, i.e.

$$k_1 = \frac{\text{AC insulation level}}{\text{rated AC voltage } (\mathbf{E}_p)} = 2.5 \tag{1.10}$$

On the other hand, with suitable converter control the corresponding HVDC transmission ratio, i.e.

$$k_2 = \frac{\text{DC insulation level}}{\text{rated DC voltage } (\mathbf{V}_d)} = 1.7 \tag{1.11}$$

Thus for a DC pole-to-earth voltage $\mathbf{V}_d$ and AC phase-to-earth voltage $\mathbf{E}_p$ the following relation exist:

$$\text{insulation ratio} = \frac{\text{insulation length required for each AC phase}}{\text{insulation length required for each DC pole}} = (k \cdot k_1/k_2)(E_p/V_d) \tag{1.12}$$

(3) The total transmission losses result from the addition of those in the converter and in DC line. For a bulk power transmission of 6 GW, Figure 1.7 compares the losses of several AC and DC voltages for various transmission distances [20]. Beyond a distance of 500 km the line losses dominate the station losses and thus DC solutions become more efficient than their AC counterparts. The comparison shows that for a distance of 1000 km (for which the various options are optimised in this example), the losses in the 800 kV DC option are about 1 % lower than for the 500 kV alternative.

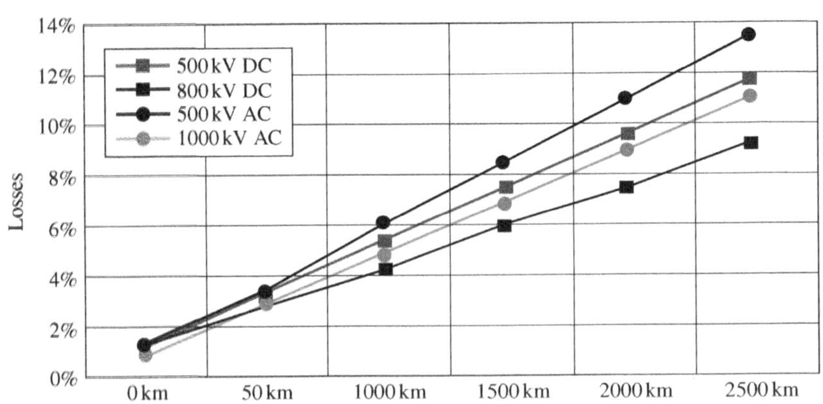

Figure 1.7 Transmission losses over transmission distance for HVAC and HVDC (Zhang, X.P. (2006), 'A grid for tomorrow', *Power Engineer*, October, reproduced by permission of the IET.)

(4) Consider a new DC transmission system to be compared with a three-phase AC system transmitting the same power and having the same percentage losses and using the same size of conductor. The DC system is considered to have two conductors at $\pm\mathbf{V}_d$ to earth.

Power in the AC system: $3\mathbf{E}_p\mathbf{I}_L$ (assuming that $\cos(\phi) = 1$)
Power in the DC system: $2\mathbf{I}_d\mathbf{V}_d$
AC losses: $3\mathbf{I}_L^2\mathbf{R}$
DC losses: $2\mathbf{I}_d^2\mathbf{R}$

Equating line losses,

$$3\mathbf{I}_L^2\mathbf{R} = 2\mathbf{I}_d^2\mathbf{R} \tag{1.13}$$

or

$$\mathbf{I}_d = (\sqrt{3}/\sqrt{2})\mathbf{I}_L \tag{1.14}$$

Equating powers,

$$3\mathbf{E}_p\mathbf{I}_L = 2\mathbf{I}_d\mathbf{V}_d \tag{1.15}$$

or

$$\mathbf{V}_d = (\sqrt{3}/\sqrt{2})\mathbf{E}_p \tag{1.16}$$

and substituting Equation (1.16) in Equation (1.12),

$$\text{insulation ratio} = (kk_1/k_2)(\sqrt{2}/\sqrt{3}) \tag{1.17}$$

For the values of k, k_1 and k_2 recommended above, the above ratio is equal to 1.2 for overhead lines and 2.4 for cables.

(5) The AC line consists of six conductors, which is equivalent to three bipolar DC circuits, each having two conductors at $\pm\mathbf{V}_d$ to earth, respectively. Thus

$$\text{power transmitted by AC}: \quad \mathbf{P}_a = 6\mathbf{E}_p\mathbf{I}_L \tag{1.18}$$

$$\text{power transmitted by DC}: \quad \mathbf{P}_d = 6\mathbf{V}_d\mathbf{I}_d \tag{1.19}$$

On the basis of equal current and insulation

$$\mathbf{I}_L = \mathbf{I}_d \tag{1.20}$$

$$\mathbf{V}_d = (kk_1/k_2)\mathbf{E}_p \quad \text{(derived from Equation (1.12) set to one)} \tag{1.21}$$

The power ratio is therefore

$$\frac{\mathbf{P}_d}{\mathbf{P}_a} = \frac{\mathbf{V}_d}{\mathbf{E}_p} = (kk_1)/k_2 \tag{1.22}$$

and since the actual losses are the same, the percentage power loss ratio will be the inverse of Equation (1.22). Thus, for the same values of k, k_1 and k_2 used above, the power transmitted by overhead lines can be increased to 147 %, with the percentage line losses reduced to 68 % and corresponding figures for cables of 294 % and 34 %, respectively.

(6) Following on from (5) above, consideration is being given to the conversion of existing AC transmission corridors for use in DC transmission, with the aim of increasing their power carrying capability. In this respect a double three-phase transmission corridor could be used as three parallel DC lines, operating at higher voltage and current levels (the latter determined purely by thermal considerations, as there is no transient stability restriction in the DC case). An interesting recent proposal for a single three-phase line is the use of a modulated control to change periodically the DC power in the three conductors to balance their thermal capability [21]; this solution requires bidirectional current carrying capability in one of the three conductors, and therefore an extra converter in opposition at each end.

1.5 The Impact of Distributed Generation

A great variety of renewable and non-renewable sources are being considered as alternative sources of power generation, closer to the points of utilisation. The most typical examples are co-generation, small hydro, wind farms, fossil-fuelled generators, photovoltaic systems, fuel cells and microturbines. These can operate as separate units or be integrated with the synchronous grid.

In addition to environmental benefits, distributed generation (DG) is also seen as offering important possibilities for improving the quality and security of power supply; it can provide improved reactive power and system voltage control, may avoid losses and user-of-system charges, as well as provide black start capability and the prospect of system islanding.

In the present state of development the proportion of DG installed varies between countries from about 10 to 30 %, and these figures are expected to increase substantially in the future.

There is an increasing tendency for dispersed generators to be connected to the distribution network through static power conversion. This is of course essential with DC sources such as fuel cells and photovoltaics, but even some large designs of wind turbines transmit their power through converters.

In the future the large number of dispersed generators may lead to alternative network designs, including for instance 'active distribution networks' where the operation of the dispersed generators is integrated with that of the network.

1.6 The Effect of Electricity Deregulation

An important factor behind the change towards greater flexibility has been the now generally accepted deregulation philosophy, which encourages the development of a competitive market, whereby energy providers and buyers have open access to the transmission services. In this new environment power generation and transmission are unbundled and the control of the grid is in the hands of an independent system operator. For competitive reasons, unbundling seems to discourage the generators from investing in equipment to provide effective reactive power control of the transmission system, and this has serious implications on voltage and power stability.

Thus, without appropriate countermeasures, the reliability and security of the power system should deteriorate. These problems can, of course, be reduced by conventional transmission system reinforcement, but the cost and the difficulty of obtaining new rights of way discourage such an approach.

The transmission network as an active market player

In the present market philosophy for the generation of electricity the network service providers play no part in the location and level of operation of the generating plants. They are therefore ill-equipped to respond with developments suited to the random needs of the market following generator bids. And yet the deregulation benefits can only be fully exploited when the network services become active players in the market, even in competition with the generators. The lack of competition has been an important factor in the gradual deterioration of national grids. Only when the transmission system becomes an active player in the market will there be sufficient incentive for suitable development of the grid.

While the owners of regulated networks receive a fixed revenue on their investment, those of unregulated networks get their income from the market players according to the services provided. The principal source of income in this case is the spot price differential between the terminals of the interconnected system. A further source of income can be the provision of ancillary services. Thus, as well as energy, there are markets developing in power reserve, frequency regulation, black start capability and voltage control.

However, to compete in the power market, the network services should be independently controllable, a condition difficult to meet by conventional synchronous interconnections, particularly when the size of the tie is small relative to the interconnected systems; in such cases the tie will be easily overloaded (and thus disconnected) even by normal operating changes in either of the interconnected systems.

A comparison of system interconnections

The liberalisation of the electricity industry relies on system interconnections to permit the exchange of power among regions or countries and to transport electrical energy more economically and (environmentally) acceptably over long distances to the load centres.

Among the advantages of interconnection are:

- the pooling of generation capability with the opportunity to utilise diverse primary energy resources;
- the creation of larger markets, which enable economies of scale to be realised in the operation of power plants and in accommodating demand growth;
- greater flexibility for the introduction of competition into electricity supply.

Traditionally such advantages have been achieved by the use of additional AC lines between the various subsystems, in order to strengthen the interconnection. However, with the increasing complexity of power systems the reliability of power supply has deteriorated and the number of blackouts in various parts of the world has increased. Thus strengthening the transmission system is essential to the reliability of block power interchanges among interconnected power systems. The probability of blackouts can be substantially decreased with the use of back-to-back HVDC interconnections due to their asynchronous character and greater controllability.

Possible additional bottlenecks inside the AC systems, resulting from the increased transmitted power, can be avoided by the use of long-distance HVDC links integrated into the system to transmit power directly from remote power generation to load centres.

The cost advantage of HVDC transmission in a multi-system environment has been quantified [22] with the help of a benchmark model of an interconnected system. The benchmark model is used in a rigorous power flow study to incorporate an extra 2000 MW transmission between two systems separated by 950 km. The AC reinforcement option required 950 km of double 400 kV and 750 km of single 400 kV lines, as well as their associated switchyards. For the HVDC alternative, two DC converter stations and a point-to-point 950 km, ±500 kV bipolar overhead line were required. When the total investment and the capitalised losses are taken into account, the DC solution was found to be 70 % cheaper.

1.7 Discussion

Both the use of FACTS and HVDC increases the power carrying capability of a transmission line. In this respect, FACTS technology has probably reached its peak with the development of the unified power flow controller (UPFC) [6, 7], where a proportion of the AC power transmission is subjected to AC–DC–AC conversion, though still has to meet all the requirements of the synchronous AC system.

Originally, the boundaries between HVDC and FACTS were determined by the type of transmission waveform (i.e. AC or DC), the solid-state devices used (which in the HVDC case had been restricted to the SCR) and the power rating of the schemes. However, as the rating and acceptability of alternative solid-state devices improve, these boundaries are gradually becoming ‘blurred’. HVDC is beginning to use the more advanced switching devices and the FACTS controllers are increasing their power rating. The general acceptance of FACTS technology has to a large extent changed the attitude towards DC transmission and the industry has become more open to alternative designs.

The asynchronous interconnector (with or without transmission distance) provides a higher degree of flexibility than the UPFC, and in many cases will be the only practical alternative for the interconnection of separate networks, by retaining (albeit increasing) the power controllability, while removing the synchronous constraints (fault-level increases, transient stability, etc.). The asynchronous link provides the perfect isolation between the interconnected parts and can be designed to provide any required level of power quality control (such as harmonic distortion, unbalance, flicker voltage, etc.). The zero-distance (or back-to-back) asynchronous interconnector has been part of HVDC transmission technology from its very beginning.

Increased transmission flexibility comes at a price in terms of either components or energy efficiency. Therefore, the concept of a new standard HVDC converter (the conventional approach) does not necessarily produce cost-effective solutions for all applications. It is thus essential to consider the degree of flexibility required for the particular transmission application (i.e. taking into account power ratings, transmission distances, extent of ancillary services expected, etc.), before deciding on the most appropriate HVDC alternative

Despite its limited controllability, the thyristor-based HVDC technology is still the preferred one for long-distance, large-power transmission. However, with the use of more advanced switching devices, such as the IGBT and IGCT (Integrated Gate Commutated Thyristor), an HVDC interconnection can now be designed to combine any or all the properties of the individual FACTS controllers: it permits maximum power transfer regardless of the AC voltage phase relationship, delivers or absorbs the required reactive power to maintain the specified voltages at the interconnected buses, contains fast emergency controls to avoid large fault current levels, is designed (if required) to control sub-synchronous resonances, etc. Moreover, the DC link permits the connection of asynchronous systems and systems of different frequencies. *A modern DC link can be the ultimate power transmission controller, i.e. the most **flexible transmission system**.*

References

1. Adamson, C. and Hingorani, N.G. (1960), *High Voltage Direct Current Power Transmission*, Garraway, London.
2. Kimbark, E.W. (1971), *Direct Current Transmission*, Wiley Interscience, New York.
3. Uhlman, E. (1975), *Power Transmission by Direct Current*, Springer-Verlag, Berlin/Heidelberg.
4. Arrillaga, J. (1983), *High Voltage Direct Current Transmission*, IEE Power Engineering Series, London.
5. Padiyar, K.R. (1990), *HVDC Power Transmission Systems – Technology and System Interactions*, New Delhi-Eastern.
6. Hingorani, N.G. and Gyugyi, L. (2000), *Understanding FACTS: Concepts and Technology of Flexible ac Transmission Systems*, IEEE Press, Piscataway, NJ.
7. Song, Y.H. and Johns, A.T. (1999), *Flexible ac transmission systems (FACTS)* IEE Power Engineering Series 30, London.
8. 'The history of high voltage direct current power transmission', *Direct Current*: Part I, December 1961, p. 260; Part II, March 1962, p. 60; Part III, September 1962, p. 228; Part IV, January 1963, p. 2; Part V, April 1963, p. 89.
9. Arrillaga, J. (1998), *High Voltage Direct Current Transmission*, Second Edition, IEE Power and Energy Series 29, London.
10. Asplund, G., Eriksson, K., Jiang, H., Lindberg, J., Palsson, R., and Svensson, K. (1998), dc transmission based on Voltage Source Converters', CIGRE SC-14, No. 14–302, Paris.
11. 'A survey of the reliability of HVDC systems throughout the world during 1995–1996', CIGRE report 14–102 (1999), Paris.
12. CIGRE WG14.20 (2001), 'Economic assessment of hvdc links', 71pp.
13. 'Determination of power losses in High Voltage Direct Current converter stations', IEC 61803-1999.
14. CIGRE B4-37/VSC document (2005), 'VSC transmission', Paris.
15. Arrillaga, J., Bowles, J., Campos Barros, J., Kanngiesser, J. and Ingram, L., (1993), 'Direct connection of generators to HVDC converters', *ELECTRA*, No. 149, August, 19–39.
16. Lamm, A.U., Uhlmann, E. and Danfors, P. (1963), 'Some aspects of tapping HVDC transmission systems', *Direct Current*, **8**(5) 124–129.
17. Arrillaga, J. (1966), 'Analytical studies on high voltage ac-dc multiterminal interconnections', PhD thesis, University of Manchester.

18. Donahue, J.A. and Bradley, D. (1993), 'Multiterminal commissioning of the Sandy Pond HVDC converter terminal', International Colloquium on HVDC and FACTS, paper 3.3, Wellington, New Zealand.
19. CIGRE SC-B4 WG-33 (2005), 'HVDC and FACTS for distribution systems', Paris.
20. Zhang, X.P., Rehtanz, C. and Song, Y. (2006), 'A grid for tomorrow', *IET Power Engineer*, October, 22–27.
21. Barthold, L.O. I-MOD Inc. (2004), US patent 6714427 B1.
22. Povh, D., Retzmann, D., Teltsch, E., Kerin, U. and Mihalic, R. (2006), 'Advantages of large ac/dc system interconnections', CIGRE B4-304, Paris.

2

Semiconductor Power Devices

2.1 Introduction

Progress in power semiconductor types and ratings is such that a review of their current state, important as it is to the subject of this book, will be very short lived, and any recommendations on their specific application must be looked at in this context.

Historically, the application of semiconductors to HVDC transmission started with the silicon-controlled rectifier (SCR) in the late 1950s. Despite its age the SCR, though with highly improved current and voltage ratings, is still the most widely used semiconductor in HVDC conversion. However, the restricted controllability and low switching frequency of the SCR has encouraged the development of alternative power semiconductors of the thyristor and transistor families.

This chapter describes the basic structure and characteristics of the power semiconductor devices currently used or with potential applicability to HVDC transmission. The physics of semiconductors, covered in many good references such as [1–3], is outside the scope of the book. However, a brief introduction to semiconductor principles is required to understand the terminology used by the industry to describe the characteristics of the devices. A thorough review of the state of power semiconductors being used and under consideration for HVDC and FACTS application was carried out by a CIGRE Working Group in 1996 [4]. Ongoing developments in power semiconductors are now an important part of the main international conferences on the subject of power electronics.

2.2 Semiconductor Principles

Semiconductors are materials with conductivities intermediate between insulators and metals, which are controllable over a wide range. These materials possess quasi-continuous bands of allowed electronic states. The energy bands are separated by forbidden bands inaccessible to electrons.

Flexible Power Transmission: The HVDC Options J. Arrillaga, Y.H. Liu and N.R. Watson

For as long as the energy bands are empty or completely filled with electrons there can be no movement of electrons, i.e. no current. If, however, there are some electrons in an otherwise empty band, these can move about in an electric field and conduct a current. Conversely, if some electrons are missing from an otherwise completely filled band, empty electron states or ***holes*** are formed; the remaining electrons may move into these holes, thus creating other holes in the states the electrons have vacated. Thus the holes can move about in the energy band. Both the electrons and holes are called ***carriers***, although conventionally the holes are considered as moving (rather than the electrons) and carrying positive charge as a current.

At low temperatures the highest band containing electrons (***valence band***) is filled completely, while the next band (***conduction band***) is empty and thus the conductivity is zero. Thermal energy or light can raise single electrons from the valence to the conduction band, a mechanism termed ***generation***. This leads to a double contribution to the conductivity, i.e. one electron in the conduction band and the hole it has left in the valence band. When an electron leaves the conduction band to fill a hole in the valence band, energy is liberated and dissipated as heat. This mechanism is called ***recombination*** and annihilates two charge carriers, i.e. one hole and one electron. The number of free electrons and holes in an ideal semiconductor is a function of temperature.

In addition to the carriers generated by thermal energy or light, further carriers can be provided by impurities (such as phosphorus or arsenic) purposely introduced into the semiconductor, a technique referred to as ***doping***. Doping can be of n-type, the species affecting it being called ***donors*** because they add a freely moving electron to the conduction band, or of p-type, which increases the conductivity by one positively charged hole. In contrast to generation, doping with impurities adds only one of the two carrier types. The carriers introduced by the impurity atoms determine the conductivity in a doped region and are called ***majority carriers***. Devices in which both types of carriers are present are called ***bipolar*** devices, as opposed to ***unipolar*** devices, in which the number of one type of carrier is much larger than the other. High-voltage devices require light doping, which is achieved by neutron irradiation of silicon.

2.3 Power Semiconductor Elements

2.3.1 The pn Rectifier

The pn diode rectifier is a two-layer bipolar device. If the p-layer voltage is positive with respect to the n-layer, a large current will flow on the application of a voltage across the device in the ***forward-biased*** direction. When the p-layer is negative with respect to the n-layer, the device is ***reversed biased*** and only a small reverse current flows through the junction even at high voltage; in this case the reverse current is almost independent of the reverse voltage applied and very dependent on temperature. The structure and current/voltage characteristics of the pn rectifier are shown in Figure 2.1.

Blocking capability

If the electric field exceeds a critical value (2×10^5 V/cm in silicon) the free carriers are accelerated to the point where they can generate further electron–hole pairs and the additional

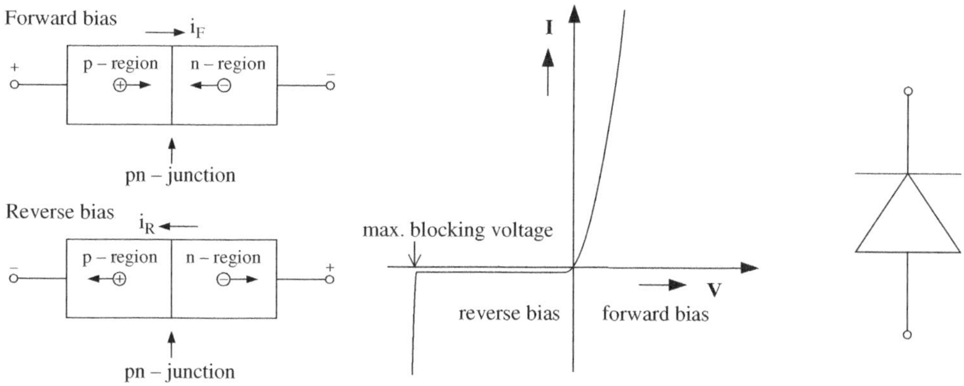

Figure 2.1 The pn structure (forward and reverse biased), current/voltage characteristic and diode symbol

free carriers are themselves accelerated and generate further electron–hole pairs. This leads to an avalanche of carriers and a large increase in reverse current which may destroy the device.

The blocking capability or ***breakover voltage***, i.e. the voltage for the onset of avalanche in silicon, is theoretically 25 kV. However, there are other limiting factors like defects, non-uniformities and shape of the doping profiles, as a result of which only 10 to 12 kV can be expected.

Switching behaviour

The main switching specification of a power diode is its rate of change of current (di/dt). When the diode is part of an inductive circuit, di/dt is controlled by the inductance; however, if the diode is used for free-wheeling, di/dt is controlled by the turn-off characteristic of the controllable solid-state device.

The switching process requires time to complete. Since the blocking action is due to the presence of a depletion zone, turn-on requires flooding the depletion zone with carriers for the device to change into its conducting state.

The time dependence of forward voltage (V_F) and forward current (I_F) during turn-on is illustrated in Figure 2.2, where $(t_1 + t_2)$ represents the forward delay (i.e. the time required for the voltage to drop to $1.1V_F$); t_1 is in the hundreds of nanoseconds and t_2 in the microseconds regions respectively.

On the other hand, turn-off requires the removal of charge carriers from the pn junction in order to build the depletion layer needed to block the voltage. At the time of commutation the central zone is flooded with carriers and in an inductive circuit the rate of change of the reverse current will be determined by the inductance; only after all the excess carriers of the central zone are removed will the reverse current reduce to a sufficiently low off-state level for the device to be able to block the applied voltage. The time dependence of the current and voltage during turn-off is shown in Figure 2.3, where t_4 is the storage time, t_5 the reverse current fall time and t_{rr} the reverse recovery time.

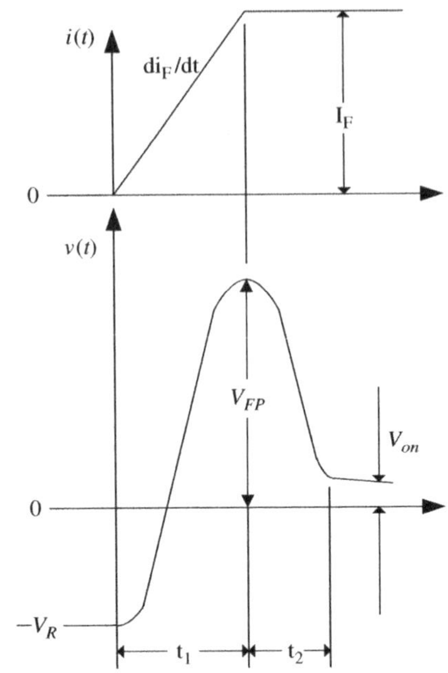

Figure 2.2 Forward voltage and current of a power diode at turn-on

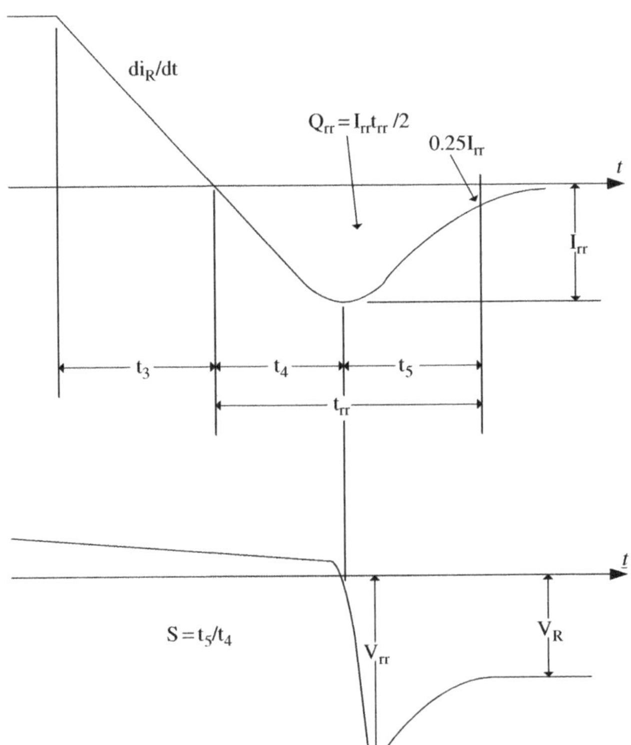

Figure 2.3 Forward voltage and current of a power diode during turn off

To reduce the switching time the carrier lifetime can be controlled by diffusion of gold or platinum and irradiation with electrons, protons or helium. However, there is a trade-off between carrier lifetime magnitudes and on-state losses.

2.3.2 The Transistor

The transistor is a three-layer bipolar device consisting of two pn junctions in series back to back as shown in Figure 2.4. Under normal operation the collector junction is reversed biased while the emitter junction is forward biased. Therefore the emitter injects electrons as minority carriers into the p-base where they drift towards the collector junction. The electric field in the collector junction space-charge region accelerates the injected electrons into the collector zone so that they can build up a current in the external circuit back to the emitter. The magnitude of this current depends on the electron concentration gradient in the base zone, and that in turn depends on the forward bias of the emitter–base junction. By varying the emitter–base current the collector current can be modulated. The collector current varies exponentially with the emitter–base voltage V_{EB} and is independent of the collector–base voltage V_{CB}; this leads to saturation, as shown by the characteristics of Figure 2.5.

2.3.3 Metal-Oxide-Semiconductor Field-Effect Transistor

The metal–oxide–semiconductor field-effect transistor (MOSFET) is a unipolar device, i.e. the charge flow consists only of majority carriers (electrons).

As shown in Figure 2.6, the gate electrode is insulated from the p-substrate by a thin layer of SiO_2, which together with the surface layer of the substrate constitutes a capacitor. Two highly doped n^+ regions at the sides of the gate area form the source (cathode) and drain (anode) which are connected to a load R_L and a voltage source V_{DS}.

In the absence of gate voltage, only a small reverse current (that of the n^+p diode) biases the drain and the device in the off-state. When a positive gate voltage is applied the capacitor is charged. The electron density at the substrate–oxide interface is enhanced by the charge stored in the capacitor and an inversion layer is formed which acts as a conducting channel between source and drain. The current is supported only by majority carriers (electrons), which have very low relaxation time and therefore permit very fast switchings. The current

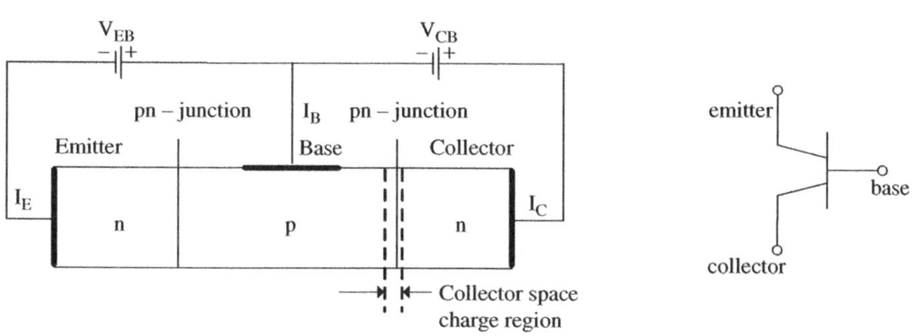

Figure 2.4 Transistor structure and symbol

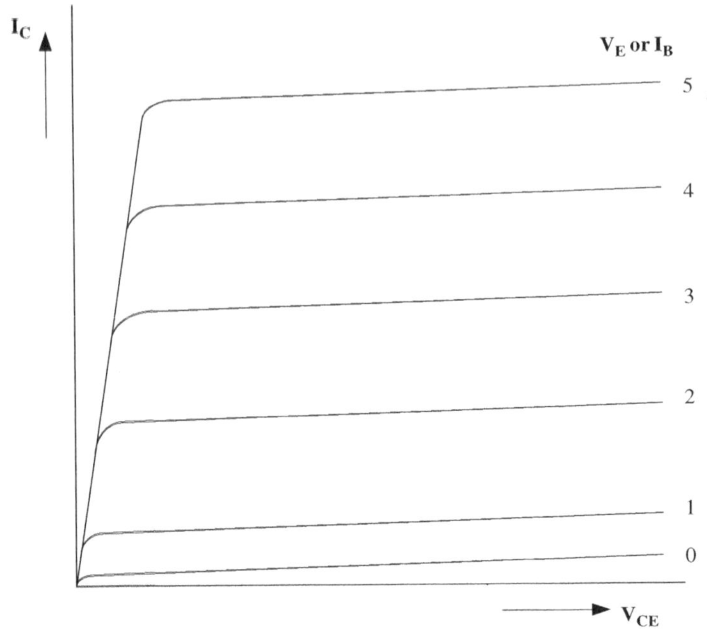

Figure 2.5 Transistor current/voltage characteristics

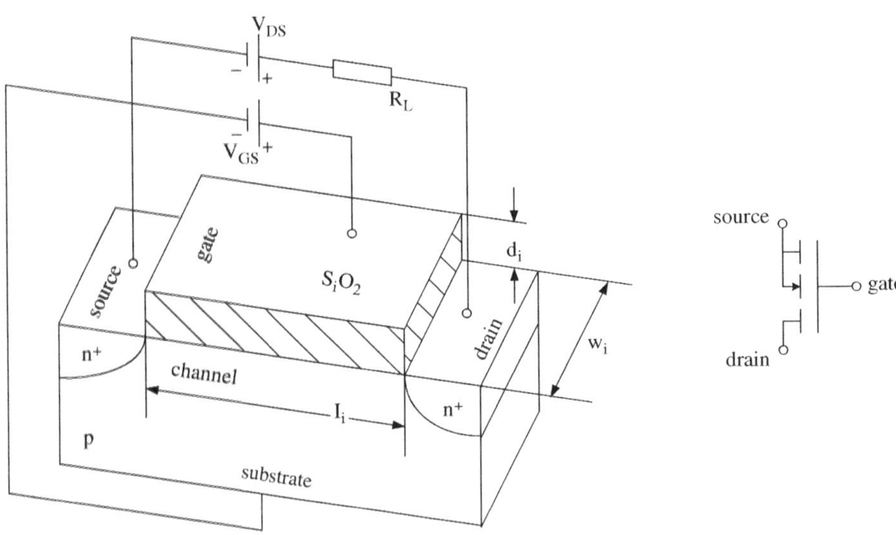

Figure 2.6 Structure and symbol of the lateral n-channel MOSFET

causes a voltage drop along the channel and, therefore, the capacitor voltage at the drain end is lower and the channel is shallower. Above a certain value of V_{DS} the channel is pinched at the drain side and becomes shorter for increasing V_{DS} levels. The voltage leads to the formation of a space-charge region at the drain n^+p junction, which takes up the additional

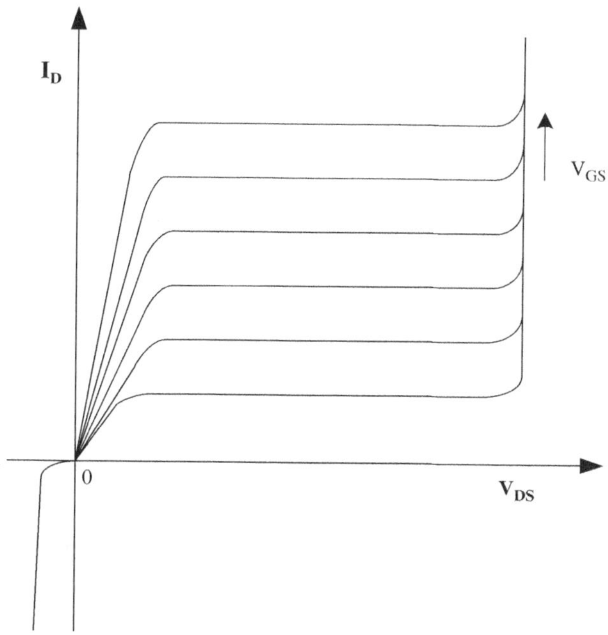

Figure 2.7 Typical MOSFET current/voltage characteristics

voltage without a further increase in current. This leads to the saturation characteristics shown in Figure 2.7.

High conductivity requires short channel lengths, which results in relatively low voltage blockings, a property that makes the MOSFET unsuitable for high-power and high-voltage applications. However, its low gate energy requirements and fast switching capability make the MOSFET an ideal auxiliary device as a gate amplifier for thyristor devices, such as the GTO (Gate Turn-Off), IGCT (Integrated Gate Commutated Thyristor), MTO (MOS Turn-Off) and ETO (Emitter Turn-Off).

2.4 Dynamic Stresses on Power Switches

2.4.1 Rate of Change of Voltage (dv/dt)

The dv/dt characteristic is the maximum rate of rise of anode to cathode voltage which will not trigger the device. Fast switching results in high dv/dt and possibly voltage overshoot at turn-off which can last for a few microseconds as shown in Figure 2.8 [5].

The dv/dt is particularly high (typically $15\,\text{kV}/\mu\text{s}$) for converters employing series devices (using synchronous switching). The stresses experienced by individual switches depend on their position within the converter arm and on the stack topologies. In a two-level converter (a subject discussed in Chapter 4) all devices are subjected to similar stresses (switching loss, conduction loss, dv/dt, di/dt and peak voltages). In multi-level arrangements (a subject discussed in Chapter 6) the outer devices are subjected to higher switching losses than

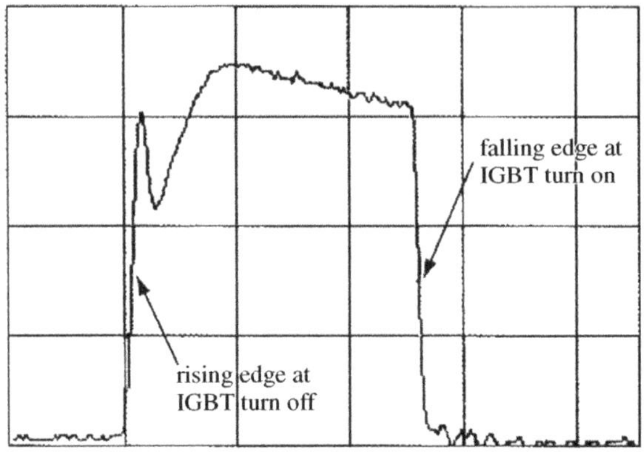

Figure 2.8 Recorded IGBT voltage waveform 500 V/div and 2S/div (Shakweh, Y. (2000), 'New breed of medium voltage converters', *Power Engineering Journal*, February, reproduced by permission of the IET.)

the inner devices, but the latter have higher conducting losses. The stress distribution is determined by the operation mode of the converter over the load duty cycle.

With reference to PWM (Pulse Width Modulation) conversion (a subject discussed in Chapter 5), in the absence of filters, dv/dt causes (i) a voltage doubling effect at the rising/falling edges of the voltage waveform, due to the wave propagation in long cables; (ii) earth currents to flow in the cable stray capacitance; and (iii) inductive and capacitive couplings between live components and earth, resulting in common mode and differential mode noise.

2.4.2 Rate of Change of Current (di/dt)

The di/dt rating indicates the maximum rate of rise of current that the device can withstand without deterioration. The latter is often caused by local 'hot spots' in the junctions of the device during turn-on. A variety of techniques are used to extend the initial conduction region, such as the use of interdigitated gates (that exploit the transversal field caused by the exciting current), the use of an amplifying gate, etc.

The admissible limit is determined by I^2t under overload conditions. The temperature of the crystal also limits the admissible di/dt. This di/dt has no unique value for all operating conditions and is thus chosen for the most severe and frequent conditions.

If the rate of rise of the current is greater than the permitted value, it must be limited by additional series inductance in the circuit.

2.4.3 Balancing Problems in Series Chains

Static balancing requires the use of large resistors in parallel with each switch in the chain (which in turn increases the power loss).

Dynamic balancing is a more difficult issue, because in its absence, the switch that turns off first (or turns on last) would have to sustain all the voltage. The problem is normally reduced by the use of snubbers; these, besides involving extra cost, slow down the switching process and reduce efficiency. Instead, the emergence of multi-level topologies and the use of asynchronous firing control within the chain can provide more cost-effective solutions.

2.5 Other Switching Issues

2.5.1 Switching Frequency

The maximum permitted switching frequency is mainly device dependent and the optimum frequency, in each case, is determined by the configuration, requirements of the load and thermal consequences. For hard-switched converters the switching frequency is often limited to below 1 kHz, whereas for topologies requiring snubbers it can be as high as 2 kHz.

2.5.2 Switching Losses

When considering static power conversion as an alternative for a particular application, the overall energy loss over the life of the plant must be present valued and taken into account in the economic comparison.

The main energy losses are caused by the switching devices, both during full conduction and under switching conditions. Apart from the energy cost, the resulting heat has to be removed from the wafer through the package into the cooling medium (normally air, gas or ionised water) at considerable extra cost.

On-state losses during full conduction are caused by the device forward voltage drop and the load current. Regarding switching losses, at turn-on the forward current rises before the voltage drops. Similarly, at turn-off, the forward voltage rises before the current falls. Thus in both cases the simultaneous presence of voltage and current represents a power loss. With high-frequency switching, as required when using PWM, the related switching losses are often higher than the on-state losses.

Also, adding to the total energy loss are the losses of the auxiliary components, such as gate driving and snubber circuitry, and the cost of the driver circuit and power supply can be higher than the device itself.

2.5.3 Soft Switching

Soft switching, a design concept used to ensure zero-voltage switching (ZVS) or zero-current switching (ZCS), is often used in static power conversion to enhance performance and reduce the energy losses. It is a difficult subject due to the bidirectional flow of power, modulation with multi-frequency involved and a wide range of load conditions. A good survey of the techniques available and related references can be found in [6].

For a given switching frequency, the device junction temperature rise is reduced significantly for the soft-switched case due to the lower semiconductor switching losses.

Many soft switching topologies have been proposed in the past 20 years for low/medium voltage. All these proposals require the use of extra inductors, capacitors, diodes and self-commutated switches. Generally, their aim is to create a resonance to provide the zero switching condition, and thus reduce di/dt in the device that is turning on and dv/dt across the device that is turning off.

Based on the location of the soft switching components, there are three main alternatives, as the component ratings and costs depend on whether or not they form part of the main power flow. These are:

1. *Load resonant converter*: An LC resonance tank is added to the AC side output power path in series, parallel or series/parallel combination to derive the ZVS or ZCS condition.
2. *Resonant DC link converter*: A resonant network is connected between the DC source and the converter bridge, and it is active for every switching action in the converter bridge to create the ZVS or ZCS condition.
3. *Transition resonant pole converter*: The resonant networks added to the converter are active in order to create the ZVS or ZCS condition only when the switching state of each main bridge switch is changing.

Solutions 1 and 2 are not appropriate for large static power converters because the soft switching components are connected in the main power transfer path. Although several solutions have been proposed for multi-level voltage source converters based on solution 3, they are complicated and costly.

Ideally, the soft switching should be synchronised by the converter firing control and should be of controllable duration so that the ZVS or ZCS conditions are established before the switching commences and terminate after the switching dynamics finish.

2.5.4 Use of Snubbers

Snubbers are used to protect the semiconductors from the voltage and current transient stresses that occur during the turn-on and turn-off switching events. They limit the voltage and current magnitudes as well as their rates of rise. The reduction of dv/dt and di/dt also lowers the electromagnetic interference (EMI) levels.

Snubbers shift the switching losses from the silicon to a low-cost passive component where it can be dissipated further away from the main switching device or recovered back into the supply or load. The reduction in losses in turn results in lower junction temperatures and lower junction temperature rise. Thus, for a given rating snubbers permit the use of higher switching frequencies.

There are three different types of snubber:

1. RC-type unpolarised snubbers, which protect diodes and thyristors from the voltages created during reverse recovery current in the presence of series inductance.
2. RC-type polarised snubbers, used with all types of switches, to provide a low voltage and control dv/dt across the switch while the current turns off.

3. *RL* polarised snubbers that reduce the voltage across a switching device as the current builds up during turn-on transients; they also limit the overcurrents due to diode reverse recovery.

2.6 Thyristor-Type Power Switches

2.6.1 The Thyristor

The thyristor, also referred to as the SCR, has existed since the mid-1950s. It is the equivalent of a binary current valve with two discrete states either conducting or blocking the current.

The thyristor consists of two transistor structures connected together such that the collector of a pnp transistor forms the base of the npn transistor, as shown in Figure 2.9. It is a four-layer, three-terminal device and its function is determined by the regenerative action of the two component transistor structures. Generally the n and p emitters are highly doped, while the n- and p-base zones have low impurity concentrations. When the cathode (the n emitter) is positive and the anode (the p emitter) negative the two emitter junctions are reversed biased and no injection of carriers occurs. In this case there is only a very small reverse current at the reverse blocking junction provided that the voltage remains below the avalanche and punch-through limits. The ***blocking state*** region is shown on the left-hand side of the *I–V* characteristic of Figure 2.10.

With the device forward biased (the anode positive) and in the absence of a gate signal, the outer pn junctions are forward biased, while at the central pn junction a space-charge region builds up that only permits a small current to flow across this junction. This is the ***off-state*** condition.

To turn the device on requires the injection of a positive current into the gate (the p base) which increases the total current flowing out of the cathode. This current must be sufficiently large (greater than the ***holding current***) for the emitters to inject sufficient numbers of carriers to keep the thyristor in conduction.

Even with zero gate current, triggering can occur under a large forward voltage (the breakover voltage), which causes a conducting state with an almost exponential on-state characteristic.

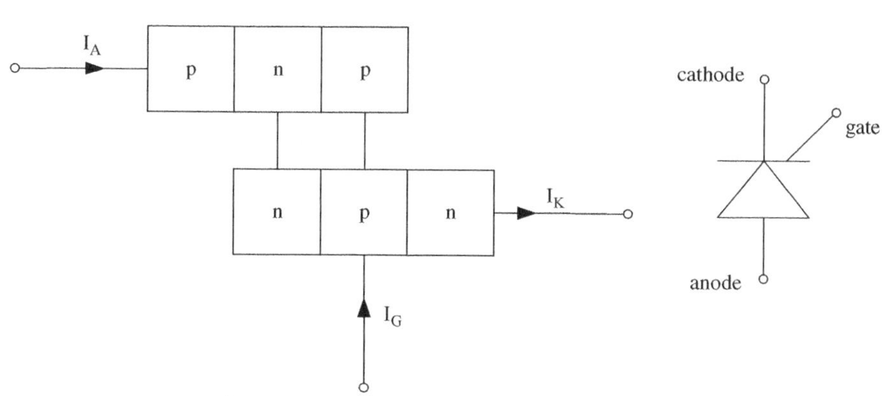

Figure 2.9 Two-transistor model and symbol of the thyristor

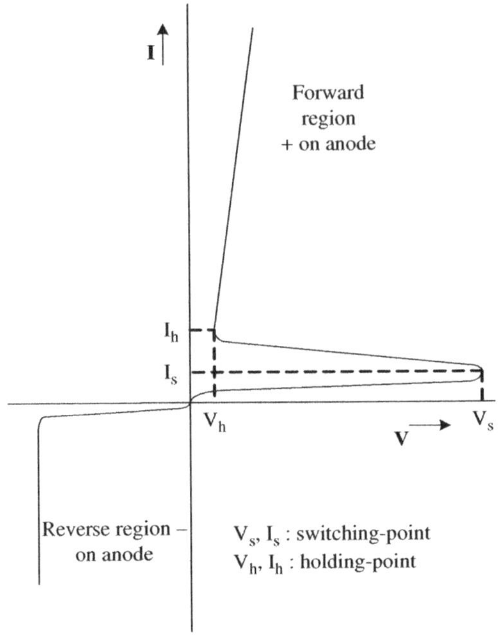

Figure 2.10 Thyristor current/voltage characteristics

Another unwanted triggering condition can occur due to high *dv/dt*. A sudden increase in forward voltage extends the space-charge region. Free carriers are then forced out of the expanding space-charge region and these create a current which, for sufficiently high *dv/dt*, will exceed the holding current of the device taking it into conduction.

The turn-on conditions described above relate to the ***electrically triggered thyristor*** (or ETT). Alternatively a light pulse can be sent to the forward blocking junction that generates sufficient carriers to turn the device on without requiring auxiliary energy. This results in the ***light-triggered thyristor*** (or LTT). An uncontrolled rise in temperature can also increase the total current to the point of turning the thyristor on.

In the ***on-state*** condition the device exhibits an almost exponential dependence of the forward current on the forward voltage, similarly to the case of a pn junction.

Turn-off in the thyristor can only be achieved by reducing the current below the holding level for a certain minimum time. This condition is achieved either by a zero-current crossing of the external current (***natural commutation***) or by ***forced commutation***.

The blocking capability of the thyristor is limited by the avalanche and punch-through effects. For high blocking voltage the resistivity and device thickness must be increased. The extra thickness, however, increases the voltage drop across the n-base region, which in turn results in higher loss and temperature. The minimum (blocking) recovery time, often given by the manufacturers as a single value, will depend on the current level used, the cooling system (which affects the junction temperature) and *di/dt* during switch-off. It is, therefore a difficult parameter to assess when considering the benefits of the ZCS condition achieved by the multi-level reinjection converters discussed in Chapter 7.

The thyristor can be designed as a symmetrical device (with both forward and reverse direction blocking capability) or as an asymmetrical device (with blocking capability in the forward direction only). The latter is a combination of a thyristor and a reverse diode in parallel on the same wafer. It is, therefore, controllable in the thyristor forward direction and always conducts in the reverse direction.

Dynamic characteristics

Thyristor turn-on is achieved by the injection of a positive current into the gate. This increases the overall current (I_T) and makes the voltage (V_T) drop, as shown in Figure 2.11. The injection starts along the gate edges and then spreads across the device face. To speed up the switching process the large devices are interdigitated, as shown in Figure 2.12 (where the white areas indicate metal and the dark areas silicon), and the device often includes several amplifier gate stages.

Initially all the current concentrates in a small area around the gate, causing high power densities; therefore, the *di/dt* rise needs to be limited by external inductance to avoid destruction. To reduce turn-on losses, the transition to the conducting state can be accelerated by using high and steep triggering pulses.

In the conducting state the base layers are flooded with carriers injected from the emitters and these carriers need to be removed before forward or reverse voltage blocking can be achieved. As this process is mostly effected by recombination, the carrier lifetime is of primary importance. A snubber circuit is normally used to protect the device from the excessive reverse voltage transients caused by the external inductance. The snubber also limits the anode *di/dt* effect, reduces the off-state and reapplied anode *dv/dt*, as well as the switching losses. The thyristor turn-off behaviour is illustrated in Figure 2.13(a).

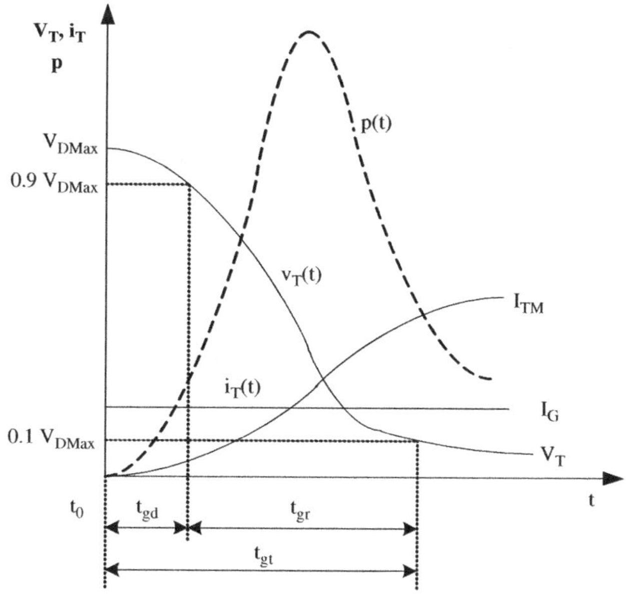

Figure 2.11 Thyristor on-state current and voltage during turn-on (t_{gd} is the delay time, t_{gr} the rise time, t_{gr} the turn-on time and $p(t)$ the turn-on loss) (Reproduced by permission of CIGRE.)

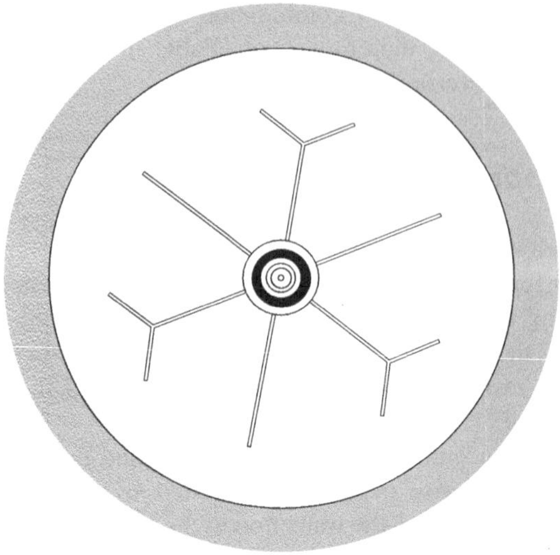

Figure 2.12 Interdigitated gate geometry of a high-power thyristor (Reproduced by permission of CIGRE.)

A critical turn-off parameter is the turn-off time t_q, i.e. the period between the current zero crossing and the point when the thyristor can take up forward voltage, because the application of forward voltage during this period would turn on the thyristor again. Figure 2.13(b) shows that the highest power loss occurs during the decrease of the reverse current. A compromise must be made between the requirements of the static and dynamic characteristics: to achieve low static losses the carrier lifetime should be large, while for fast switching and low dynamic losses the carrier lifetime needs to be small.

A number of advanced thyristors, developed to permit turn-off control, are considered in the following sections. The main types are the GTO, the IGCT, the MTO, the MOS controlled thyristor (MCT) and the field controlled thyristor (FCT). The last, sometimes referred to as the static induction thyristor (SITh), has so far been an unsuccessful competitor in high-voltage and high-power applications because of the low turn-off gain and special manufacturing difficulties.

Series operation

When the switching devices are to be operated in series, each unit must share equally the total forward and reverse voltage, both during steady-state operation and under transient conditions.

The build-up of anode current of the individual thyristor units during turn-on can be purposely delayed (by circuit elements) to ensure that all the series units have turned on. This is necessary to prevent a voltage spike from appearing on the unit with the longest turn-on time.

The structure of the series thyristor chains used in HVDC conversion is described in detail in Chapter 8.

In some medium-voltage applications, only one of the units in series needs to be turned on by the external signal, the remaining units being slave fired or triggered by the decaying

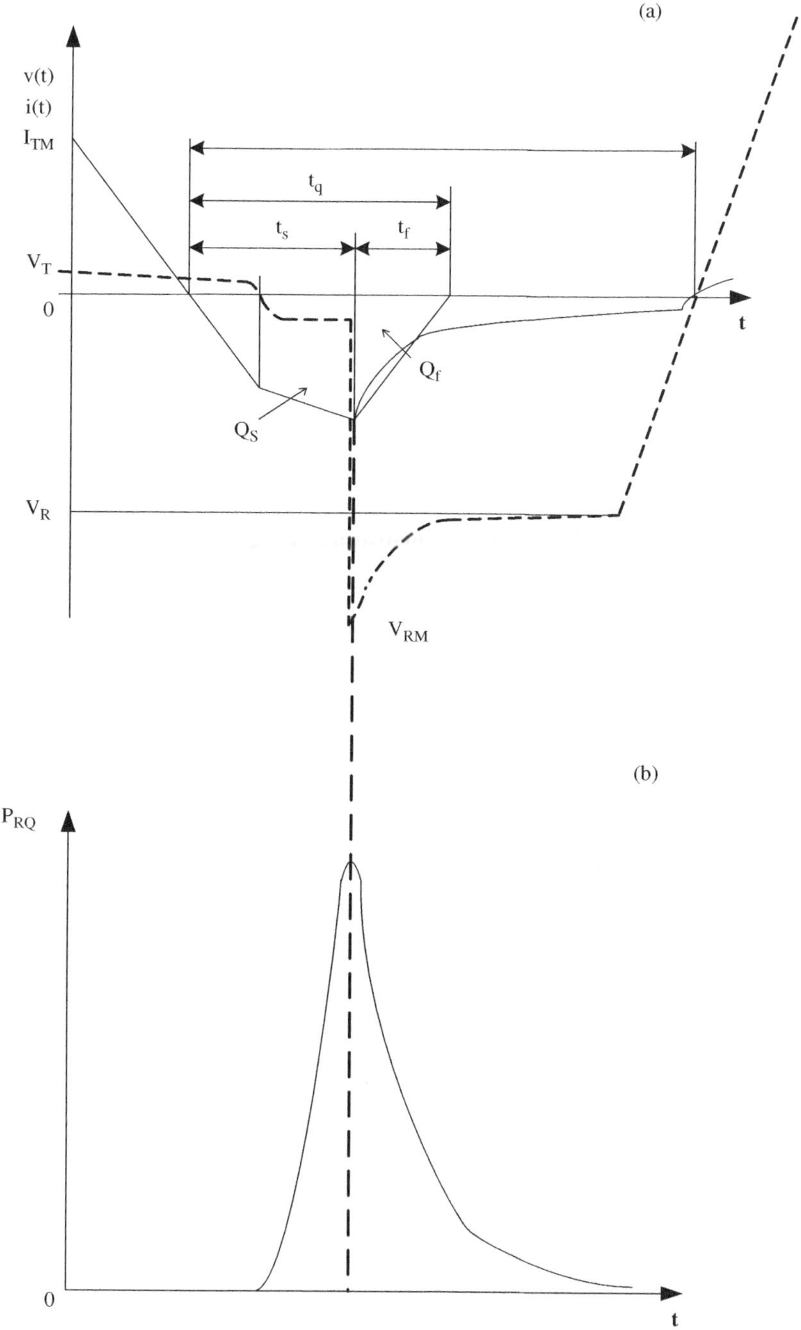

Figure 2.13 Thyristor turn-off: (a) voltage and current versus time; (b) power loss

anode to cathode voltage of the externally triggered unit. However, in general, each unit in a series string requires an isolated gate to cathode turn-on signal to ensure a reliable turn-on with minimal transients.

Parallel operation

When the SICs are connected in parallel they must share equally the total anode current. As the forward voltage drop of the SCR is almost independent of the load current, wide variations in the division of the current are possible. Thus, even when matched cells are used, their current rating must be reduced depending on the number of parallel cells. Also, to keep the junction temperature of each cell approximately equal, all parallel cells should be mounted on a common heat sink.

The gate to cathode signal should remain in all the cells until the anode current has reached the holding current. Even after one of the parallel cells has decreased the anode to cathode voltage of the other parallel cells to a few volts, the gate signal must be maintained to ensure that the remaining units turn on. Current balancing reactors (Figure 2.14) are usually connected in the anode circuit to equalise the ampere-turns of the parallel circuits. Finally, the conductor layout must be arranged to equalise the inductance and resistance of the individual devices placed in parallel.

2.6.2 Gate Turn-Off Thyristor (GTO)

Gate-controlled thyristor turn-off, introduced in the late 1970s, made it possible to design self-commutating converters of larger power ratings. Like the conventional thyristor, the GTO, shown in Figure 2.15 [4], is a three-terminal device.

The GTO switching, however, requires more complex gating and snubber circuits. Unlike the conventional device, the GTO allows the extraction of holes from the p-base by means of a high negative current on the gate, which stops the n^+ emitter injection and turns the device off. However, since in the thyristor the lateral dimensions are bigger than its thickness, a

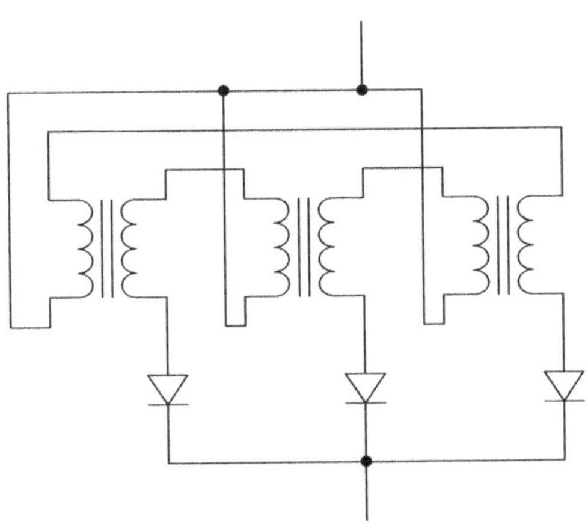

Figure 2.14 Current balancing reactors

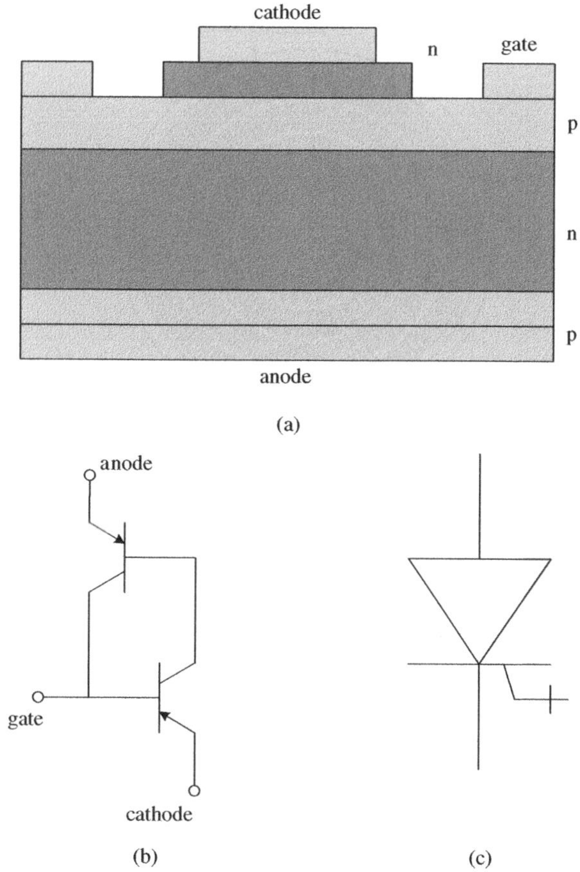

Figure 2.15 GTO structure, circuit diagram and symbol

high lateral voltage drop under the n^+ emitter would occur and a large part of it would continue injecting electrons. Therefore the GTO n^+ emitter is configured as an array of many single and narrow stripes, as shown in Figure 2.16, and to reduce the voltage drop in the gate, the p-base layer (which constitutes the gate) is metallised outside of the n emitters.

The GTO is manufactured as a whole wafer or as a die and is available in module, plastic and capsule packages. In the on-state the GTO behaves exactly as a thyristor, and, therefore, turn-on requires the application of a positive triggering current to the gate; di/dt is less critical, though, because all the emitter area is turned on quickly due to its multi-cell structure. As the n emitters are not shorted, a negative gate voltage must be applied continually during the off-state to improve the blocking behaviour at high temperatures. Shorts are introduced in the p emitter to improve the turn-off capability. Also the carrier lifetime is kept low to provide higher forward voltage drop and holding current. To counteract this effect, current needs to be fed continuously into the gate during the on-state.

Due to the cellular structure of the n emitters constituting the cathode, as well as its thyristor-like characteristics, the GTO is prone to current filamentation during turn-off. This

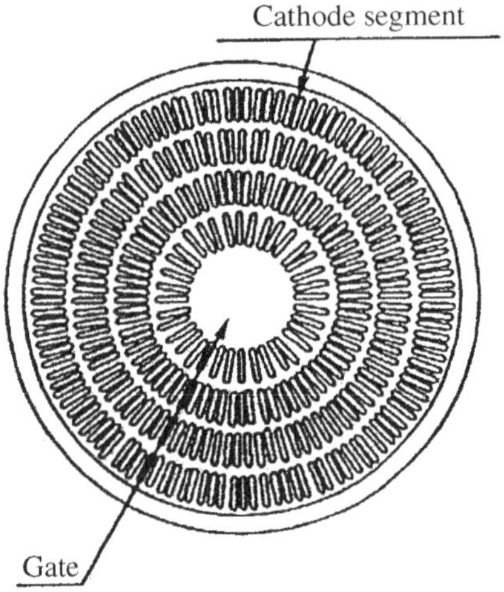

Figure 2.16 GTO emitter layout (Reproduced by permission of CIGRE.)

effect is due to its exponential (thyristor-type) forward current characteristic and to a negative temperature characteristic of the on-state resistance (caused by the temperature dependence of the carrier concentration). Therefore the turn-off losses, which occur in a small volume (that of the last few emitters turning off), are high (about 10 to 20 % greater than an equivalent rated conventional thyristor); thus GTOs require special snubber arrangements, as shown in Figure 2.17.

The resistor R_S limits the capacitor discharge current during turn-on and the R_S, C_S combination determines the minimum time required by the GTO to turn off again. For instance, for the case of a GTO with a peak voltage rating of 4.5 kV and a peak turn-off current of 4 kA [7], to turn the GTO safely the snubber capacitor C_S must be at least 6 mF to be able to dissipate the energy stored in the capacitor after the switching. Also the energy stored in the di/dt limiting inductor L_S at turn-off is dissipated via the discharge resistor and diode.

Figure 2.18 displays the current/voltage characteristics during turn-on and turn-off, where:

- t_s is the storage time in which the carriers are removed from under the cathode emitters, while the load current is maintained by its inductance.
- t_f is the fall time in which the current reduces very quickly.
- i_{tail} is the tail current; in this region the remaining carriers are removed via recombination across the anode emitter shorts.
- V_{DSP} is a forward voltage spike caused by stray inductance in the snubber circuit (a possible source of destruction that must be avoided).

Also the turn-on delay of the free-wheeling diode D_F, together with the stray inductances of the diode and circuit, cause an overvoltage V_{DM} that must not exceed the maximum rating.

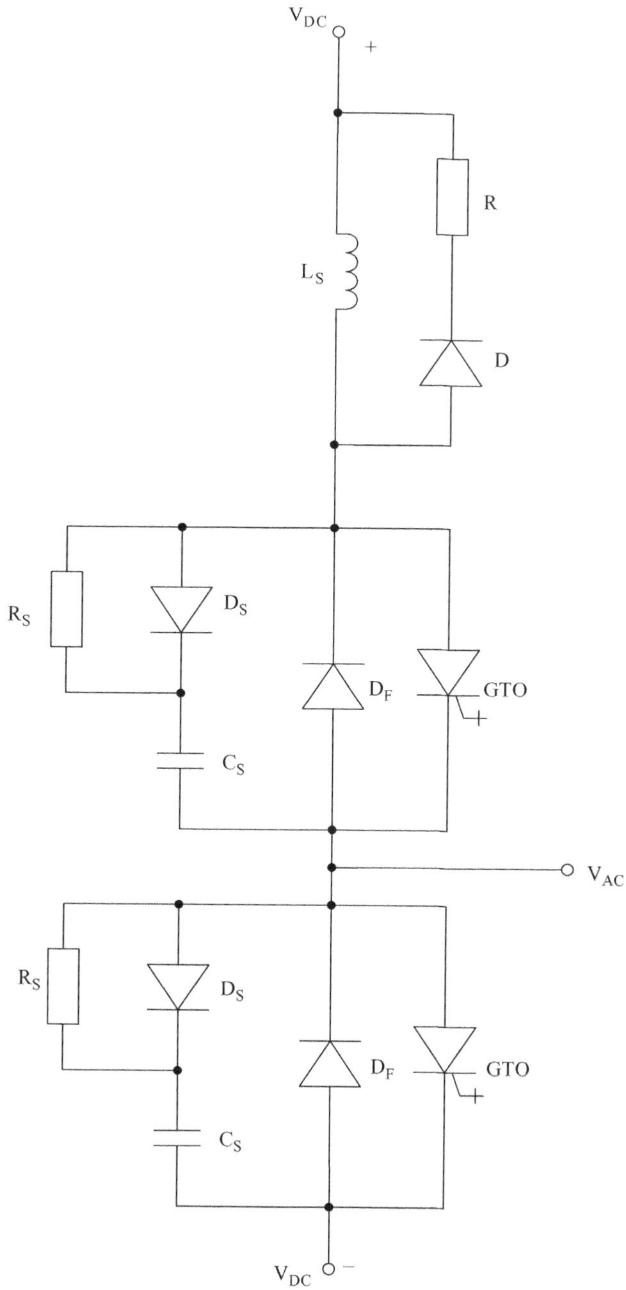

Figure 2.17 Snubber circuit in a GTO chain

The GTO can be designed for symmetrical or asymmetrical operation. A symmetrical GTO (with forward and backward blocking voltage capability) is needed when the DC voltage can reverse its polarity, as in the case of current source conversion. The asymmetrical GTO is used in voltage source conversion and permits larger power

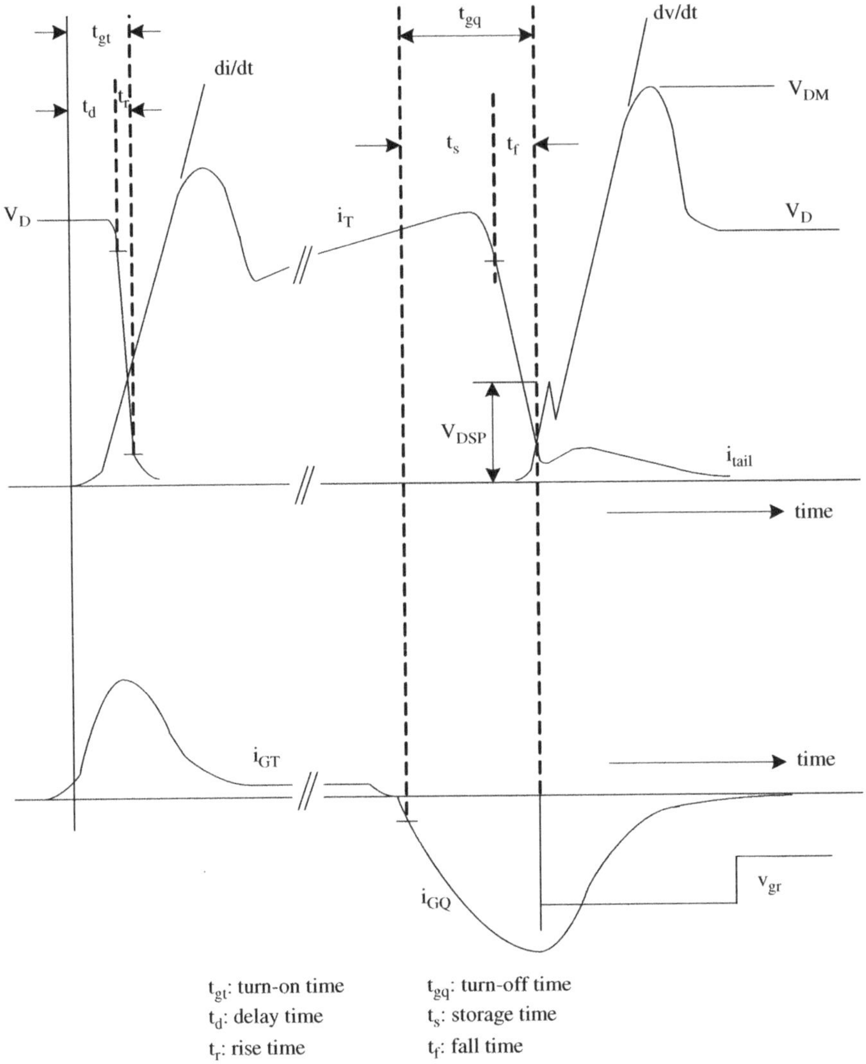

Figure 2.18 Effect of snubber on switching behaviour (for the circuit of Figure 2.17) (Reproduced by permission of CIGRE.)

ratings. In this case the reverse voltage blocking capability is eliminated on purpose by means of internal anode to cathode shorts; these decrease the forward voltage drop and improve the switching characteristics (two important factors influencing the converter losses).

The GTO is turned off by gate current and the turn-off gain is between 3 and 5. Series connection of devices is possible by matching the turn-off storage time of the devices. Presently a 6 kV, 6 kA GTO is commercially available, with a maximum operating frequency of 500 Hz.

2.6.3 Insulated Gate-Commutated Thyristor (IGCT)

The need to simplify the GTO snubber circuit requirements and improve its switching performance has led to the development of the IGCT. It is a hard-driven GTO, in which during turn-off the full main current is commutated to the gate. However, the pulse is of a short duration and the energy dissipated in the gate drive is very small. As a result, there is no current concentration during turn-off, a problem of the conventional GTO commutation. Consequently, turn-off capability is greatly improved and snubberless turn-off is possible. The turn-off storage time is decreased to about one-tenth when compared with that of a GTO and its operating frequency can be increased. The small storage time also simplifies the series connection of IGCTs. The gain is only 1, but the gate charge is only half that of the GTO and the gate power is substantially decreased. An integrated annular gate unit on a printed circuit board achieves a very fast gate drive, enabling a current rise as high as 4 kA/μs with a gate cathode voltage of about 20 V. Thus the di/dt withstand can be increased by two to three times that of the GTO and, therefore, the anode reactor size can be reduced considerably.

The symbol and device structure, as shown in Figure 2.19, normally include a reverse diode. In common with other GTO devices, the design minimises the inductance of the gate drive and cathode loop, an important parameter to achieve the required fast rising gate current. Thus high dv/dt is achieved, as well as a greatly increased current turn-off capability.

The storage time, which for the conventional GTO is of the order of 15–20 μs, is reduced to about 1 μs.

During conduction the minimum current needed to switch on the IGCT at low di/dt is typically under 2A (significantly lower than for a conventional GTO).

Any adjustment of the speeds at which anode current and voltage vary during switching, i.e. di/dt and dv/dt, must be either achieved by external components or included in the design of the device. In the IGCT the turn-off dv/dt capability is easily achieved through anode design, but the turn-on di/dt capability is more difficult to achieve.

Figure 2.20 [7] shows a typical IGCT inverter with a di/dt limiting circuit inserted between the quasi-zero impedance supply (DC link) and the inverter. An inductor L limits

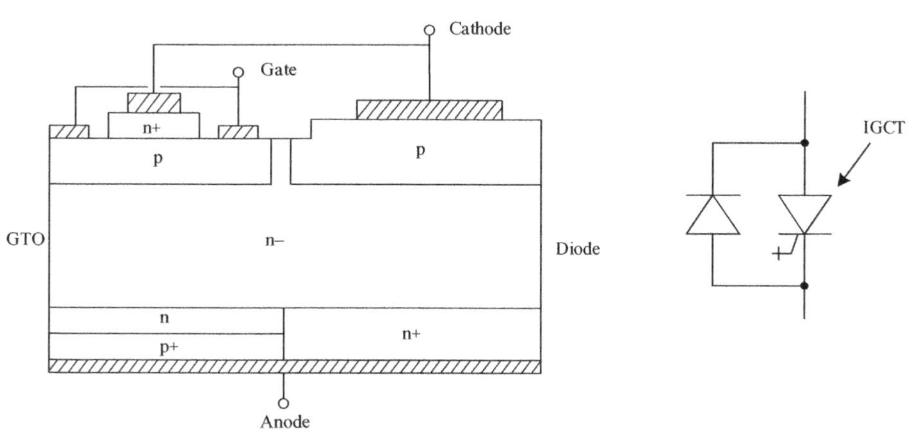

Figure 2.19 IGCT structure including a GCT and a reversed diode

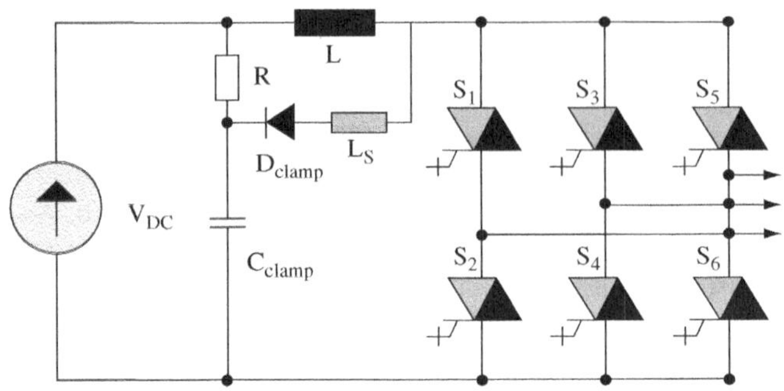

Figure 2.20 A typical IGCT inverter

the rate of rise of current in the inverter when one of the IGCTs is turned on. The limitation of current rise is determined by the allowable rate of fall of current in the associated diode, which in most cases is integrated onto the same wafer. The energy stored in the inductor L, once the diode is commutated, is dissipated in a resistance R and a clamp capacitor (C_{clamp}) can be used to minimise the voltage overshoot. In the event of failure of two devices in one phase, the inductance limits the resulting current to a value $I_{fault} = V_{DC}\sqrt{C/L}$, where C is the capacitance of the DC link.

Electrolytic capacitors and MOSFET switches are used by the gate drivers to provide the required peak gate current at turn-off. As these components reduce the reliability and increase the size of the device, a practical gate driver limits the switching frequency of the IGCT to about 500 Hz.

Like the GTO, the IGCT has high current and voltage ratings; devices with 6 kV/6 kA and 6.5 kV capabilities are already commercially available. As compared with other devices, the IGCT has low on-state voltage and low total power losses (about one-half of those of the conventional GTO) as shown in Figure 2.21 [8]. Although the IGCT has overcome many of the problems of the conventional GTO, the gate driver is still complex and a large linear di/dt limiting inductor is needed for the anti-parallel diode.

Due to its very short storage time (of about 1 μs), which permits small tolerances (under 0.2 μs) in turn-off times of the different devices, the IGCT provides very good voltage sharing and is perfectly suited for the series connection, as required in high-voltage applications.

2.6.4 MOS Turn-Off Thyristor (MTO)

The main aim in the development of the MTO, as compared with the previous thyristor devices, was to achieve a switch with similar power handling but faster turn-off capability, lower turn-off switching losses and a simpler gate drive. The structure of the device is shown in Figure 2.22 [5]. The small turn-off gate drive required by the MTO reduces the cost and improves the reliability of the device, as compared with the conventional GTO and IGCT. The storage time is comparable with that of the IGCT (1–2 μs), which makes it suitable for series operation. Moreover, the MTO structure permits double-sided cooling.

The MTO gate drive requirement for turn-on is similar to that of the GTO. A peak current of 30–60 A is applied from gate to cathode for 5–10 μs; a few microseconds are needed

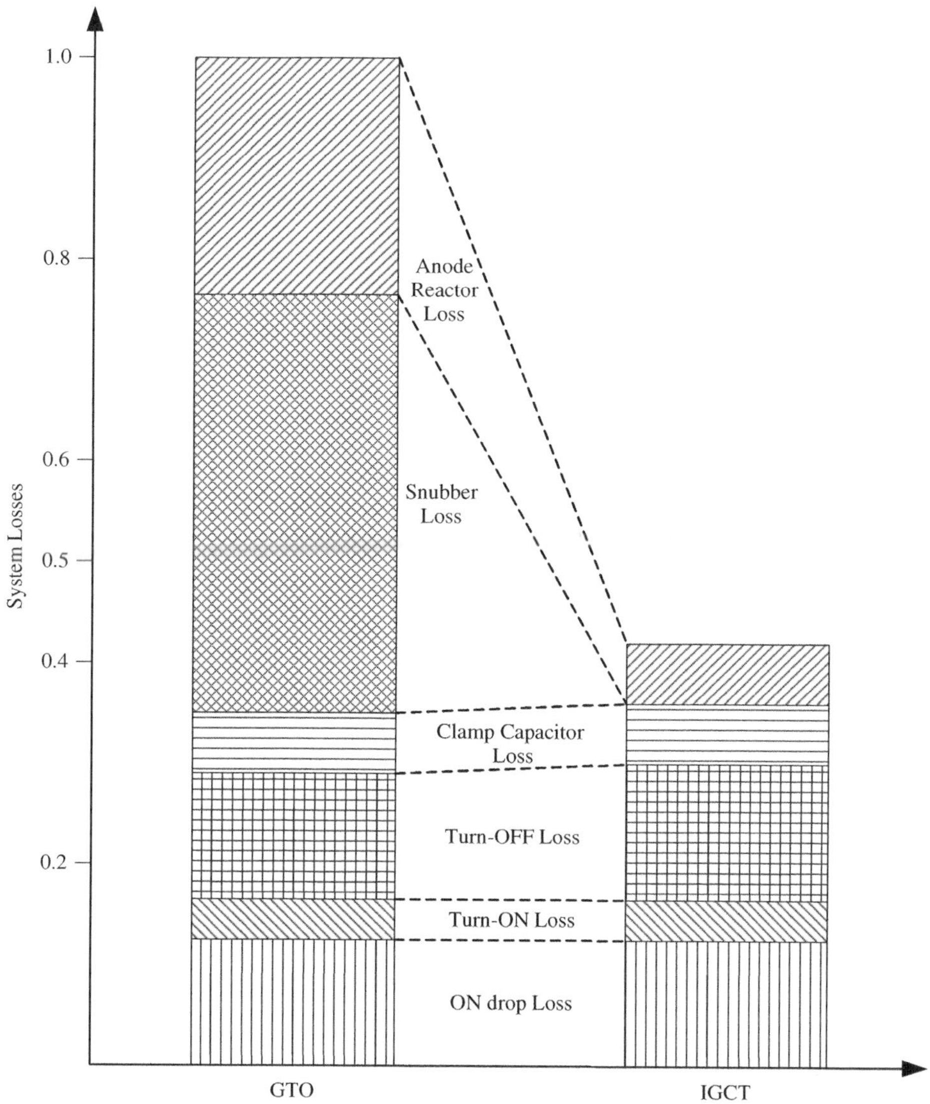

Figure 2.21 Comparison of system loss for a converter using GTO or IGCT (GCT)

before the anode – cathode current begins to rise and the voltage starts to fall. Following full turn-on, some gate current (about 0.5 %) must be maintained to ensure that the gate does not unlatch.

Turn-off requires the application of about 15 V on a second gate terminal which is held continuously during the off-state period.

In the double-gated MTO the MOSFETs are bonded in order to provide high-impedance turn-off gates on both the anode and the cathode sides. This permits the device to operate at much higher frequencies than other power devices (up to 10 kHz).

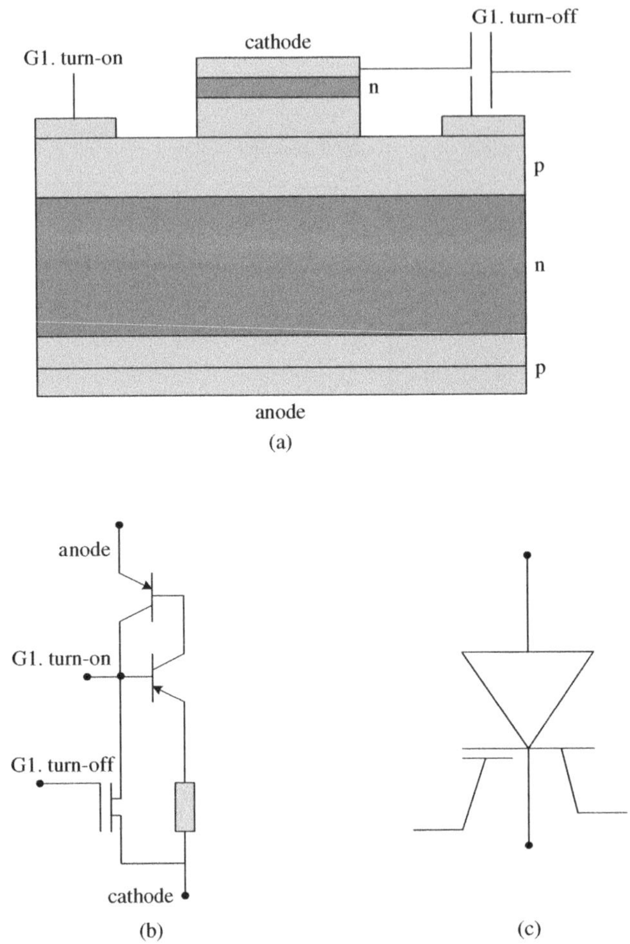

Figure 2.22 MTO structure, circuit and symbol

2.6.5 MOS Controlled Thyristor (MCT)

Again, the purpose of the MCT is to obtain faster turn-off capability and lower turn-off switching losses. The distinguishing feature of the MCT is the incorporation of MOSFETs into a cellular thyristor. Most of these are used to switch emitter shorts to block the forward voltage. These MOSFETs must, therefore, be switched off when the MCT is to be turned on. At the same time a smaller number of MOSFETs of opposite type are switched on.

There are two types of MCT, i.e. the p-MCT, which has a p-type lightly doped layer to take up the off-stage voltage, and the n-type MCT, with an n-type layer as in the conventional thyristor. These are shown in Figure 2.23.

In the case of the n-type MCT, an n-type MOSFET is connected across the cathode of the npn transistor. When this MOSFET is turned on by the application of a positive gate to cathode voltage, a current flows from anode to base in the lower npn transistor; as a result the latter turns on and produces the latched turn-on of the MCT. Similarly, a p-type MOSFET

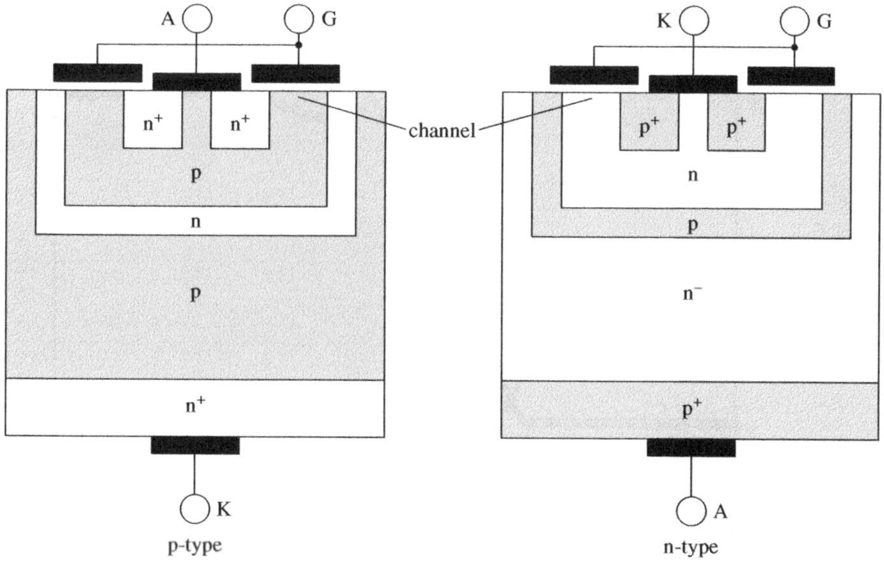

Figure 2.23 MCT structures

connected across the gate–cathode of the cathode side npn transistor is activated to turn the device off; in this case a gate voltage must be applied to the base of the p-MOSFET to turn it off.

The MCT structure uses many single emitter cells which can be turned on very fast. Turn-off, on the other hand, is less straightforward and requires imposing limits for the maximum controllable current. This is due to the relatively low efficiency of the MOSFET shorts; a voltage limit is also imposed due to the device tendency to filamentation. Both these limits lead to amplification of the inevitable lack of current homogeneity effect, which restricts the device application to high powers. The MCT is switched either against high voltage, with the result of a reduction in the controllable current, or at high current, with the voltage level well below that of a comparable thyristor or GTO.

Thus, further development of the MCT is needed before its ideal characteristics, i.e. fast switching with low on-state and switching losses, can be exploited in high-power static conversion.

2.6.6 Emitter Turn-Off Thyristor (ETO)

The ETO, a promising new device for high-power applications, is reported to combine the high-voltage and high-current capability of the traditional thyristor and the easier gate controllability of the IGBT. The present voltage and current ratings are claimed to be 4500 V and 4000 A respectively. As compared with present competing technologies, the ETO offers improved performance in current conduction and turn-off speed, as well as snubberless turn-off current capability. These are shown in Figures 2.24 and 2.25. The switching speed capability is of the order of 1 kHz and the total losses for switching frequencies of 500 and

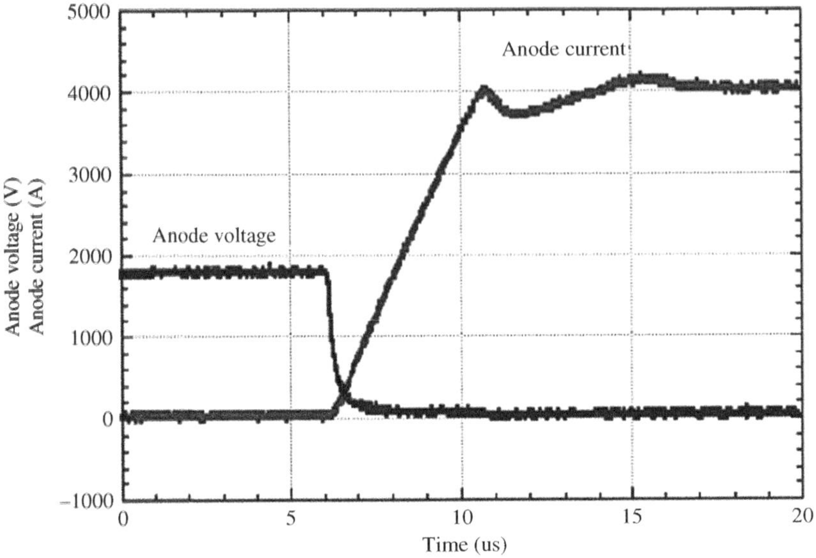

Figure 2.24 ETO turn-on characteristics (Reproduced by permission of CIGRE)

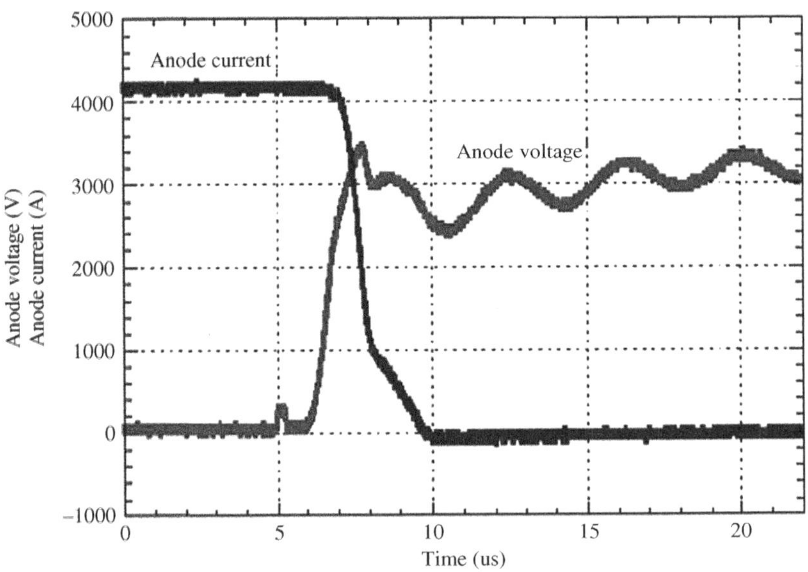

Figure 2.25 ETO snubberless turn-off characteristics (Reproduced by permission of CIGRE)

1000 Hz are similar to those of the IGCT. A modular voltage source converter has been developed to demonstrate these benefits [9].

Due to very high silicon utilisation and use of conventional thyristor technology the ETO is a very cost-effective device. Further development in the device is expected to include the following capabilities: (i) built-in voltage sensing; (ii) current sensing; (iii) temperature sensing; (iv) control power self-generation: (v) high-voltage current saturation.

2.7 Insulated Gate Bipolar Transistor (IGBT)

A development from MOSFET technology for high power ratings, the IGBT is designed with high switching speed and low conduction loss. Its feasibility for high-voltage application was realised in the 1990s, and had already reached 6.5 kV by 2002. Although the IGBT behaves as a bipolar transistor, it has a double transistor structure, like the thyristor. The IGBT combines a high-impedance and low-power gate input with the power handling capacity of the bipolar transistors and thyristors.

As shown in the cross-section of Figure 2.26, the IGBT is achieved by adding a p^+ layer; this layer is referred to as the collector of the IGBT but acts as a p emitter of a pnp bipolar transistor to the MOSFET drain. The bipolar transistor is driven by the electron current from the MOSFET portion, which provides the base current, and holes are injected from the p^+ emitter.

A condition referred to as latch-up can occur at high voltage, when substantial injection of electrons from the source will occur, turning on the two transistor components (npn and pnp). Then the thyristor composed of these transistors will latch on and the IGBT will latch up. For each IGBT geometry there is a critical value of drain current that will activate this condition. Once latch-up occurs, the gate has no control of the drain current and the IGBT turn-off will have to be forced as in a conventional thyristor. However, unlike in the conventional thyristor, this will have to be done quickly to avoid device destruction by excessive power dissipation. The risk of latch-up can be reduced by controlling the doping profiles to ensure that the npn transistor does not turn on.

Control is accomplished by using a pattern of MOS transistors distributed on the surface of the device. As a MOS gated device, the IGBT only needs a small amount of power to be supplied (via a voltage source) to the control gate for full turn-on and turn-off control

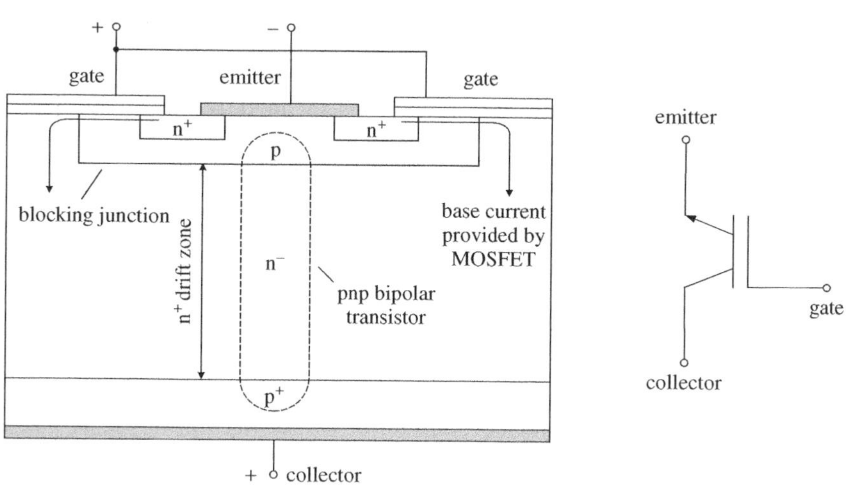

Figure 2.26 IGBT cell structure and symbol

(typically 10 to 15 V). Although the drive circuit still needs enough power to charge and discharge the gate capacitance, the circuitry required for this purpose is small compared with that of other power devices.

The IGBT is turned on by applying a positive base voltage. When the gate is made positive with respect to the emitter, n carriers are drawn into the p channel near the gate region, which forward biases the base of the npn transistor and thus causes the turn-on. Being a cellular device, the IGBT avoids the problem of slow current spreading that occurs in the conventional thyristor.

The peak current in the on-state is limited by the overall transistor gain, i.e. for a given base current the collector current is largely independent of the collector–emitter voltage. This is the case even under short-circuit conditions, provided that the IGBT is switched off within about 10 μs after the occurrence of a fault.

When the gate voltage is removed, the MOSFET is turned off and the base drive current cut off. A space-charge region builds up quickly at the emitter pn junction; therefore, the IGBT can pick up voltage immediately and its turn-off delay is much shorter than the storage time in bipolar transistors.

An attractive feature of the IGBT is the ability to exercise linear control through the gate, i.e. di/dt and dv/dt can be controlled during the commutations. So, unlike the GTO, a di/dt limiting reactor is not required in the IGBT case.

During switching the IGBT must be capable of turning off the peak current, including ripple. Moreover, some additional current rating is needed for current control regulation and protection during transient conditions; in particular, the valve should be able to turn off the current following short circuits. The IGBT has a square turn-off (or true reverse-biased) safe operating area and, therefore, snubbers are not normally required for inductive switching.

The ability to withstand high voltages and currents is provided by the vertical transistor component, which therefore needs to be of sufficient thickness to withstand high voltages. This vertical transistor is the major determining factor in the conductivity of the semiconductor material and therefore the voltage drop over the device in the conducting state.

Recently a new type of IGBT, referred to as the IEGT, has become available that takes advantage of the effect of electron injection from the emitter to achieve a low saturation voltage similar to that of a GTO. Therefore this device achieves a low forward voltage drop during conduction (like the thyristor). The maximum present ratings of the IEGT are 4.5 kV and 2.1 kA.

Figure 2.27 shows the semiconductor components of an IGBT inverter bridge, which for large power ratings may consist of several modules, each comprising an IGBT and anti-parallel free-wheeling diode (FWD). These two components are integrated in the same semiconductor package to provide current capability in the reverse direction and to prevent the application of reverse voltage. Both the IGBT and the diode will normally consist of a number of chips in parallel to obtain the rated current capability. Moreover, the IGBT and the diode must have the same voltage capability. In this arrangement, no ancillary components are needed, the di/dt controlling function being implemented by gate control, whereby the turn-on of the IGBT is slowed to match the allowable turn-off speed of the diode. The absence of the inductance between the inverter and the DC link results in extremely high fault currents in the event of simultaneous failure of two devices in the same phase. Thus the applicability of the topology shown in Figure 2.27 is limited to low voltages (of about 1.5 kV). At higher voltages decoupling of link and inverter via inductance or resistance is required to limit the collateral explosion damage due to the electromagnetic force caused by the fault current.

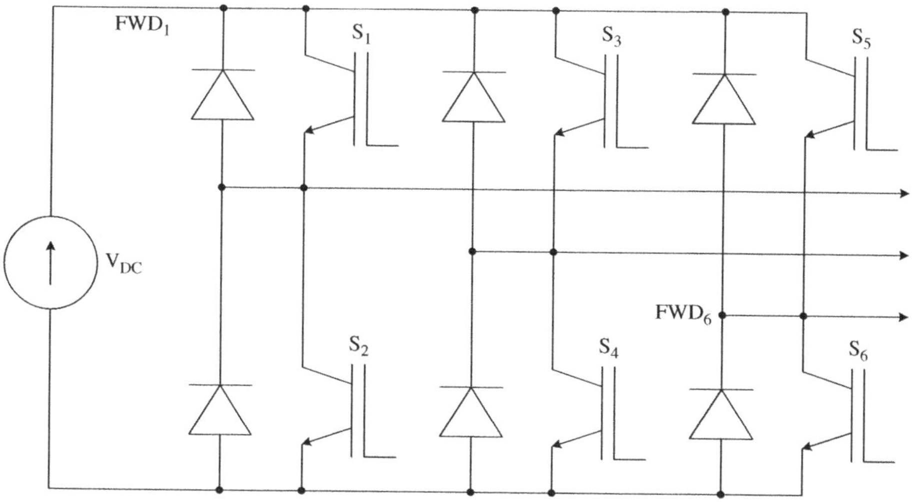

Figure 2.27 Typical IGBT inverter

2.7.1 IGBT (Series) Chains

As an MOS device with a high-impedance gate, the IGBT requires low energy to switch the device and maintain power flow. This facilitates the series connection by providing good voltage distribution even at high switching frequencies (in kilohertz). Moreover, the gate power for the series-connected semiconductors can be provided (as in the conventional HVDC valves) via the voltage dividing circuits across the individual IGBTs.

In high-voltage applications, many IGBTs are connected in series and redundant devices are included. Valves with up to 200 series-connected IGBTs have already been used in HVDC installations. By appropriate servicing at regular intervals only a short circuit caused by a direct lightning strike would cause simultaneous failure of all redundant devices of a phase leg, and, therefore, the link inductance can be omitted. A faulty IGBT cannot be allowed to create an open circuit, since the valve must continue operating. Instead, the faulty device must be capable of conducting current until the next scheduled maintenance period. However, standard IGBT modules do not have the required short-circuit capability, as they use bond wires that open the circuit upon failure. Therefore specially developed press pack designs need to be used for DC transmission valves [10–12].

In the series stacks, the parallel-connected IGBT and diode chips are pressure contacted, with each contact designed to take full load current should a chip fail.

Static and dynamic voltage sharing

Unbalanced voltage sharing, particularly during switching transitions, leads to higher switching losses. Adequate static voltage sharing of series IGBTs is achieved by employing a parallel-connected resistor per IGBT module. The resistor function is to provide a leakage current which is dominant over the IGBT's, thus forcing sharing to occur by potential divider action.

All IGBTs experience a change in the turn-on delay time with temperature, which varies from device to device. The turn-on time variation is also affected by rise time, i.e. the slower turning-on devices will have a greater share of the total arm voltage. IGBTs also experience turn-off variation times, depending on temperature and commutated current levels, which vary from device to device.

In the IGBT series chain, small differences in the region of 10 nanoseconds are critical to voltage sharing and, thus, active voltage control of the gate driver circuit is often used to ensure voltage balancing, both statically and during switchings. However, in addition to active voltage control, a small *RC* snubber for damping oscillations is normally employed, as shown in Figure 2.28. The use of a snubber capacitor helps to achieve dynamic voltage sharing (due to spread in turn-off delays) while a di/dt inductor helps the IGBTs to share

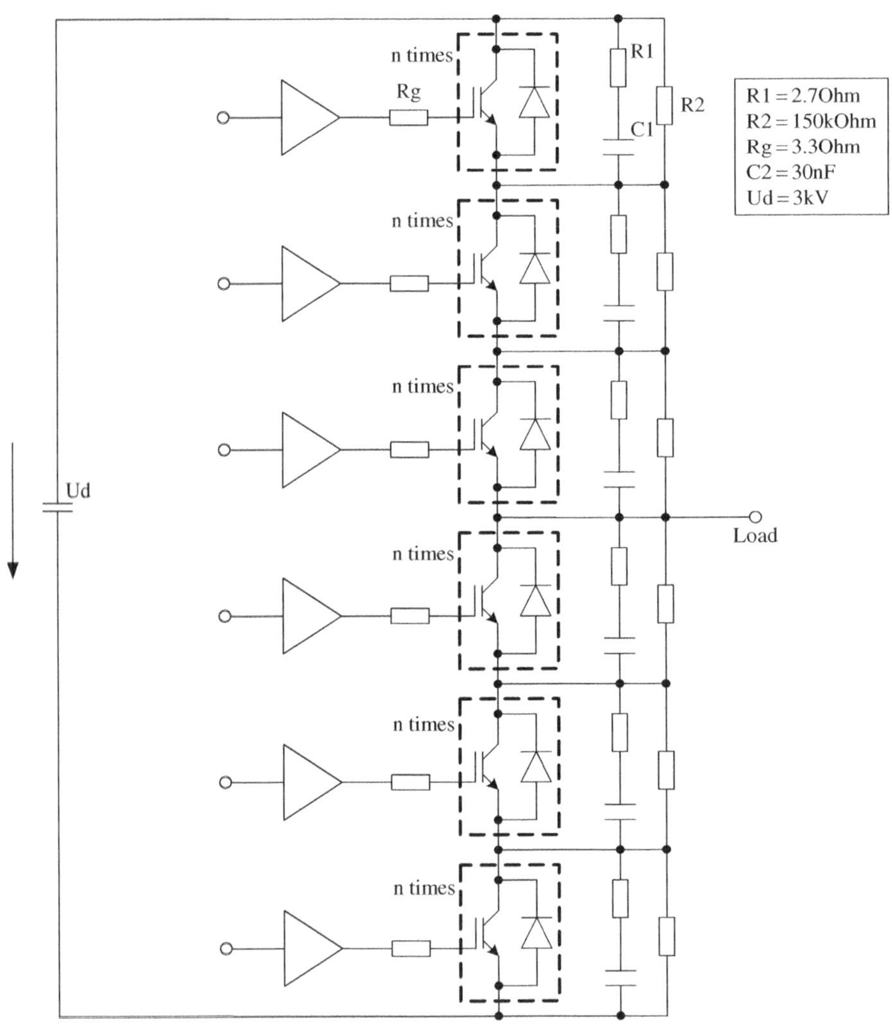

Figure 2.28 Circuit diagram of an IGBT chain phase module

voltage (due to spread in turn-on delays) and diode voltage sharing under reverse recovery conditions. A *di/dt* limiting inductor reduces the turn-on losses substantially.

2.8 Diodes

Diodes are extensively used as part of the static conversion process, in most cases in an auxiliary capacity. Their main application is across the turn-off devices in voltage source conversion; they are also used as clamping devices in multi-level conversion and in snubber and gate drive circuits.

As indicated earlier, the diode plays an important part in the topology chosen, by determining the allowable turn-on *di/dt* of a voltage source inverter and, hence, constituting the principal part of the turn-on losses, which will be dissipated in the silicon or in the circuit resistance. Its effect is illustrated in Figure 2.29, which shows the general turn-on characteristics for the switches of the IGCT and IGBT inverters of Figures 2.20 and 2.27. To be effective as part of a static converter circuit, the diode's turn-off, i.e. the current following commutation, must decrease smoothly to avoid overvoltages in the inductive circuit. Also the turn-on voltage should be low, to ensure a fast flooding of the depletion zone with carriers, so that the diode begins conduction before the voltage builds up too much.

Finally, the static and dynamic losses should be low. In each application, a compromise needs to be made between these contradictory requirements. A low-voltage turn-on requires a high carrier lifetime and a thin, lightly doped centre region. A high carrier lifetime provides soft recovery, because the reverse current does not break down sharply; this, however, produces high switching losses as these increase with the carrier lifetime. Conversely, the on-state losses reduce with increased carrier lifetimes. Thus the trade-off depends on the particular application; for instance, when high switching frequencies are used, it is advisable to reduce the switching losses at the expense of the on-state losses.

An ideal device for fast switching applications is the Schottky diode, obtained by depositing a metal film onto a lightly doped semiconductor. In this device the turn-on and turn-off are much faster, due to the lack of minority carriers. As the forward voltage drop of

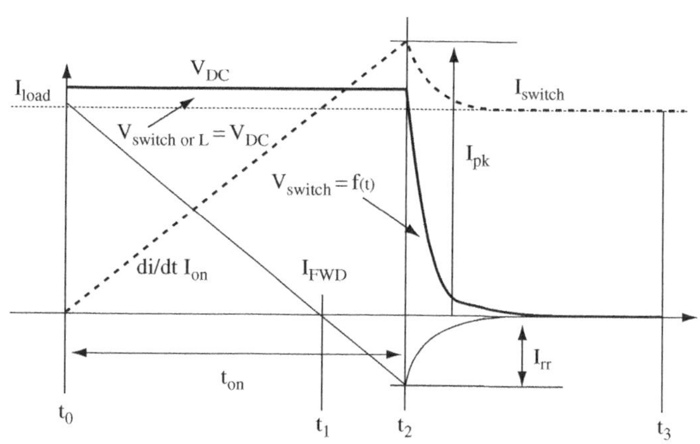

Figure 2.29 Turn-on characteristics for the inverters of Figures 2.20 and 2.25

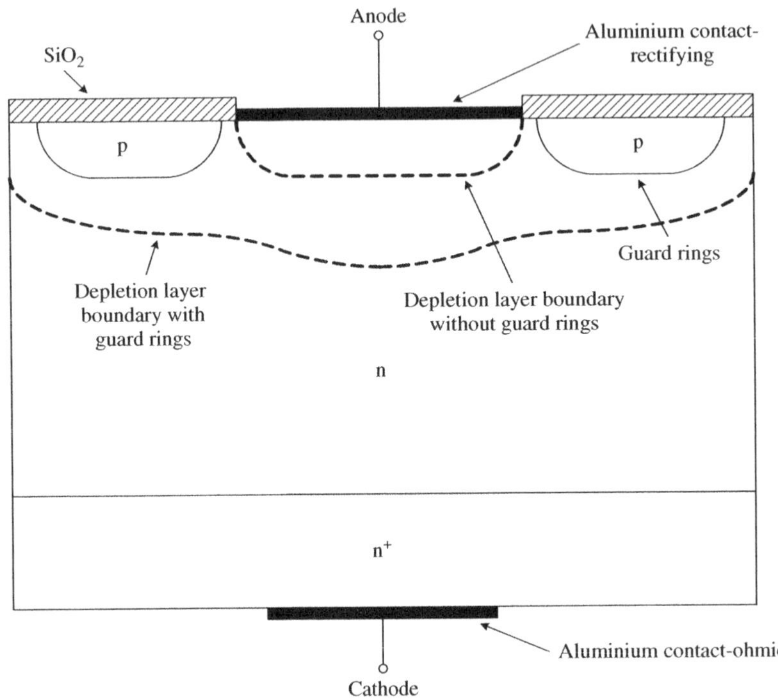

Figure 2.30 Cross-section of a Schottky diode

a Schottky contact is very low (typically 0.3 V), the forward losses are substantially reduced as compared with those of the pn diode. A cross-section of the Schottky diode is shown in Figure 2.30, which illustrates (dashed line) the use of guards rings to improve the breakdown voltage capability of the device. However, at present the blocking voltages are still too low (e.g. under 200 V for the silicon Schottky diode).

Usually, as the current falls to zero following conduction, the voltage across the diode jumps to a negative value; this causes a temporary flow of reverse current to remove the excess of internal charge and restore the depletion layer, as required by the applied reverse voltage. The reverse current flow in the diode increases the current that the turn-off devices have to turn on in voltage source converters and this results in increased turn-on losses. The eventual development of silicon carbide diodes (with 100 times lower switching losses) should overcome the turn-on loss limitations of choke-less topologies and make this device a practical proposition for IGBT applications.

For HVDC application, power diodes must be adequately sized to support the maximum current following short circuits and withstand the related stresses because, unlike the IGBTs, they cannot be switched off. In the case of a DC side short circuit there is a path through the diodes and the current is only limited by the short-circuit impedance of the AC network and the reactance of the converter (normally the leakage reactance of the converter transformer). The diodes will also experience a surge inrush current during energisation through the AC breaker; in this case an overvoltage will occur on the DC bus, which has to be limited by external components, such as pre-insertion breaker resistors.

Power diodes are currently produced commercially for typical voltage and currents of 6 kV and 6 kA.

2.9 Prognostic Assessment

2.9.1 Ratings and Applicability

At present the power semiconductor devices available for high-power applications are based on silicon technology and they can be broadly classified in two groups.

(1) One group includes devices with four-layer, three-junction monolithic structures, the two early devices in this category being the SCR and GTO. The devices in this group have low conduction losses and high surge and current carrying capabilities; they operate only as on/off switches.

Recent developments in this group are the MCT, ETO, MTO and GCT or IGCT. These recent devices were developed to provide fast turn-off capability and low turn-off switching losses.

The IGCT has the lowest total loss (including both device and peripheral circuits) of all present power semiconductors. On the other hand, the SCR remains by far the most cost-efficient device for very high-power application. The SCR power ratings presently available are typically 12 kV/1.5 kA and 8.5 kV/5 kA.

The majority of commercially available GTOs are of the asymmetrical type; they are reversed connected to a fast recovery diode, such that the GTO does not require reverse voltage capability. Asymmetrical GTOs have been used extensively in PWM two- and three-level VSCs, active filters and custom power supplies. However, current development is focusing on higher blocking voltages, hard switching (also called snubberless) ring gate (to reduce the size of the snubber and switching losses), anode buffer layer, thinner wafer, absence of emitter shorts and improved switching performance.

The ratings are changing fast and therefore any comparisons made must indicate the date and source of the information. For instance, a 1999 published IEE review [5] of typical maximum ratings (Figure 2.31) shows that the GTO offered the best maximum blocking voltage and turn-off current ratings; these were 6 kV and 4 kA, respectively, the switching frequency being typically under 1 kHz. More recently, however, the industry has concentrated on the IGCT developments, for which the maximum ratings quoted at the time of writing are 6 kV/6 kA.

(2) The second group contains devices of three-layer, two-junction structure which operate in switching and linear modes; they have good turn-off capability. These are: the bipolar junction transistor (BJT), Darlington transistor, MOSFET, injection enhanced transistor (IEGT), carrier stored trench-gate bipolar transistor (CSTBT), static induction transistor (SIT), field controlled transistor (FCT) and IGBT.

There is little manufacturing enthusiasm for developing further some of the devices in the second group, because of the perceived advantages of the IGBT; at present the voltage and peak turn-off currents of the silicon-based IGBTs are 6.5 kV and 2 kA. A new type of IGBT (referred to as IEGT) has become available that takes advantage of the effect of electron injection from the emitter to achieve a low saturation voltage similar to that of the GTO.

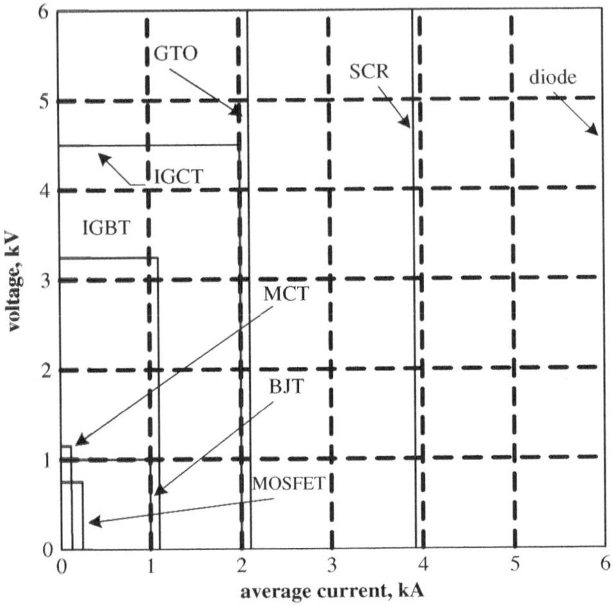

Figure 2.31 Voltage and current ratings of the main power semiconductors (Shakweh, Y. (1999), 'Power devices for medium voltage PWM converters', *Power Engineering Journal*, December, reproduced by permission of the IET.)

The main candidates for high-power conversion (as required in HVDC applications) appear to be the IGCT and IGBT. Compared with the GTO and IGCT types, the IGBT requires much less gate power and has considerably superior switching speed capability, therefore permitting the use of higher switching frequencies (these can be typically 3 kHz for soft-switched devices compared with 500 Hz for the IGCT). While high switching frequencies have some disadvantages (like switching and possibly snubber losses), they help to reduce the harmonic content and therefore filtering arrangements, reduce machine losses and improve the converter dynamic performance.

The IGBT is more reliable than the IGCT under short-circuit conditions. It is designed to sustain a current surge during conduction and also during turn-on. However, short-circuit faults need to be detected quickly so that turn-off is achieved within 10 μs. The IGCT on the other hand has no inherent current limiting capability and must be protected externally. The overall reliability of the IGCT for high-power and high-voltage application is very impressive, given the reduced voltage stress achieved by the series connection and its potential to eliminate the need for snubbers.

HVDC converters require high blocking voltage switches, which can only be achieved by series connection of individual units. Both the IGBT and IGCT are suitable for series connection as the turn-on and turn-off times are relatively small and the switching speeds high. Reliable static and dynamic voltage sharing techniques are now available for thyristor and transistor types of switching devices. Parallel sharing resistors are perfectly adequate for static balancing. In the case of the IGBT good dynamic voltage sharing can also be achieved by means of adaptive gate control of the individual units.

The current ratings of present devices are sufficient large for high-voltage applications and the use of device paralleling is rarely required. The IGBT technology is suited, however, for parallel operation since the high-current IGBT modules themselves consist of many parallel chips.

Heavy investment in IGBT technology is favouring this device at the expense of the GTO and IGCT alternatives. The availability of press pack IGBTs at high voltages and currents is strengthening their position in high-voltage applications, where series operation and redundancy of power switches are required. For high-voltage applications, however, there are still problems with the IGBT due to stray inductances and diode reverse recoveries and, thus, at this stage it is not clear to what extent the fast switching capability should be exploited, as the resulting voltage spikes may exceed the allowable limits.

In the future, new wide-band-gap (WBG) semiconductor materials such as silicon carbide (SiC) and gallium nitride (GaN), instead of silicon, should increase the power handling capability and switching speed. The best candidate at the moment seems to be SiC, a material that provides low on-state voltage, low recovery charge, fast turn-on and turn-off, high blocking voltage, higher junction temperature and high power density. In particular this material permits a substantial increase in the allowable peak junction temperature, thus improving the device surge capability and reducing the complexity of the cooling system. With the utilisation of SiC unipolar power switches it is possible to reduce the power losses by a factor of 10. A recent forecast of the future voltage ratings achievable with SiC switching devices (in relation to those of present silicon technology) is shown in Figure 2.32.

2.9.2 Relative Losses

The IGBT turn-off losses are lower than those of the SCR and IGCT and so are the turn-on losses in the case of soft-switched IGBTs. The forward voltage drop of the IGBT is, however, much higher than that of a thyristor of comparable voltage rating.

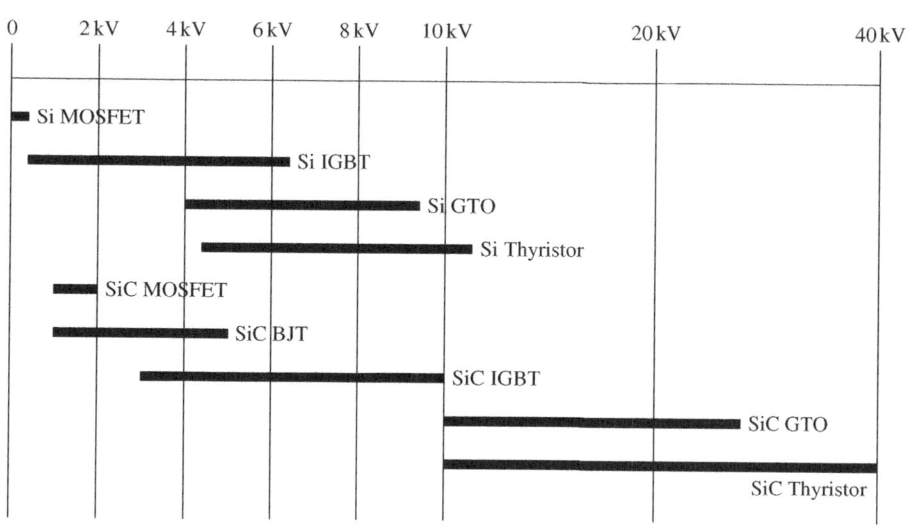

Figure 2.32 Silicon carbide (SiC) switches compared with silicon (Si) Switches (Reproduced by permission of CIGRE.)

In the assessment of the energy loss of a converter, the most important factor is the frequency used for the switching of the valves, which depends on the type of configuration and control. To illustrate this point a comparison between different alternative converters of the LCC and VSC types has produced the following figures for the power loss of the complete converter station [13]:

- An IGBT-based two-level VSC with a PWM frequency of 1950 Hz has a power loss of approximately 3 %.
- An IGBT-based three-level VSC with a PWM frequency of 1260 Hz has a power loss of approximately 1.8 %.
- The loss figure for an SCR-based LCC station (including valves, filters and transformers) is 0.8 %.

References

1. Ghandhi, S.K. (1983), *Semiconductor Power Devices*, John Wiley & Sons, Inc., New York.
2. Baliga, B. (1996), *Power Semiconductor Devices*, PWS, Boston.
3. Blicher, A. (1976), *Thyristor Physics*, Springer-Verlag, New York.
4. Lips, P. (1996), 'Semiconductor power devices for use in HVDC and FACTS controllers', CIGRE WG 14.17.
5. Shakweh, Y. 'New breed of medium voltage converters', *IEE Power Engineering Journal*, (1999), December, 297–307 and (2000), February, 12–20.
6. Mohan, N., Undeland, T.M. and Robbins, W.P. (2003), *Power Electronics, Converters, Applications and Design*, Third Edition, John Wiley & Sons Inc., New York.
7. Carroll, E.I. (1999), 'Power electronics for very high power applications', *IEE Power Engineering Journal*, April, 81–87.
8. CIGRE Working Group B4-33 (2005), 'HVDC and FACTS for distributions systems', Paris.
9. Huang, A.Q., Chen, B., Han, C., Du, Z., Batthacharya, S., Baran, M., Edris, A., Ingram, M. and Atcitty, S. (2006), 'Emitter turn-off (ETO) thyristor: an emerging power semiconductor switch with lower cost, improved reliability and scalability of FACTS controllers', CIGRE paper B4-107, Paris.
10. Kaufmann, S. and Zwick, F. (2002), '10 kV IGBT press pack modules with series connected chips', Proceedings of the 14th International Symposium on Power Semiconductor Devices and ICs, IEEE ISPSD'02, 89–92.
11. Kaufmann, S., Lang, T. and Chokhawala, R. (2001), 'Innovative press pack modules for high power IGBTs', Proceedings of the 13th International Symposium on Power Semiconductor Devices and ICs, IEEE ISPSD'01, 59–62.
12. Fujii, T., Yoshikawa, K., Koga, T., Takahashi, Y., Kakiki, H., Ichijyou, M. and Seki, Y. (2000), '4.5 kV-2000 A power pack IGBT (ultra high power flat-packaged PT type RC-IGBT)', Proceedings of the 12th International Symposium on Power Semiconductor Devices and ICs, IEEE ISPSD'00.
13. CIGRE Study Committee B4-WG 37 (2005), 'VSC transmission', Paris.

3

Line-Commutated HVDC Conversion

3.1 Introduction

The use of thyristor control in VSC would need very large series impedance to accommodate rapid changes in the AC supply voltage. This solution, however, is uneconomical in terms of reactive power compensation requirements and therefore CSC is, instead, the only practical option for conventional LCC technology.

LCC, although still the most widely used HVDC transmission alternative, has very limited flexibility due to the use of thyristor switching. The lack of turn-off controllability of the conventional thyristor results in poor power factors and considerable waveform distortion. Thus, while the LCC configuration is very simple, the external plant required for reactive power compensation and filtering is elaborate and expensive.

The three-phase bridge converter is the basic configuration adopted for large-power applications. Compared with other possible alternatives, such as the three-phase double star or the six-phase diametrical connections, the bridge configuration provides better utilisation of the converter transformer and a lower peak inverse voltage across the converter valves. In the bridge converter two valves are connected to each phase terminal, one with the anode and the other with the cathode connected to it. Although the conducting path of the bridge configuration uses two valves in series, this does not increase the power loss because, due to the high voltage involved, the valves of HVDC converters consists of many series-connected switches.

While the main object of this chapter is the structure and operating characteristics of the three-phase AC–DC converter, special attention is given to the subject of converter/AC system harmonic interaction, a problem of particular importance in LCC.

3.2 Three-Phase AC–DC Conversion [1]

The main component of an AC–DC converter for HVDC application is the line-commutated three-phase (six-pulse) thyristor bridge, which, as explained in the previous section, is

Flexible Power Transmission: The HVDC Options J. Arrillaga, Y.H. Liu and N.R. Watson

designed to operate in CSC. This implies a current stiff DC side, achieved by the connection of a large series reactor. As the AC system has substantial inductance, the interface requires the provision of shunt capacitors at the converter AC side terminals; the installed capacitance is normally in the form of passive filters (which for all the frequencies below the tuned frequency will be capacitive), but often some extra capacitor banks have to be added to provide the required reactive power compensation. The bridge thyristors must block voltages of both polarities but only conduct current in one direction.

The structure of the high-voltage large-power converter required for use in HVDC transmission is described in greater detail in Chapter 8.

3.2.1 Basic CSC Operating Principles

The operation of the three-phase bridge rectifier is first explained for the ideal case, where the bridge is connected to an infinitely strong AC side (i.e. of zero source impedance) and to a DC side with infinite smoothing reactance. Under these conditions, the transfer of DC (commutation) between valves on the same side of the bridge takes place instantaneously. The switching sequence and the rectified voltage waveform are illustrated in Figure 3.1 for the case of an uncontrolled bridge rectifier (i.e. on diode operation); valves 1, 3, 5 at the top and 4, 6, 2 at the bottom are connected to phases red, yellow and blue, respectively.

With reference to Figures 3.1(a) and (g), at instant A phases R and Y are involved through conducting valves 1 and 6. This operating state continues up to point B, after which valve 2 becomes forward biased, since its anode, directly connected to that of valve 6, is positive with respect to its cathode (connected to phase blue); therefore at point B the current commutates naturally from valve 6 to valve 2 (Figure 3.1(b)).

A similar argument applies at point C, with reference to valves 1 and 3 on the upper half of the bridge. The anode of valve 3 (connected to Y phase) begins to be positive with respect to its cathode (connected to phase R through the conducting valve 1) and a commutation takes place from valve 1 to valve 3 (Figure 3.1(c)). This is followed by commutations from valve 2 to valve 4 at point D, valve 3 to valve 5 at point E, valve 4 to valve 6 at point F, and valve 5 to valve 1 at point G. This completes the switching cycle sequence.

The DC output waveform in Figure 3.1(g) shows the voltage variation of the positive (common cathode) and negative (common anode) poles with respect to the transformer neutral. Figure 3.1(h) shows the DC output voltage, i.e. the voltage of the positive pole with respect to the negative pole. It is seen that the output voltage has a ripple varying at six times the main frequency.

Each valve carries the full value of DC for one-third of the cycle, and there are always two valves conducting in series.

3.2.2 Effect of Delaying the Firing Instant

On the assumption that the converter terminal voltage is balanced and undistorted (i.e. perfectly filtered), by delaying the firing instants of the valves with respect to the voltage crossings, the commencement of the natural commutations described in Section 3.2.1 can be delayed by a definite time interval. The effect of this action on the direct voltage waveforms is illustrated in Figures 3.2(a) and (b). It is noticeable that the voltage area, and therefore the mean direct voltage, are reduced proportionally to the magnitude of the delay.

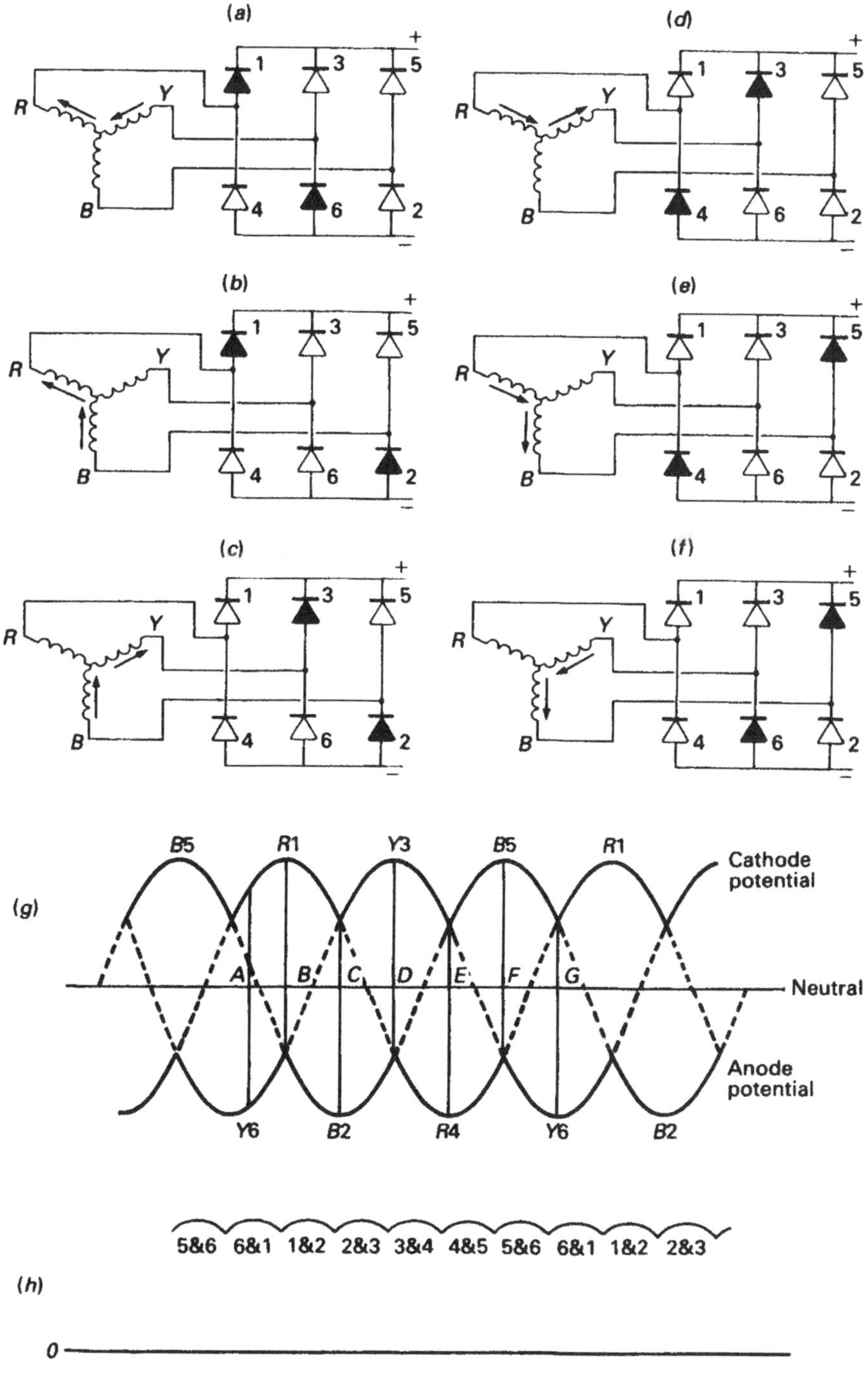

Figure 3.1 Bridge conducting sequence and DC voltage waveforms

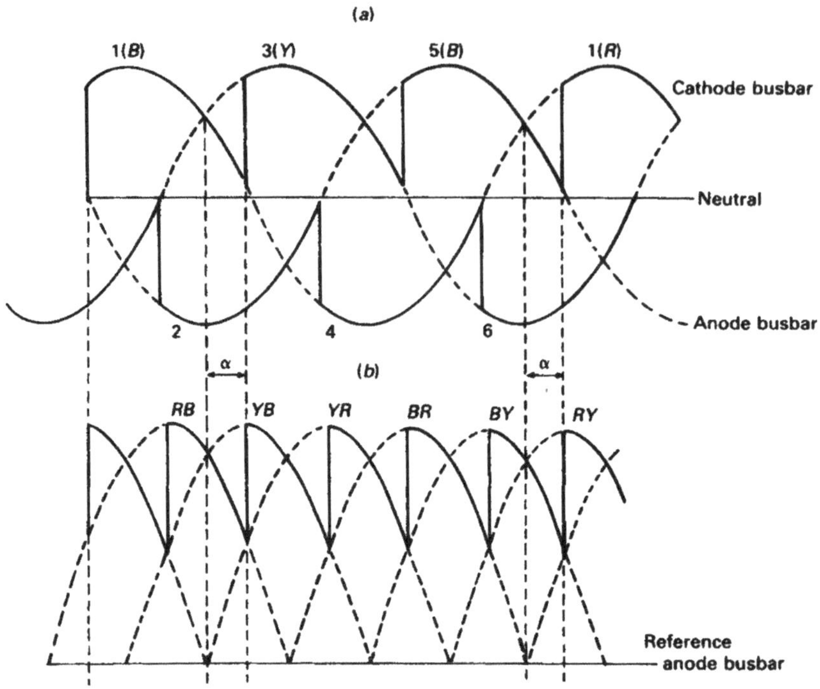

Figure 3.2 Effect of firing delay on voltage waveforms: (a) common anode and common cathode voltages; (b) direct voltage

In the presence of a large smoothing reactor on the DC side, the voltage waveform of Figure 3.2(b) will produce constant DC, the level of which depends on the mean voltages at both ends of the link and the link resistance. For the idealised commutating conditions described in Section 3.2.1, the valve current will be a rectangular pulse lasting 120°, its relative position with reference to the corresponding voltage waveform being determined by the firing delay angle (α).

For delay angles above 60° some negative voltage periods begin to appear. If the bridge output were connected to a pure resistance, the bridge unidirectional current conduction property would prevent reverse current flow during these negative voltage periods, and the operation would then be intermittent. However, the provision of a large smoothing reactor maintains positive current flow during the negative periods, and energy is transferred from the reactor magnetic field to the AC system. The voltage waveforms for a delay of 90°, illustrated in Figures 3.3(a) and (b), show equal positive and negative voltage regions (indicated by horizontal and vertical shaded areas, respectively); the mean direct voltage is therefore zero with a 90° delay.

When the delay angle is larger than 90° the mean voltage is negative and the bridge operation can only be maintained in the presence of a DC power supply. This supply overcomes the negative voltage and forces the current to conduct in the same direction (i.e. from anode to cathode), in opposition to the induced emf in the converter transformer. This indicates that power is being supplied to the AC system, i.e. the converter is inverting. Figures 3.4(a) and (b) illustrate the (ideal) limiting case of full inversion, which would require a delay angle of 180°.

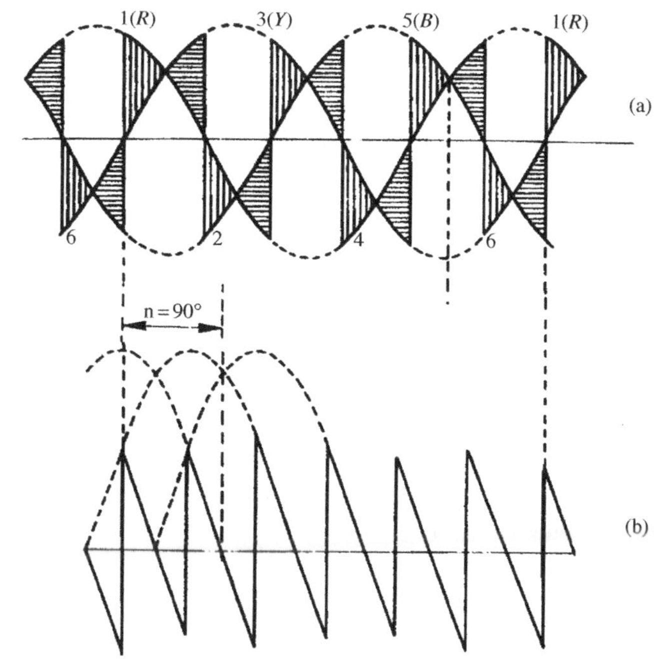

Figure 3.3 Voltage waveforms with 90° firing delay: (a) common anode and common cathode voltages; (b) direct voltage

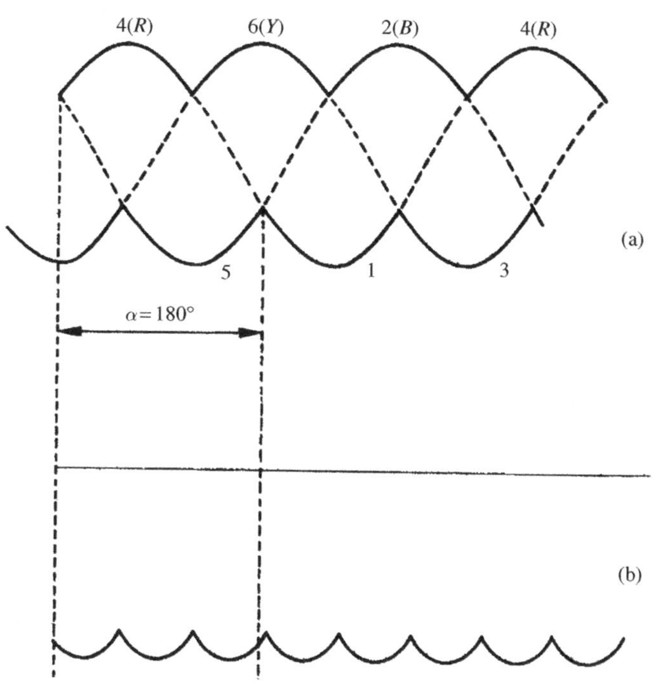

Figure 3.4 Voltage waveforms on full inversion: (a) common anode and common cathode voltages; (b) direct voltage

3.3 The Commutation Process

The zero-impedance supply required to produce the voltage and current waveforms described in Section 3.2 never exist in practice. Even if the AC system impedance were negligible, there would always be some impedance between the converter and the AC system due to the converter transformer leakage reactance. Although in theory converter transformers are not essential to the static power conversion process, there are practical reasons for their presence; among them, the possibility of phase shifting multiple bridges and the availability of on-load tap-changing, which reduces the reactive power compensation requirements. The main effect of AC system reactance is to reduce the rate of change of current or, in other words, to lengthen the commutating time.

During the commutation, the magnetic energy stored in the reactance of the previously conducting phase has to be transferred to the reactance of the incoming phase. That energy only depends on the DC current level and the inductance per phase. The speed of the commutation process, and therefore the rate of change of current, are also affected by two other parameters, i.e. the supply voltage and the firing delay angle.

Before analysing the commutation process, it is necessary to define the terms commutating voltage and commutation reactance.

The commutating voltage waveforms are those appearing on the DC line during the periods when no commutations are taking place. In these operating regions only DC current flows through the AC system impedance, and the voltage waveform is therefore sinusoidal.

Considering the AC waveform distortion produced by the converter, it will be necessary to go back to the system source voltage to find an undistorted supply to the converter. In practice, however, converter transformer phase shifting and filtering are provided with every converter station and the voltage waveform at the filter busbar is reasonably sinusoidal (under steady-state and normal operating conditions). Such voltage can therefore be used as the commutating voltage. The commutation reactance is defined as the reactance between the commutating voltage and the converter valves.

3.3.1 Analysis of the Commutation Circuit

Figure 3.5 illustrates the commutation process between valves 1 and 3 of a converter bridge, connected to a system with a source voltage v_c, a commutation reactance per phase X_c and negligible source resistance.

With reference to Figure 3.5(a), the commutation from valve 1 to valve 3 can start (by the firing of value 3) any time after the upper voltage crossing between v_{CR} and v_{CY} and must be completed before the lower (negative side) crossing of these two voltage waveforms. Beyond that point $v_{CY} > v_{CR}$ and therefore a commutating current i_c $(= i_3)$ builds up at the expense of i_1 so that at all times $i_1 + i_3 = I_d$, as shown in Figure 3.5(c).

If the commutation reactances are balanced, the rates of change of i_3 and $-i_1$ are equal; therefore, the voltage drops across X_{CR} and X_{CY} are the same and thus, during the overlap period, the direct voltage v_d is the mean value of v_{CY} and v_{CR}. The voltage waveforms for early (rectification) and late (inversion) commutations are shown in Figure 3.5(b).

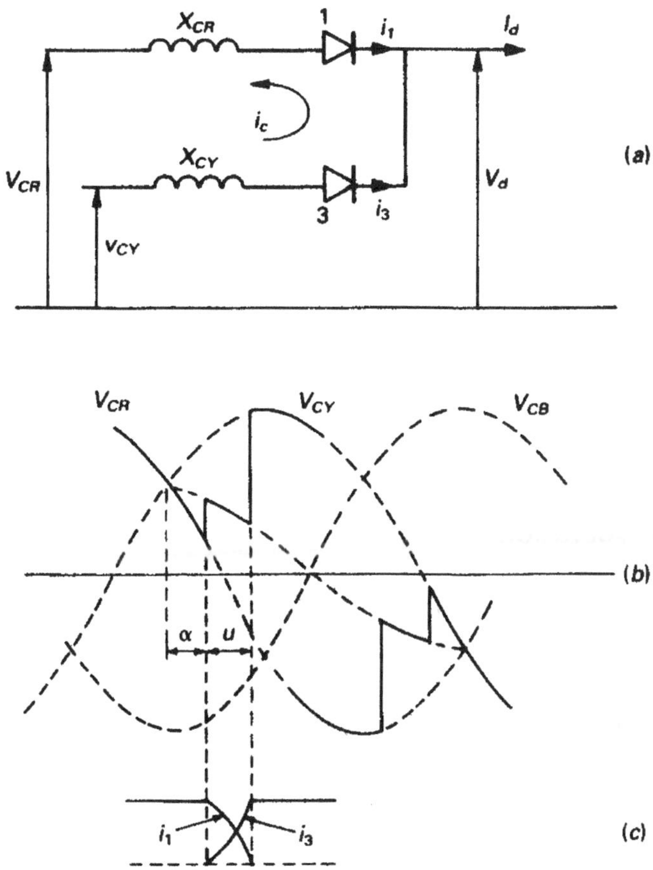

Figure 3.5 The commutation process: (a) equivalent circuit; (b) voltage waveforms showing an early (rectification) and late (inversion) commutations; (c) current waveforms

From the circuit of Figure 3.5(a), and assuming $X_{CR} = X_{CY} = X_C$, the following analysis applies:

$$v_{CY} - v_{CR} = 2(X_c/\omega)d(i_c)/dt \tag{3.1}$$

and using as a reference the voltage crossing between phases R and Y,

$$v_{CY} - v_{CR} = \sqrt{2}V_c \sin(\omega t)$$

where V_c is the phase-to-phase rms voltage.

Thus (3.1) can be written as

$$\frac{1}{\sqrt{2}}V_c \sin(\omega t)d(\omega t) = X_c di_c \tag{3.2}$$

Integrating from $\omega t = \alpha$,

$$\frac{1}{\sqrt{2}}\int_{\alpha}^{\omega t} V_c \sin(\omega t)d(\omega t) = X_c \int_0^{i_c} d(i_c) \tag{3.3}$$

the instantaneous expression for the commutating current becomes

$$i_c = \frac{V_c}{\sqrt{2}X_c}[\cos(\alpha) - \cos(\omega t)] \tag{3.4}$$

and substituting the final condition, i.e. $i_c = I_d$ at $\omega t = \alpha + u$, yields

$$I_d = \frac{V_c}{\sqrt{2}X_c}[\cos(\alpha) - \cos(\alpha + u)] \tag{3.5}$$

3.4 Rectifier Operation

The description and analysis of the conversion process (both for rectifier and inverter operation) are made here with reference to a perfectly symmetrical system voltage source. A more detailed analysis under unbalanced conditions can be found in [2].

Typical voltage and current waveforms of a bridge operating as a rectifier, with the commutation effect included, are shown in Figure 3.6, where P indicates a firing instant (e.g. P_1 is the firing instant of valve 1), S indicates the end of a commutation (e.g. at S5 valve 5 stops conducting), and C is a voltage crossing (e.g. C1 indicates the positive crossing between phases blue and red).

Figure 3.6(a) illustrates the positive (determined by the conduction of valves 1, 3, 5) and negative (determined by the conduction of valves 2, 4, 6) potentials with respect to the transformer neutral. Figure 3.6(b) shows the direct voltage output waveform.

The potential across valve 1, also shown in Figure 3.6(b), depends on the conducting valves. When valve 1 completes the commutation to valve 3 (at S1) the voltage across will follow the red–yellow potential difference until P4. Between P4 and S2 the commutation from valve 2 to valve 4 (shown in Figure 3.6(a)) reduces the negative potential of phase red and causes the first voltage dent. The firing of valve 5 (at P5) increases the potential of the common cathode to the average of phases yellow and blue; this causes a second commutation dent, at the end of which (at S3) the common cathode follows the potential of phase blue (due to the conduction of valve 5). Finally, the commutation from valve 4 to valve 6 (between P6 and S4) increases the negative potential of valve 1 anode and produces another voltage dent.

Figures 3.6(c) and (d) illustrate the individual valves (1 and 4) and Figure 3.6(e) the phase (red) currents, respectively.

A number of reasonable approximations have to be made to simplify the derivation of the steady-state equations that follow: the converter valves are treated as ideal switches; when calculating the power loss, the valve resistance can be added to that of the DC transmission line; the AC systems consist of perfectly balanced and sinusoidal emfs; and the commutation reactances are equal in each phase and their resistive components are ignored; the DC current is constant and ripple-free, i.e. the presence of a very large smoothing reactor is assumed.

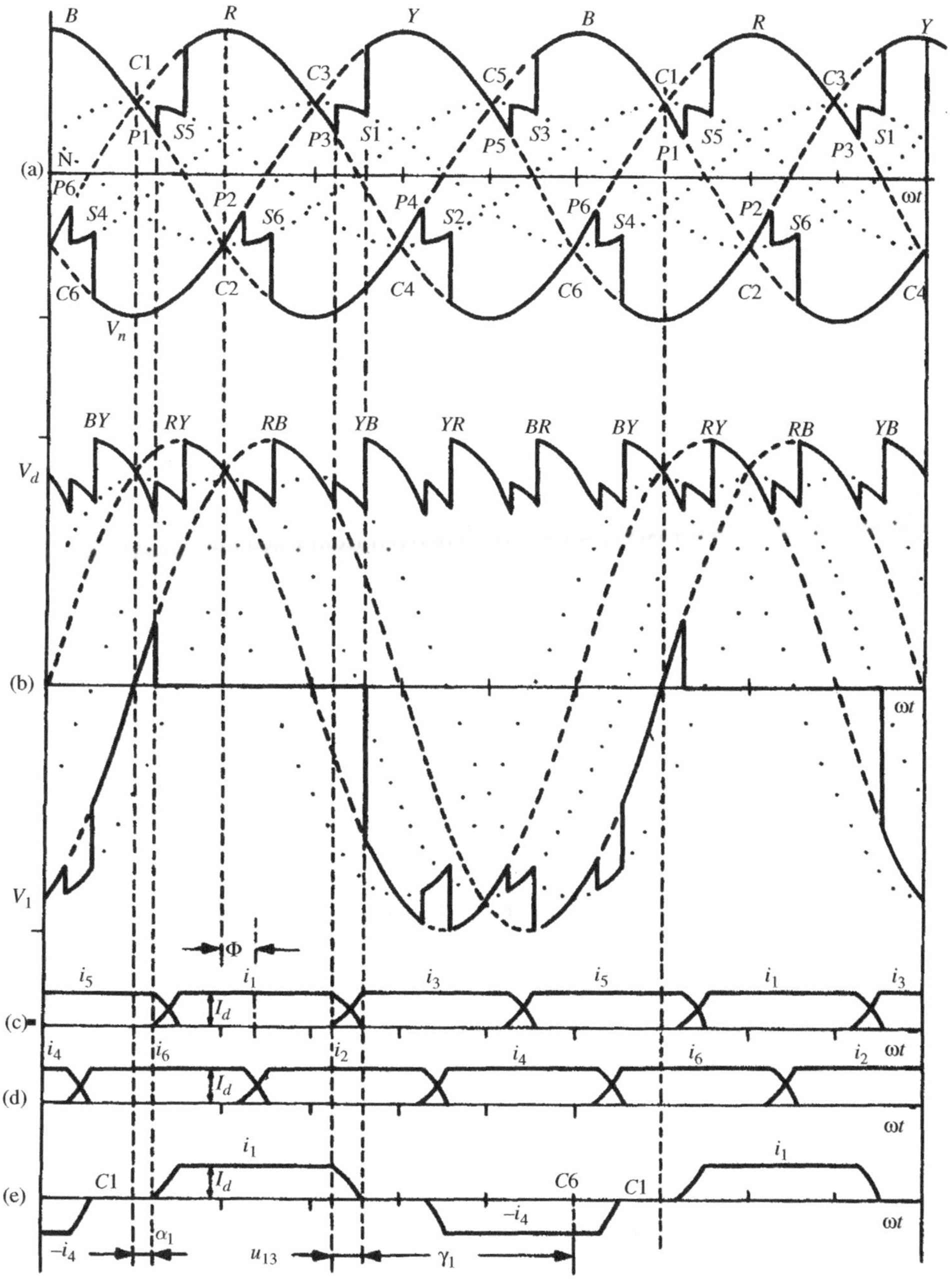

Figure 3.6 Typical six-pulse rectifier operation: (a) positive and negative direct voltages with respect to the transformer neutral; (b) direct bridge voltage $\mathbf{V}_d$ and voltage across valve 1; (c), (d) valve currents i_1 to i_6; (e) AC line current of phase R

Mean direct voltage

The following expression can be easily derived for the average output voltage with reference to the waveforms of Figure 3.6:

$$V_d = (1/2)V_{c0}[\cos(\alpha) + \cos(\alpha + u)] \tag{3.6}$$

where V_{c0} is the maximum average DC voltage (i.e. at no-load and without firing delay), which for the three-phase bridge configuration is

$$V_{c0} = (3\sqrt{2}/\pi)V_c$$

and V_c is the phase-to-phase rms commutating voltage referred to the secondary (or valve side) of the converter transformer.

Equation (3.6) specifies the DC voltage in terms of V_c, α and u. However, the value of the commutation angle (u) is not normally available and a more useful expression for the DC voltage, as a function of DC current, can be derived from Equations (3.5) and (3.6), i.e.

$$V_d = V_{c0}\cos(\alpha) - \frac{3X_c}{\pi}I_d \tag{3.7}$$

AC

The rms magnitude of a rectangular current waveform (neglecting the commutation overlap) is often used to define the converter transformer MVA, i.e.

$$I_{rms} = \sqrt{\left[(1/\pi)\int_{-\pi/3}^{\pi/3} I^2 d(\omega t)\right]} = \frac{\sqrt{2}}{\sqrt{3}}I_d \tag{3.8}$$

Since harmonic filters are normally provided at the converter terminals, the current flowing in the AC system contains (ideally) only fundamental component frequency and its rms magnitude (obtained from Fourier analysis) is

$$I_1 = I_d\sqrt{6}/\pi \tag{3.9}$$

If the effect of commutation reactance is taken into account, the current waveform for a star/star transformer connection is shown in Figure 3.6(e). Using Equations (3.4) and (3.5) the currents of the incoming and outgoing valve during the commutation are defined by Equations (3.10) and (3.11), respectively:

$$i = \frac{I_d[\cos(\alpha) - \cos(\omega t)]}{\cos(\alpha) - \cos(\alpha + u)} \quad \text{for } \alpha < \omega t < \alpha + u \tag{3.10}$$

$$i = I_d - I_d\frac{\cos(\alpha) - \cos(\omega t - 2\pi/3)}{\cos(\alpha) - \cos(\alpha + u)} \quad \text{for } \alpha + \frac{2\pi}{3} < \omega t < \alpha + \frac{2\pi}{3} + u \tag{3.11}$$

In between commutations the current is

$$i = I_d \quad \text{for} \quad \alpha + u < \omega t < (2\pi/3) + \alpha \tag{3.12}$$

The fundamental component of the current waveform, defined by Equations (3.10), (3.11) and (3.12), is

$$I = \frac{\sqrt{6}}{\pi} I_d \sqrt{\left\{[\cos(2\alpha) - \cos\ 2(\alpha + u)]^2 + [2u + \sin(2\alpha) - \sin 2(\alpha + u)]^2\right\} / \{4\,[\cos(\alpha) - \cos(\alpha + u)]\}} \tag{3.13}$$

3.5 Inverter Operation

In an LCC the following three conditions must be met to achieve power inversion, i.e. a supply of power from the DC to the AC side:

1. Presence of an active AC voltage source which provides the commutating voltage waveforms.
2. Provision of firing angle control to delay the commutations beyond $\alpha = 90°$.
3. A source of DC power supply.

The process of inversion has been described in Section 3.2.1 with reference to an ideal system without commutation reactance. In practice, the presence of commutation reactance prevents full inversion because the delay angle must be considerably lower than 180° to take into account the commutation angle and ensure that the outgoing valve recovers its capability to withstand reverse voltage before the voltage crossing.

As shown in Figures 3.7(a) and (e), the commutation from valve 1 to valve 3 (at P3) requires that phase Y is positive with respect to phase R. The commutation must not only be completed before C6, but some extinction angle γ_1 ($> \gamma_0$) must be left for valve 1, which has just stopped conducting, to re-establish its blocking ability. This puts a limit on the maximum angle of firing $\alpha = \pi - (u + \gamma_0)$ for successful inverter operation. If this limit were exceeded, valve 1 would pick up the current again, causing a commutation failure, which results in a short circuit on the DC side and a momentary loss of power transmission. Commutation failures are mainly caused by events in the AC system. Typical causes are a voltage drop or a phase angle change in the AC system resulting from a switching event or system fault. Commutation failures are random and statistical in nature and, therefore, the design objective is to try and minimise their rate and consequences. Unlike the process of rectification, where the delay angle α can be chosen accurately to satisfy a particular control constraint, the same is not possible with respect to the inverter firing angle, because of the uncertainty of the overlap angle (u). Events taking place after the instant of firing are beyond predictability and, therefore, the minimum extinction angle γ_0 must contain a margin of safety to cope with reasonable uncertainties (values between 15° and 20° are typically used). However, the greater the extinction angle, the greater the DC system voltage rating

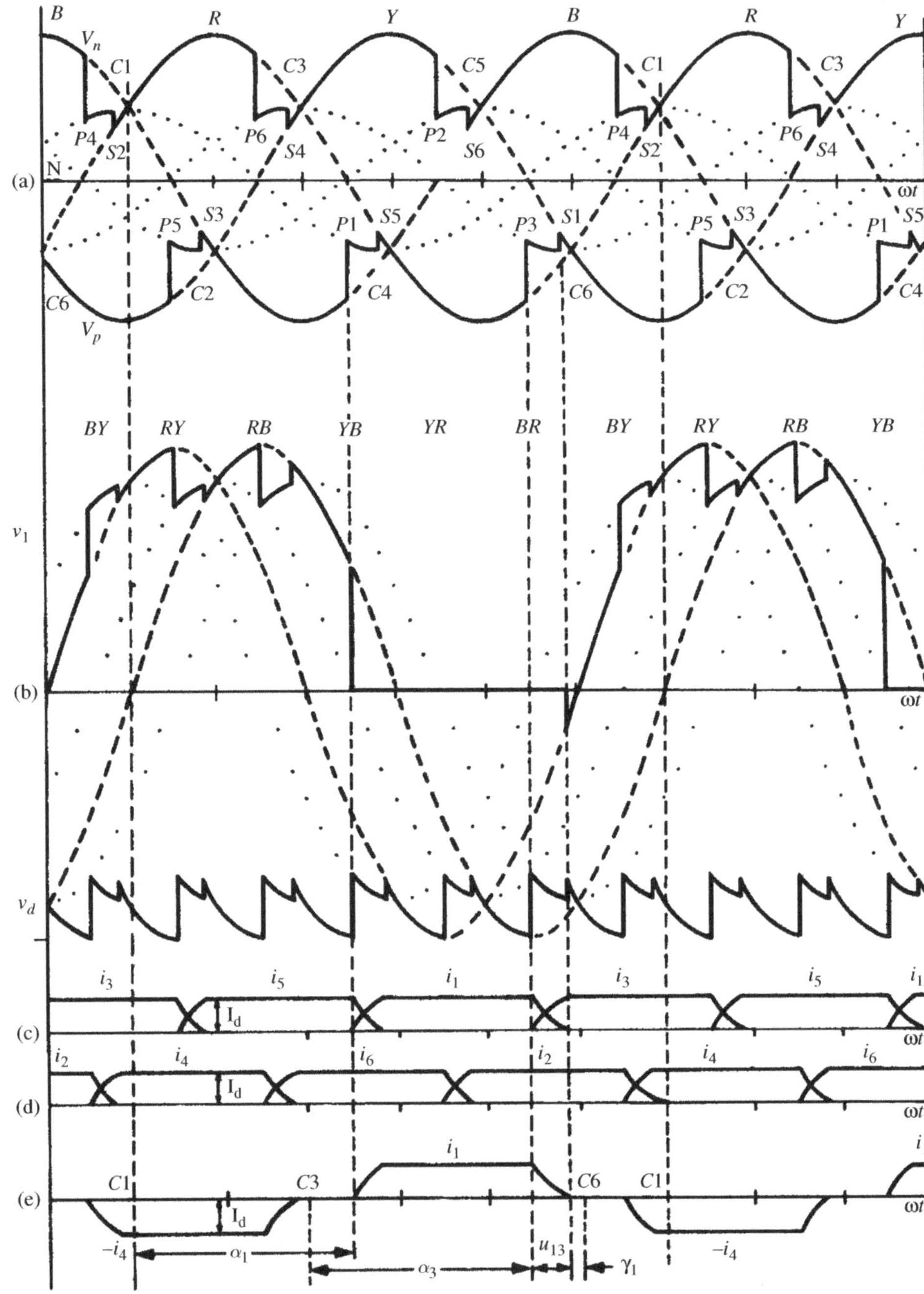

Figure 3.7 Typical six-pulse inverter operation: (a) positive and negative direct voltage with respect to the transformer neutral; (b) voltage across valve 1 and direct bridge voltage $\mathbf{V}_d$; (c), (d) valve currents i_1 to i_6; (e) AC line current of phase R

and cost; moreover, the overall system losses are increased with higher extinction margins, because the reactive power absorption of the converter plant is correspondingly increased.

Although the analysis carried out for the rectification process is equally applicable to inverter operation, for convenience, the inverter equations are often expressed in terms of the angle of advance $\beta(=\pi-\alpha)$ or the extinction angle $\gamma(=\beta-u)$.

Thus, omitting the negative sign of the inverter DC voltage, the following expressions apply:

$$V_d = \frac{3\sqrt{2}}{\pi} V_{term} \cos(\gamma) - \frac{3X_c}{\pi} I_d \tag{3.14}$$

or

$$V_d = \frac{3\sqrt{2}}{\pi} V_{term} \cos(\beta) + \frac{3X_c}{\pi} I_d \tag{3.15}$$

or

$$V_d = \frac{3}{\sqrt{2}\pi} V_{term} [\cos(\beta) + \cos(\gamma)] \tag{3.16}$$

The expression for the DC current is now

$$I_d = \frac{V_c}{\sqrt{2}X_c} [\cos(\gamma) - \cos(\beta)] \tag{3.17}$$

3.6 Power Factor and Reactive Power

Due to the firing delay and commutation angles, the converter current in each phase always lags its voltage, as shown in Figure 3.6(c). The rectifier therefore absorbs lagging current (i.e. consumes reactive power).

With perfect filtering no distorting current flows into the AC system, and the power factor can be approximated by the displacement factor $\cos(\phi)$, where ϕ is the phase difference between the fundamental frequency voltage and current components.

Thus under ideal conditions, with losses neglected, the active fundamental AC power (P) is the same as the DC power, i.e.

$$P = \sqrt{3} V_c\, I \cos(\phi) = V_d I_d \tag{3.18}$$

and

$$\cos(\phi) = V_d I_d / (\sqrt{3}\ V_c I) \tag{3.19}$$

Substituting Equation (3.9) in (3.18) yields the following expression for the DC voltage in terms of the AC side variables:

$$V_d = \frac{3\sqrt{2}}{\pi} V_c \cos(\phi) \tag{3.20}$$

Also, substituting V_d and I_d from Equations (3.6) and (3.9) in (3.19), the following approximate expression results for the power factor caused by the rectification process:

$$\cos(\phi) = {}^1/_2[\cos(\alpha) + \cos(\alpha + u)] \tag{3.21}$$

The reactive power expressed in terms of the active power, is

$$Q = P\tan(\phi) \tag{3.22}$$

where $\tan(\phi)$, derived from Equations (3.13) and (3.19), is

$$\tan(\phi) = \frac{\sin(2\alpha + 2u) - \sin(2\alpha) - 2u}{\cos(2\alpha) - \cos(2\alpha + 2u)} \tag{3.23}$$

Similarly to Equation (3.21), the following approximate expression can be written for the power factor in the inversion process:

$$\cos(\phi) = {}^1/_2[\cos(\gamma) + \cos(\beta)] \tag{3.24}$$

Referring to the AC voltage and valve current waveforms in Figures 3.7(a) and (e), it is clear that the current supplied by the inverter to the AC system lags the positive half of the corresponding phase voltage waveform by more than 90°, or leads the negative half of the same voltage by less than 90°. It can be said that the inverter either 'absorbs lagging current' or 'provides leading current', both concepts indicating that the inverter, like the rectifier, acts as a sink of reactive power. This point is made clearer in the vector diagram of Figure 3.8.

Equations (3.18), (3.21) and (3.22) show that the active and reactive powers of a controlled rectifier vary with the cosine and sine of the control angle respectively. Thus, when operating

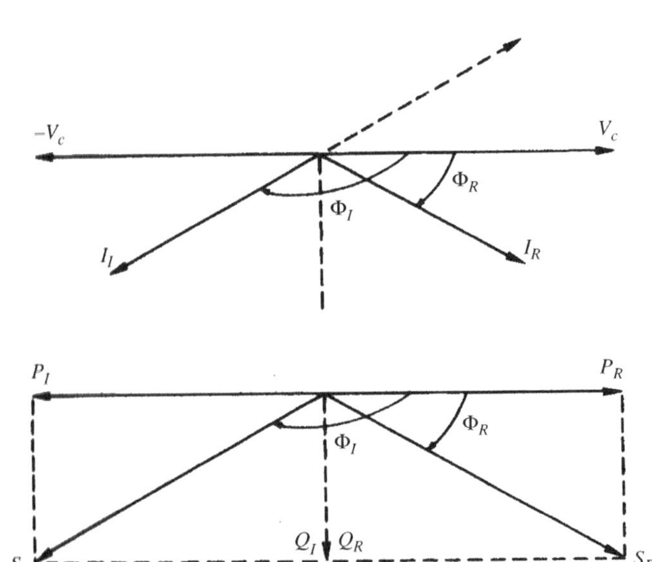

Figure 3.8 Vector diagrams of current and power (suffix R for rectification and I for inversion)

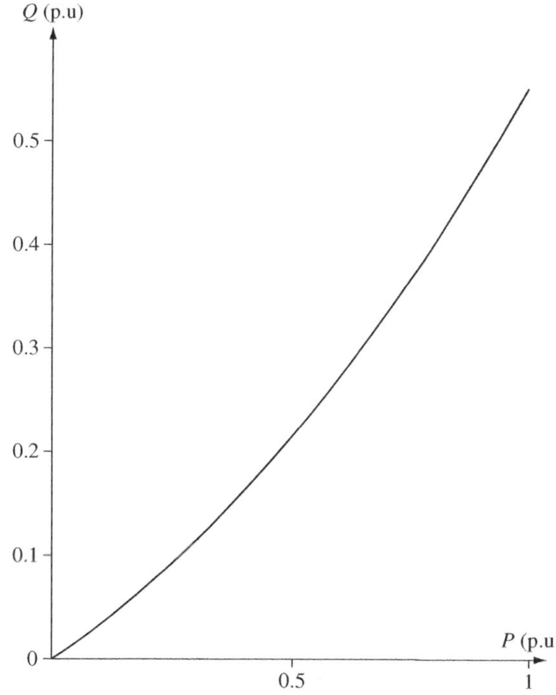

Figure 3.9 Variation of reactive power with active power

on constant current, the reactive power demand at low powers ($\phi \approx 90°$) can be very high. Such operating condition is prevented in the case of an HVDC converter by the use of on-load transformer tap-changers, which try and reduce the steady-state control angle (or the extinction angle) to the minimum specified. Under such controlled conditions, Figure 3.9 shows a typical variation of the reactive power demand versus active power of an HVDC converter; the reactive power demand is shown to be approximately 60 % of the power transmitted at full load.

3.7 Characteristic Harmonics [3]

Line-commutated static converters generate characteristic harmonic voltages and currents on the DC and AC sides, respectively. The orders of these harmonics are related to the pulse number of the converter configuration, defined as the number of non-simultaneous commutations per cycle of the fundamental frequency.

A converter of pulse number p ideally generates only characteristic voltage harmonics of orders pk on the DC side and current harmonics of orders $pk \pm 1$ on the AC side (where k is any integer).

The derivation of the characteristic harmonics is based on the following assumptions:

1. The supply voltages are displaced exactly by one-third of a cycle in time from each other and consist only of fundamental frequency.

2. The DC current is perfectly constant (i.e. has no frequency components). This can only be achieved if the DC smoothing reactor has infinite inductance.
3. The valves begin conducting at equal time intervals.
4. The commutation impedances are the same in the three phases (i.e. all the overlap angles are the same).

3.7.1 DC Side Harmonics

For the three-phase bridge configuration (i.e. $p = 6$) the order of harmonics is $n = 6k$. The repetition interval (see Figure 3.6(c)) is $\pi/3$ and it contains three different functions which, using as a time reference the voltage crossings, are expressed as follows:

$$V_d = \sqrt{2}V_c \cos(\omega t + \pi/6) \quad \text{for } 0 < \omega t < \alpha \tag{3.25}$$

$$V_d = \sqrt{2}V_c \cos(\omega t + \pi/6) + \frac{1}{2}\sqrt{2}V_c \sin(\omega t) = \frac{\sqrt{6}}{2}V_c \cos(\omega t) \quad \text{for } \alpha < \omega t < \alpha + u \tag{3.26}$$

$$V_d = \sqrt{2}V_c \cos(\omega t - \pi/6) \quad \text{for } \alpha + u < \omega t < \pi/3 \tag{3.27}$$

From Fourier analysis, the rms magnitudes of the harmonic voltages are determined from the expression

$$V_n = \frac{V_{c0}}{\sqrt{2}(n^2-1)}\left\{(n-1)^2\cos^2\left[(n+1)\frac{u}{2}\right] + (n+1)^2\cos^2\left[(n-1)\frac{u}{2}\right]\right.$$
$$\left. -2(n-1)(n+1)\cos\left[(n+1)\frac{u}{2}\right]\cos\left[(n-1)\frac{u}{2}\right]\cos(2\alpha+u)\right\}^{1/2} \tag{3.28}$$

Equation (3.28) shows some interesting facts. First, for $\alpha = 0$ and $u = 0$, the relative value of the nth harmonic reduces to

$$\frac{V_{n0}}{V_{c0}} = \sqrt{2}/(n^2-1) \approx \sqrt{2}/n^2 \tag{3.29}$$

giving 4.04, 0.99 and 0.44 % for the 6th, 12th and 18th harmonics, respectively. Generally, as α increases, harmonics increase as well, and for $\alpha = (\pi/2)$ and $u = 0$ the value of the nth harmonic becomes

$$\frac{V_n}{V_{c0}} = \sqrt{2}n/(n^2-1) \approx \sqrt{2}/n \tag{3.30}$$

By way of example, Figure 3.10 illustrates the variation of the sixth harmonic with firing and overlap angles.

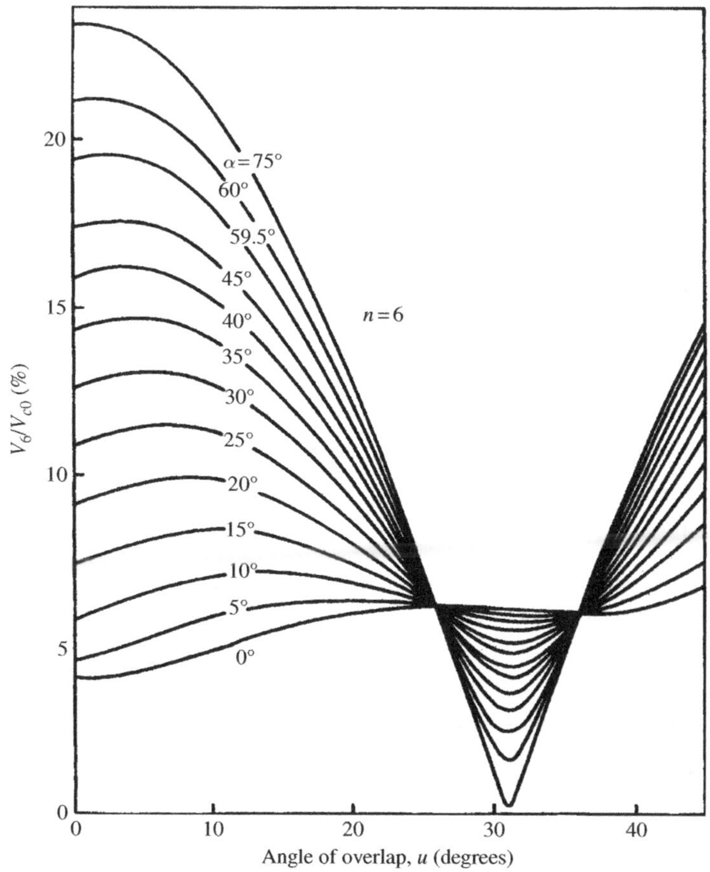

Figure 3.10 Variation of sixth harmonic on the DC voltage in relation to angles of delay and overlap

3.7.2 AC Side Harmonics

With zero commutation reactance the ideal current waveform for a star/star-connected converter transformer, shown in Figure 3.11(a), is defined as follows:

$$\begin{aligned} &i = I_d \text{ for } -\frac{\pi}{3} < \omega t < \frac{\pi}{3} \\ &i = 0 \text{ for } -\frac{2\pi}{3} < \omega t < -\frac{\pi}{3} \text{ and } \frac{\pi}{3} < \omega t < \frac{2\pi}{3} \\ &i = -I_d \text{ for } -\pi < \omega t < -\frac{2\pi}{3} \text{ and } \frac{2\pi}{3} < \omega t < \pi \end{aligned} \tag{3.31}$$

The Fourier series for this current waveform is

$$i = \frac{2\sqrt{3}}{\pi} I_d \left[\cos(\omega t) - \frac{1}{5}\cos(5\omega t) + \frac{1}{7}\cos(7\omega t) - \frac{1}{11}\cos(11\omega t) + \ldots \right] \tag{3.32}$$

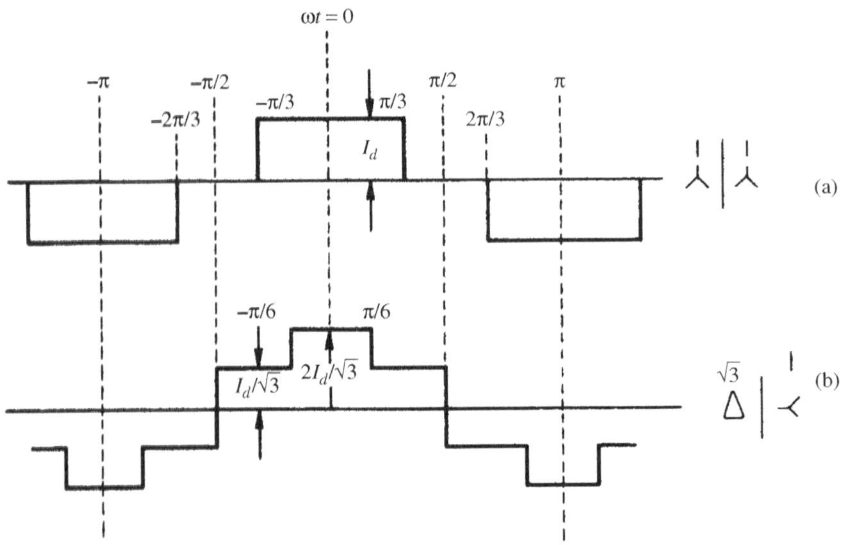

Figure 3.11 Idealised phase current waveforms on the primary side: (a) star–star transformer connection; (b) delta–star transformer connection

which shows that the harmonics of orders $(6k-1)$ rotate in the positive sequence (i.e. in the same sense as the fundamental), whereas those of orders $(6k+1)$ rotate in the negative sequence.

The magnitude of the nth harmonic is given by

$$I_n = \frac{\sqrt{6}}{n\pi} I_d \tag{3.33}$$

and that of the fundamental

$$I_1 = \frac{\sqrt{6}}{\pi} I_d \tag{3.34}$$

3.8 Multi-Pulse Conversion

3.8.1 Transformer Phase Shifting

When the power rating justifies the use of more than one bridge, the need for external filtering can be reduced by appropriate phase shifting of the individual converter transformers, which can be achieved by the use of different transformer connections.

The most common transformer connections are the star and delta types. A transformer consisting of star-connected primary and secondary windings does not alter the phase relationship of the input and output line-to-line voltages. However, the use of either delta–star or star–delta primary to secondary connections introduces a 30° phase shift between the input and output line-to-line voltages. The resulting current waveforms are shown in Figure 3.11(b). The Fourier series for the delta–star transformer connection is

$$i = \frac{2\sqrt{3}}{\pi} I_d \left[\cos(\omega t) + {}^1/_5 \cos(5\omega t) - {}^1/_7 \cos(7\omega) - {}^1/_{11} \cos(11\omega t) + \ldots\right] \tag{3.35}$$

Equation (3.35) is the same as (3.32), the star–star Fourier series, with the exception that the harmonics 5, 7 (k = odd numbers) are of opposite sequence.

A detailed description of alternative converter transformer and autotransformer connections providing different phase shifts can be found in [4]. These include the zigzag, polygon, differential delta, centre-tapped delta, fork and differential fork connections, all of which can be used to provide high-pulse converter configurations. While these connections can be considered for low-voltage applications, such as the metal reduction industry, they are usually found uneconomical in high-voltage power conversion.

Double bridge configuration

Equations (3.32) and (3.35) indicate that the use of two bridges (either in parallel or series on the DC side), one with a star–star and the other with delta–star or star–delta transformer (i.e. with a 30° phase shift between them), will inject only harmonic currents of orders $n = 12k \pm 1$ into the AC system.

The series-connected double bridge converter, shown in Figure 3.12, has become the standard configuration in HVDC transmission. The current waveform and harmonic spectrum of the double bridge 12-pulse configuration are illustrated in Figure 3.13 (with the overlap angle ignored). The characteristic harmonics decrease with increases in commutation angle (u), the rate of decrease being greater for higher harmonics. By way of example, Figure 3.14 shows the variation of the fifth harmonic with commutation angle for different values of the firing delay.

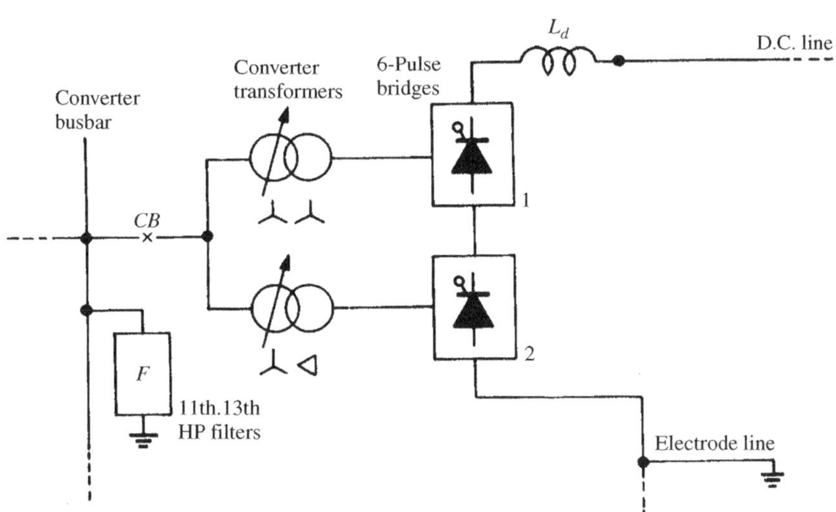

Figure 3.12 The 12-pulse configuration

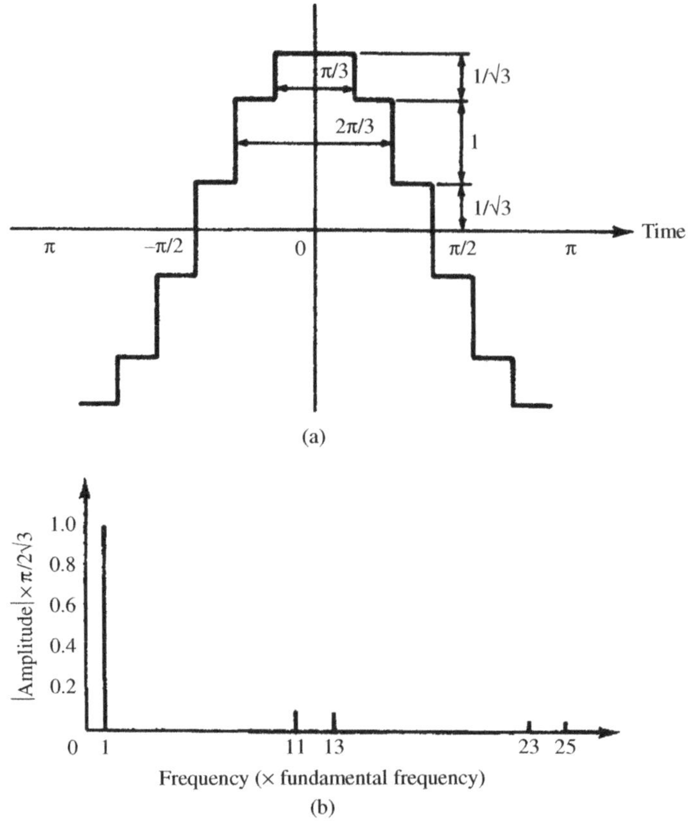

Figure 3.13 Idealised phase current waveform with 12-pulse operation: (a) current waveform; (b) frequency domain representation

During normal operation α is small (say up to 10°) and u is large (say 20°), whereas during disturbances, when α is nearly 90°, u is very small and the harmonics approach their maximum.

For inverter operation α and $(\alpha + u)$ should be replaced by the extinction angle γ and the angle of advance β, respectively.

The DC side voltages are also 30° out of phase and so are their harmonic components. Since 30° of main frequency corresponds to a half-cycle of sixth harmonics, this harmonic will be in phase opposition in the two bridges. Similarly for the 12th harmonic, 30° corresponds to one cycle, giving harmonics in phase; for the 18th, 30° corresponds to one and a half cycles, giving harmonics in opposition, and so on.

The addition of further appropriately shifted transformers connected in parallel on the AC side provides the basis for increasing pulse configurations. For instance, 24-pulse operation is achieved by means of four transformers with 15° phase shifts and 48-pulse operation requires eight transformers with 7.5° phase shifts. Although theoretically possible, pulse numbers above 48 are rarely justified due to the practical levels of imbalance and distortion found in the supply voltage waveforms, which can have as much influence on the voltage crossings as the theoretical phase shifts.

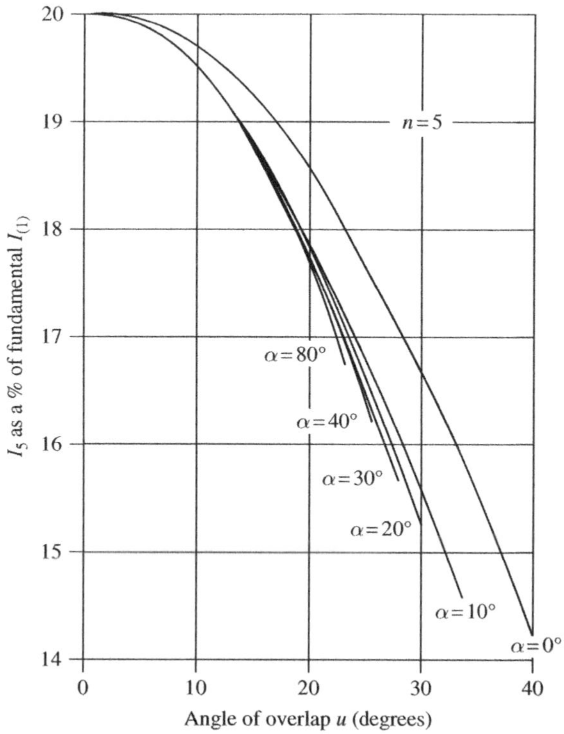

Figure 3.14 Variation of fifth harmonic with the angle of commutation overlap

As in the case of the 12-pulse connection, the alternative phase shifts involved in higher pulse configurations require the use of appropriate factors in the parallel transformer ratios to achieve common fundamental frequency voltages on their primary and secondary sides.

The theoretical harmonic currents are related to the pulse number (p) by the general expression $pk \pm 1$ and their magnitudes decrease in inverse proportion to the harmonic order.

Although the multi-bridge concept has been extensively used for high-power, low-voltage applications, it is not found to be cost effective for DC transmission and the 12-pulse series configuration has become the standard configuration.

3.8.2 DC Ripple Reinjection [5]

Single bridge pulse multiplication

As shown in Figure 3.1(g), static converters produce a ripple voltage at their DC output. With six-pulse rectification the ripple has a period of $(1/6)T$, where $T = 1/f$. However, with respect to the star point of the converter side transformer windings, each DC pole has non-sinusoidal ripple voltage of period $(1/3)T$, i.e. a triple frequency voltage. This voltage has the same phase relationship on each DC pole and is referred to as the common mode DC ripple voltage.

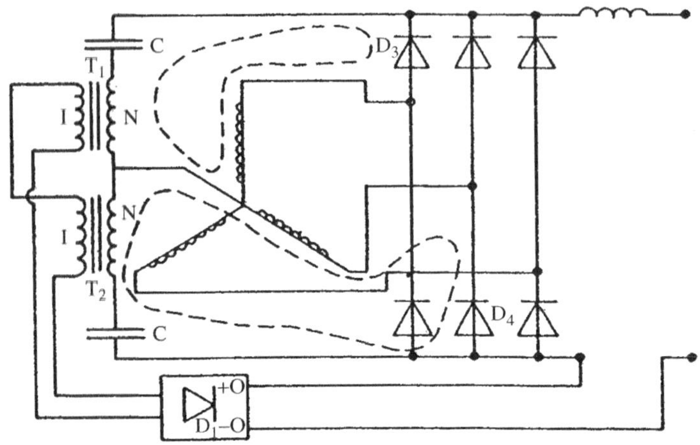

Figure 3.15 Bridge rectifier with ripple reinjection. T_1, T_2, feedback transformers; N, turns ratio; C, blocking capacitor; D_1,feedback rectifier. (The dashed lines indicate the paths of injected current, while D_3 and D_4 are on)

The principle of DC ripple reinjection is applicable to the line-commutated three-phase current source converter bridge and the circuit configuration is shown in Figure 3.15. The converter transformer is star connected on the converter side and must have either a delta primary or delta tertiary winding. It requires an auxiliary single phase transformer with two primary windings connected to the common mode DC ripple voltage via blocking capacitors. This transformer provides the commutating voltage for a single phase full-wave rectifier (or feedback converter) connected to the secondary windings. The output of the feedback converter is connected in series with the DC output of the six-pulse converter bridge. The frequency of the harmonic injection is determined by the supply frequency, and therefore the harmonic source is always synchronised with the mains frequency.

The required phase adjustment of the injected current is achieved by thyristor control in the feedback converter, and the firing angle control of the feedback converter is locked to that of the main converter. When the thyristors of the feedback converter are fired 30° after the firings of the main converter valves, the waveforms in Figure 3.16 result. A Fourier analysis of waveform E shows that, for a particular ratio of the injected current to rectifier current ratio, all harmonics of orders $6n \pm 1$ (where $n = 1, 3, 5, \ldots$) are zero, while those of the remaining orders (i.e. for $n = 2, 4, 6, \ldots$) retain the same relationship with the fundamental as before. The result is that the original six-pulse converter configuration has been converted to a 12-pulse converter system from the point of view of AC and DC system harmonics.

Double bridge pulse multiplication [6]

The reinjection principle is even more effective when the 12-pulse (instead of the six-pulse) converter is used as a basis for the pulse transformation. In this case the centre point between the two bridges is used, instead of the transformer neutral, as the reinjection point and the DC ripple across the bridge (which in this case is $1/6T$) provides the commutating voltage source for the reinjection bridge.

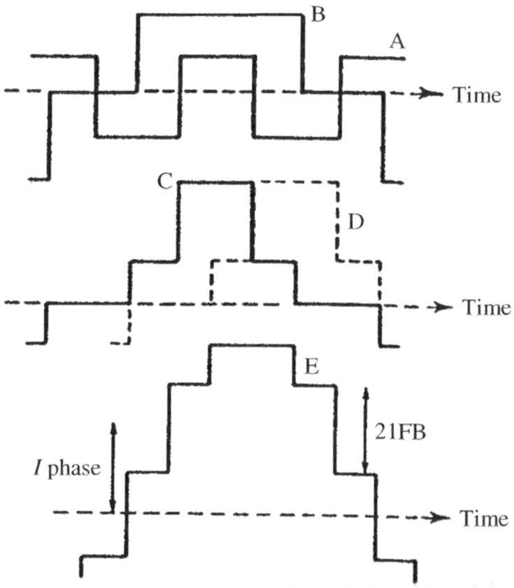

Figure 3.16 Current waveforms for the converter of Figure 3.15: (A) reinjection injected current; (B) rectifier current before modification; (C) modified phase current, (D) second-phase current displaced 120°; (E) resultant phase current on delta primary

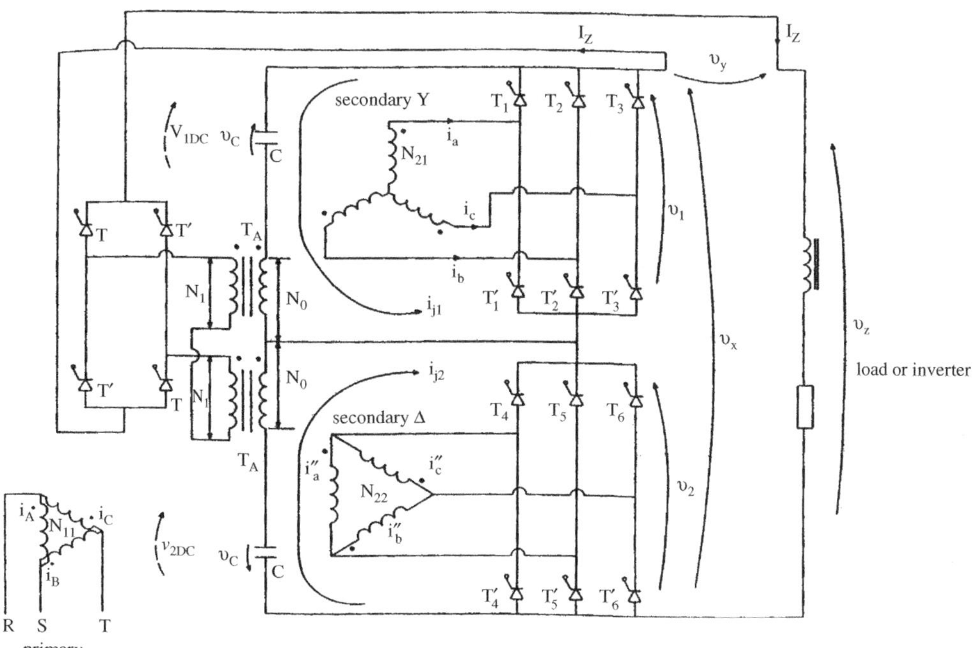

Figure 3.17 The 24-pulse series bridge configuration

For the series configuration, Figure 3.17 illustrates the modified circuit required to implement the reinjection concept; the additional components are two capacitors C, needed to block the DC component of voltages v_1 and v_2, two transformers T_A (operating at the ripple frequency, which is six time the fundamental frequency) and a single phase converter bridge.

The commutating voltage of the ripple reinjection thyristors is the common mode ripple frequency voltage of the 12-pulse converter group. The voltage waveforms associated with the pulse doubling process of the series connection are shown in Figure 3.18 for any pulse number p of the main converters. The individual voltages of the converter bridges, v_1 and v_2, are shown in graph (a). The output of the reinjection circuit (v_y) is a 12-pulse voltage

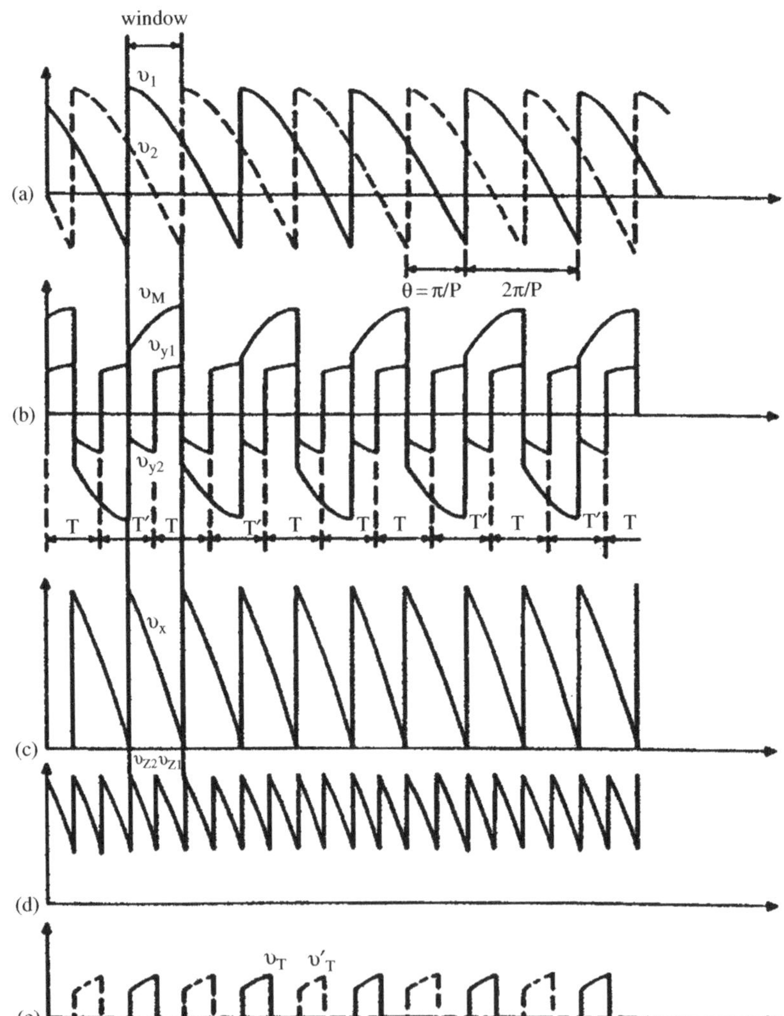

Figure 3.18 Voltage waveforms on the DC side of the 24-pulse reinjection configurations

waveform (graph (b)). Its level is adjusted by the transformer (or autotransformer) turns ratio (N_1/N_0) and its phase is shifted by firing angle control. Finally, v_y is added to the converter terminal voltage v_x (graph (c)) to double the number of pulses of the combined (output) voltage waveform v_z (graph (d)).

Clearly an appropriate selection of firing delay for the reinjection bridge valves T, T' (in Figure 3.17) and a suitable turns ratio of the feedback transformer T_A produce a voltage waveform v_Y, which when added to v_x doubles the pulse number of the output voltage v_z, regardless of the value of the firing angle.

Figure 3.19 shows the AC waveforms associated with the circuit of Figure 3.17. In this figure, graphs (c), (f) and (h) are the conventional current waveforms in the absence of the reinjection circuit. Graph (a) shows the valve conducting sequences of the main bridges. Using the 12-pulse ripple voltage as commutating voltage for the reinjection bridge, a six-pulse reinjection current (i_j) is produced on the secondary side of the reinjection transformers (shown in graph (b)).

On the phases of the delta-connected bridge the reinjection current ($i''_{\Delta 0}$) is shown in graph (d) and the addition of (d) and (c) produces the delta winding current (graph (e)). Similarly, in the star-connected bridge the reinjection current (i_j) is added to the basic waveform (f) to produce $i''_a/\sqrt{3}$ (graph (g)). Finally the addition of (e) and (g) results in the 24-pulse output current waveform (i_A) shown in graph (j).

However, the ripple reinjection concept has not been found competitive with the conventional converter configuration supplemented by filters. This is mainly due to the part played by the filters in the provision of reactive power requirements of the LCC process. It will be shown in Chapters 4 and 7 that the use of self-commutating reinjection switches can make this concept far more attractive.

3.9 Uncharacteristic Harmonics and Interharmonics

The harmonics caused by imperfect system conditions encountered in practice cannot be predicted with the idealised models described in previous sections. In general each of the three main parts of the system is always in error to some extent:

1. The AC system voltages are never perfectly balanced and undistorted and the system impedances, in particular those of the converter transformer, are not exactly equal in the three phases.

2. DC may be modulated from another converter station in the case of a rectifier–inverter link.

3. The firing angle control systems often gives rise to substantial errors in their implementation.

As a result the static converters generally produce harmonic orders and magnitudes not predicted by the Fourier series of the idealised waveforms.

The uncertain nature of the 'uncharacteristic' harmonics makes it difficult to prevent them at the design stage. Filters are not normally provided for uncharacteristic harmonics and, as a result, their presence often causes more problems than the characteristic harmonics.

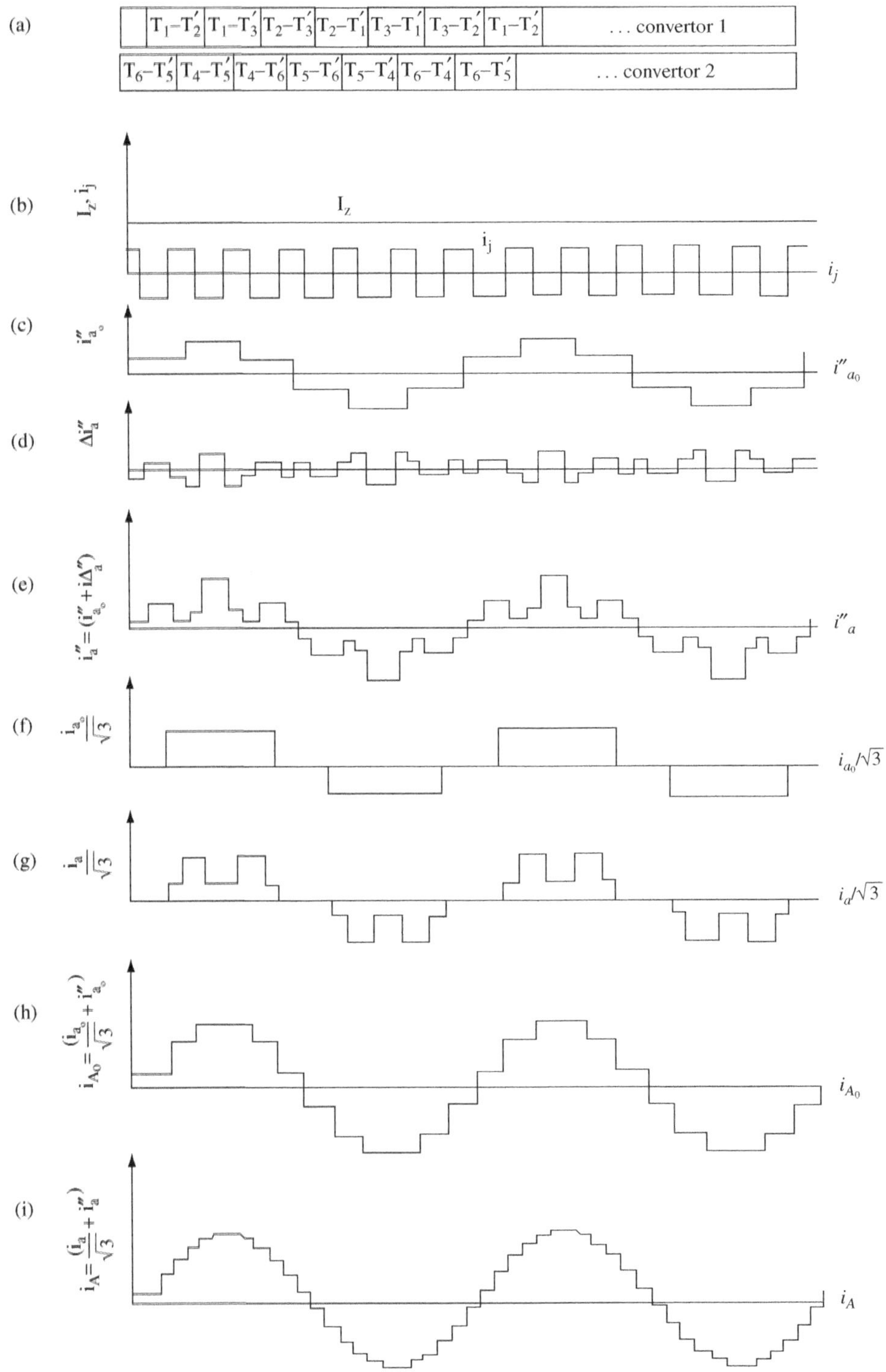

Figure 3.19 Current waveforms for the 24-pulse reinjection configurations

Table 3.1 Harmonic measurements during back-to-back testing of the New Zealand HVDC converters

Harmonic	400 A DC (one-third full load) current; phase-to-neutral voltages at Benmore on 220 kV		
	Red phase (%)	Yellow phase (%)	Blue phase (%)
1	100	100	100
2	0.5	0.7	1.0
3	2.9	0.3	1.0
4	0.6	0.3	0.4
5	0.25	0.15	0.25
6	0.25	0.30	0.35
7	0.15	0.15	0.1
8	0	0.05	0.1
9	0.05	0.05	0.15
10	0.05	0.05	0.05
11	0.1	0.15	0.1
12	0.15	0.05	0.15
13	0.05	0.05	0.05
14	0.05	0.05	0.05
15	0.15	0	0.2
16	0	0.1	0.15
17	0.3	0.3	0.3
18	0	0.05	0.1
19	0.3	0.3	0.7
20	—	—	—
21	—	—	—
22	0.2	0.2	0.5
23	0.4	0.2	0.3
24	0.2	0.2	0.15

By way of example, Table 3.1 shows the results of harmonic measurements during back-to-back testing of the New Zealand HVDC station at Benmore. All the harmonic voltages are unbalanced, particularly the third and ninth. The table also illustrates the presence of all current harmonic orders, odd and even, with the uncharacteristic orders causing higher voltage distortion than the characteristic ones. A realistic quantitative analysis of the uncharacteristic harmonic components can only be achieved by a complete three-phase computer model of the system behaviour with detailed representation of the converter controls. A mostly qualitative assessment of the main problem areas and the sensitivity of the system to small deviations from the ideal conditions are considered in this section.

3.9.1 Imperfect AC Source

Deviations from the perfectly balanced sinusoidal supply can be caused by (i) the presence of negative sequence fundamental frequency in the commutating voltage; (ii) harmonic

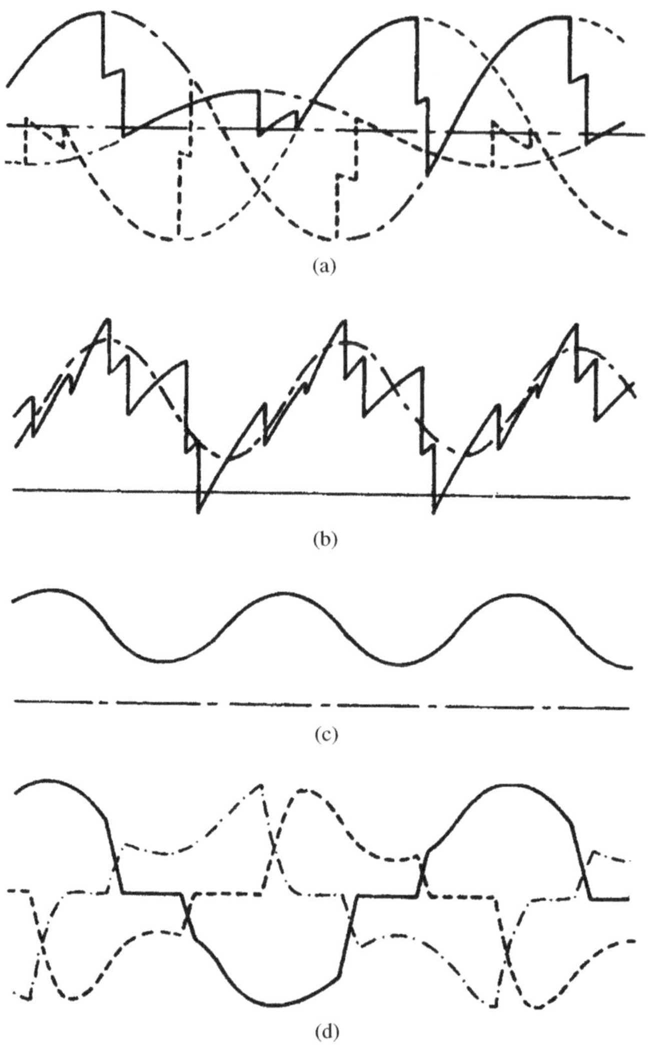

Figure 3.20 Sustained unbalanced voltage on single converter bridge: (a) three-phase voltages; (b) direct voltage; (c) direct current; (d) three-phase currents

voltage distortion of positive or negative sequence; (iii) asymmetries in the commutation reactances. In general an imperfect AC source produces asymmetrical firing references and DC modulation. The first problem can be eliminated by using equidistant firing control [7], but the second problem still remains. This effect, illustrated in Figure 3.20 for the case of an unrealistically high level of fundamental voltage asymmetry, produces considerable second-harmonic content on the DC side and third-harmonic on the AC side [8].

Under normal operating conditions the expected levels of asymmetry and distortion are small and their effects can be approximated with reasonable accuracy.

With reference to unbalanced conditions, a CIGRE Working Group [9] examined the effect of different factors of asymmetry (one at a time) on an otherwise standard 12-pulse converter. The AC system was assumed to be of negligible impedance (i.e. the terminal voltage was perfectly sinusoidal) and the DC ripple was ignored. In practice, of course, the result of imbalance is a combination of each effect, which may either increase or decrease the harmonic content of the individual cases.

Five cases were considered as follows:

1. The voltage in one phase is reduced by 1 % with respect to the other two.
2. The transformer's leakage reactances are 20 % (for five of the phases) and 21 % (for the sixth phase).
3. Different transformer reactances are considered, i.e. 20 % (for the star–star) and 21 % (for the star–delta).
4. Different transformer turns ratios are considered, i.e. 100 % (for the star–star) and 100.5 % (for the star–delta).
5. The firing angles are 15° (for five valves) and 15.2° (for the remaining valve).

The effect of each of these cases on the first 10 harmonic currents generated by the converter are shown in Table 3.2. These results show that the effects of phase reactance imbalance and firing angle imbalance are different for the two types of converter transformer. These are more severe for the star–star transformer (i.e. in the absence of a delta winding).

With reference to supply distortion, if a small positive or negative sequence signal V_n (per unit of the normal fundamental voltage) is added to the otherwise ideal three-phase supply of a 12-pulse converter configuration, the order and maximum level (V_k) of uncharacteristic harmonic voltage at the rectified output come under one of the following categories [10]:

Table 3.2 Harmonic currents (in % of the fundamental) generated by different unbalanced types

Harmonic order	Unbalanced type				
	1	2	3	4	5
1	100.00	100.00	100.00	100.00	100.00
2					0.0279
3	0.069	0.130			0.0268
4					0.0253
5		0.208	0.414	0.0666	0.0234
6					0.0213
7		0.0184	0.366	0.0516	0.0189
8					0.0163
9	0.043	0.0904			0.0136
10					0.0109

Case 1: If $n+k=12p_1+1$ and $n-k=12p_2+1$, where p_1 and p_2 are any integers, then

$$V_k = \begin{cases} V_n\left(\dfrac{n\sqrt{2}}{n^2-k^2}\right) & \text{if } n^2 > k^2 \\ V_n\left(\dfrac{k\sqrt{2}}{k^2-n^2}\right) & \text{if } k^2 > n^2 \end{cases} \tag{3.36}$$

Case 2: If $n+k=12p_1+1$ but $n-k \neq 12p_2+1$, then

$$V_k = V_n \frac{1}{\sqrt{2}(n+k)} \tag{3.37}$$

Case 3: If $n-k=12p_1+1$ but $n+k \neq 12p_2+1$, then

$$V_k = V_n \frac{1}{\sqrt{2}(n-k)} \tag{3.38}$$

Case 4: If $n+k \neq 12p_1+1$ and $n-k \neq 12p_2+1$, then

$$V_k = 0 \tag{3.39}$$

An interfering AC side voltage harmonic (V_n) of order (−1, +2, −2, +3, −3, −4, −5) will cause a corresponding DC side voltage harmonic (V_k) of order (2,1,3,2,4,5,6) respectively, with a magnitude of 0.707 (as a ratio of the interfering harmonic magnitude). V_k is expressed as per unit of the maximum average DC voltage and V_n as per unit of the normal AC voltage.

The values given above are in practice applicable also to converters of higher pulse number. Thus, while characteristic harmonics can be reduced by using high pulse numbers, the uncharacteristic harmonics due to AC imbalance cannot.

It should be noted that the above is an approximation due to neglect of commutation reactance but is generally sufficiently valid at low harmonic orders (below the fifth).

The cause of the imperfection may also be some asymmetry in the commutation reactances, i.e.

$$X_a = X_0(1+g_a)$$
$$X_b = X_0(1+g_b)$$
$$X_c = X_0(1+g_c)$$

where X_0 is the mean reactance and each value of g can vary between $\pm g_0$. In this case the maximum level of uncharacteristic AC harmonic currents I_n of order n (in phase a) for the case of a six-pulse bridge occurs when

$$g_a = 0 \quad g_b = \pm g_0 \quad g_c = \mp g_0 \quad \text{for } n = 3,\ 9,\ 15,\ \text{etc.}$$

or

$$g_a = \pm g_0 \quad g_b = \mp g_0 \quad g_c = 0 \quad \text{for } n = 5,\ 7,\ 11,\ 13,\ \text{etc.}$$

The maximum value of I_n neglecting changes of DC and AC voltage is obtained from the expression

$$I_n = \frac{I_1 g_0}{n(n^2-1)i_d X_0 \sqrt{3}} \times [n^4[\cos(\alpha+u)-\cos(\alpha)]^2 + 2n^3 \sin(\alpha)\sin\{nu[\cos(\alpha+u)-\cos(\alpha)]\} + n^2\{\sin^2(\alpha)+\sin^2(\alpha+u)+2\cos(nu)[\cos^2(\alpha)-\cos(u)]+2\cos(\alpha)[\cos(\alpha+u) - \cos(\alpha)]\} + 2n\cos(\alpha)\sin(nu)[\sin(\alpha)+\sin(\alpha+u)]+2\cos^2\{\alpha[1-\cos(nu)]\}]^{1/2} \tag{3.40}$$

for $n = 3, 9, 15$, etc., where I_1 is the fundamental rms current, i_d is the DC current per unit, X_0 is the mean commutation reactance per unit, α is the firing angle and u is the overlap (or commutation) angle. For $n = 5, 7, 11, 13$, etc., the above expression should be divided by 2.

Table 3.3 gives values for a typical case of $X_0 = 0.2$ per unit, $\alpha = 15°$, $g_0 = 0.075$.

Unequal commutation reactances also cause uncharacteristic voltages on the DC side. The highest magnitude of these occurs when $g_a = 0$, $g_b = +g_0$ and $g_c = -g_0$. Only even harmonics occur, given by

$$V_n(\max) = \frac{i_d X_0 g_0 V_{dio}}{2\sqrt{6}} \tag{3.41}$$

where V_{dio} is the theoretical no-load DC voltage.

As an example, for $i_d = 1$, $X_0 = 20\,\%$ and $g_0 = 0.075$, $V_n(\max)$ is 0.31 % of V_{dio} for $n = 2, 4, 8, 10, 14, 16$, etc., independently of harmonic order or of firing angle.

3.9.2 DC Modulation

If we now assume a perfect three-phase supply and equidistant firing, the addition of a small current harmonic component I_k of low order k (say between 1 and 4) on the DC side will generate a component I_n of different order, but of the same sequence, on the AC side, the maximum level of which, in % of I_1, is approximately 0.707.

Table 3.3 Typical harmonic currents (in % of the fundamental) for a 0.075 per unit asymmetry in the commutation reactances

n	I_n (% of I_1)
3	0.70
5	0.33
7	0.29
9	0.50
11	0.22
13	0.19
15	0.31

The amplitude I_n is in multiples of $I_1 I_k / I_d$, where I_1 is the rms fundamental current at the AC busbar, I_k is the rms interfering current on the DC side at order k and I_d is the DC.

This effect (and magnitude) is independent of the prime cause of the DC modulation. It is again approximate, and valid only at low frequencies, because the commutation reactance has been neglected.

3.9.3 Control System Imperfections

No general rules can be given in this case. By way of example, Ainsworth [10] describes the effect of modulating the harmonic content of the voltage applied to the oscillator of the DC control system using the phase-locked oscillator principle [7]. Assuming constant DC current and AC voltages, a V_c modulating harmonic signal of order n (per unit, referred to the normal steady–state control voltage), causes DC voltage components of orders $n_1 = \pm n \pm 12p$ in a 12-pulse converter, where p is any integer.

The magnitude (per unit of the maximum average rectified voltage) of the DC voltage modulation is

$$V = \frac{V_c \sin(\alpha_0)\cos(n_1 u_0/2)}{n_1} \tag{3.42}$$

where α_0 and u_0 are the mean firing and overlap angles respectively.

The total AC magnitude referred to one phase of one valve winding at harmonic order n_2 due to a similar excitation is

$$I_A(n_2) = \frac{I_d V_c 2\sqrt{3}\sin(n_2 u_0/2)}{n_2[\cos(\alpha_0) - \cos(\alpha_0 + u_0)]} \tag{3.43}$$

for $n_2 = \pm n \pm (1, 11, 13, \ldots)$ only.

3.9.4 Firing Asymmetry

AC system imperfections or firing errors result in pulse width deviations from the characteristic quasi-rectangular current waveform. Kimbark [11] described the effect of late and early firings with reference to a six-pulse bridge converter.

If the positive current pulses start early by an angle ε and the negative ones start late by the same angle, the non-conductive intervals are increased by 2ε. The even symmetry, which eliminates the even-ordered harmonics, is now lost and the even harmonics for small overlap angles are given by the expression

$$\frac{I_n}{I_1} = \frac{2\sin(n\varepsilon)}{2n\cos(\varepsilon)} \approx \varepsilon \tag{3.44}$$

For example, for $\varepsilon = 1°$, the second and fourth harmonics are each approximately 1.74 % of the fundamental current.

If the firings of the two valves connected to the same phase are late by ε, then the positive and negative current pulses of that phase are $\varepsilon°$ shorter than normal. Moreover, the current pulses of one of the remaining phases (the leading phase) are increased by ε, while those of the lagging phase remain unaltered. This produces triplen harmonic currents. On the assumption of zero overlap angle, the ratio of the triplen harmonics ($h = 3q$) to the fundamental current is expressed by

$$\frac{I_n}{I_1} = \frac{\sin(q\pi \pm 1.5q\varepsilon)}{3q\sin(\pi/3 \pm \varepsilon/2)} \tag{3.45}$$

For small values of ε the approximate levels of third harmonic are given by

$$\frac{I_3}{I_1} \approx \frac{1.5q\varepsilon}{3q\sqrt{3}/2} = 0.577\varepsilon \tag{3.46}$$

That is, for $\varepsilon = 1°$, $I_3 = 1\,\%$ of the fundamental.

3.9.5 Magnification of Low-Order Harmonics

The following simplified reasoning explains the harmonic magnification phenomena. Let us consider the case of a static converter (represented as a harmonic current injector) fed from an AC system of internal impedance Z_r at the h harmonic.

The AC system admittance at the fundamental and low-order harmonics is predominantly inductive, i.e.

$$Y_r = G_r - jB_r \tag{3.47}$$

In the absence of filters or compensation, the harmonic current (I_h) generates at the point of connection a harmonic voltage of amplitude

$$V_h = Z_r I_h = I_h / Y_r \tag{3.48}$$

When a capacitor bank or filters of admittance Y_f are present, the harmonic voltage at the point of connection becomes

$$V_h' = I_h/(Y_r + Y_f) \tag{3.49}$$

and, since the admittance of a filter bank is predominantly capacitive ($Y_f = j\ B_f$),

$$V_h' = I_h/(G_r - jB_r + jB_f) \tag{3.50}$$

When $B_r = B_f$ the harmonic voltage is only limited by the system resistance, which is generally very small. Thus when $Y_r + Y_f < Y_r$ the harmonic distortion is magnified, the magnification factor being

$$V_h'/V_h = Y_r/(Y_r + Y_f) \tag{3.51}$$

Consequently, a low-order non-characteristic harmonic current, which has no practical adverse effect in the absence of the capacitor or filter banks, can be amplified to give a voltage greater than the filtered harmonics.

3.10 Harmonic Reduction by Filters

3.10.1 AC Side Filters

When planning the installation of large converter plant, the decision has to be made between using a converter configuration with low levels of waveform distortion or installing harmonic compensation equipment at the terminals. External harmonic compensation is achieved by means of filters. In each case the decision will depend on such factors as the power and voltage rating of the equipment to be installed and the effect of the local (internal) waveform distortion on the rest of the plant.

The size of a filter is defined as the reactive power that the filter supplies at fundamental frequency. It is substantially equal to the fundamental reactive power supplied by the capacitors. The total size of all the branches of a filter is determined by the reactive power requirements of the harmonic source and by how much this requirement can be supplied by the AC network and/or by extra shunt capacitors. The amount of filter capacity needed to limit harmonics will depend on the system impedances and will typically be of the order of 20 to 30 % of the converter rating.

The ideal criterion of filter design is the elimination of all detrimental effects caused by waveform distortion, including telephone interference, which is the most difficult effect to eliminate completely. However, the ideal criterion is unrealistic for technical and economic reasons. A more practical approach is to try and reduce the problem to an acceptable level at the point of common coupling with other consumers, the problem being expressed in terms of harmonic current, harmonic voltage, or both. A criterion based on harmonic voltage is more convenient for filter design, because it is easier to guarantee staying within a reasonable voltage limit than to limit the current level as the AC network impedance changes.

Typical specified factors to be taken into account in filter design are the voltage distortion caused by individual harmonics (V_n), the total voltage distortion defined as

$$V_{TD} = \sqrt{\sum_{n=2}^{\infty} V_n^2} \tag{3.52}$$

and the telephone influence factor (TIF).

The TIF gives an approximation of the effect of the distorted voltage or current waveform of a power line on telephone noise, without considering the geometrical aspects of coupling. The harmonic frequencies which are sensitive to the ear are given high weighting factors, since even if the harmonic magnitudes are small, these harmonics may result in unacceptable telephone noise. The TIF is defined as

$$\mathrm{TIF} = \frac{1}{V}\left[\sum_{f=0}^{\infty} (K_f P_f V_f)^2\right]^{1/2} \tag{3.53}$$

where

K_f = 5000(f/1000)=5f

P_f = **C**-message weighting

V_f = rms voltage of frequency f on the power line

and

$$V = \left[\sum_{f=0}^{\infty} V_f^2 \right]^{1/2} \tag{3.54}$$

The quality of a filter (**Q**) expresses the sharpness of tuning and is therefore defined differently for tuned and high-pass filters. The sharpness of tuning of a resonant filter branch increases with the ratio of its resonance inductance or capacitance to its resistance, whereas in the case of a high-pass filter, the sharpness increases in inverse proportion to that ratio.

The high-**Q** or tuned filter is sharply tuned to one or two of the lower harmonic frequencies such as the fifth and seventh. The low-**Q** or damped filter provides a low impedance over a broad band of frequencies and is often used to eliminate the higher order harmonics, e.g. 17th up. It is normally referred to as a high-pass filter.

To comply with the required harmonic limitations the design of filters involves the following steps:

1. The harmonic current spectrum produced by the converter is injected into a circuit consisting of filters in parallel with the AC system, as shown in Figure 3.21, and the harmonic voltages are calculated.

2. The results of step 1 are used to determine the individual and total harmonic distortion levels.

3. The stresses in the filter components, i.e. capacitors, inductors and resistors, are then calculated and with them their ratings and losses.

Three components require detailed consideration in filter design, i.e. the harmonic current source and the filter and system admittances.

The harmonic current source content should be varied through the range of converter operation. As far as the system and filter admittances are concerned, it is essential to calculate the minimum total equivalent admittance at each harmonic frequency, which will result in maximum voltage distortion.

The obvious filter design is a single broad band-pass configuration capable of attenuating the whole spectrum of injected harmonics (e.g. from the fifth up in the case of a six-pulse

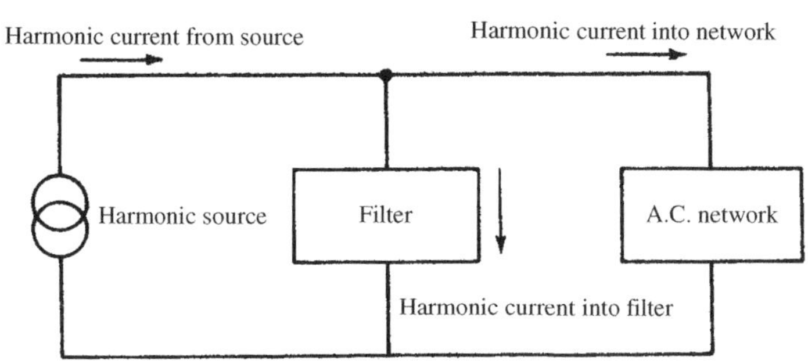

Figure 3.21 Simplified circuit for the harmonic source, filter and AC network impedance

converter). However, the capacitance required to meet such a target is too large, and it is usually more economical to attenuate the lower harmonic orders by means of single arm tuned filters, which for the double bridge HVDC converter are the 11 and 13, as shown in Figure 3.13.

The derivation of AC system harmonic impedances, their incorporation in AC filter design and a thorough description of the alternative configurations used in LCC conversion can be found in [3].

3.10.2 DC Side Filters

On the DC side of HVDC converters the voltage harmonics generate harmonic currents, whose amplitudes depend on known elements such as the delay and extinction angles, the overlap angle and the impedance of DC circuits (i.e. smoothing reactors, damping circuits, surge capacitors and the line itself).

In contrast to the AC filters discussed above, the DC filters:

- do not carry fundamental frequency power and therefore have substantially lower losses;
- do not need to provide reactive power, their only function being harmonic mitigation;
- have a main capacitor that has to withstand the full pole-to-neutral DC voltage.

Also, the harmonic impedances do not change with the operating conditions and it is therefore possible to use tuned filters with higher **Q** factors and thus smaller capacitors and reactors.

The criteria to be met by the filters relate mainly to telephone interference from the DC line and must comply with the CCITT directives:

- the induced voltage measured at subscribers' sets, for safety reasons, should not normally exceed 60 volts rms;
- the induced psophometric emf should not exceed 1 mV, if psophometric weightings are used for noise measurements; or
- 20 dB_{rnc} of noise when using the **C**-message weighting.

The performance requirements are normally assessed by the equivalent disturbing current, whereby all harmonic currents on the DC line are reduced to an 'equivalent disturbing current' (I_{eq}) at a single frequency, under the assumption that this current should cause the same interference effect on the telecom lines, i.e.

$$I_{eq} = \sqrt{\sum_{n=1}^{n=m} (I_n.H_n.C_n)^2} \tag{3.55}$$

where

n = harmonic order: the upper limit usually considered ranges between 2.5 and 5 kHz

C_n = normalised weighting factors at harmonic n, referred to 800 Hz or 1000 Hz for the psophometric and **C**-message weights, respectively

H_n = frequency-dependent factor, taking into account variations (if significant) of the mutual impedance among the HVDC line and the telecom lines, of shielding and telecom circuit balance.

The assessment of interference levels requires detailed information on the harmonic voltage and current profiles along the HVDC line; electromagnetic induction from harmonic currents is normally the main problem. Moreover, since both ends of the link contribute to the disturbance, it is necessary to obtain the profiles from each end and add their effects.

3.11 Frequency Cross-Modulation Across the LCC

The mechanism of converter cross-modulation at harmonic and interharmonic frequencies is summarised in Figure 3.22. The presence of a harmonic at k times fundamental frequency on the DC side of a 12-pulse HVDC converter will produce, on the AC side, positive sequence harmonics of orders $k+1$, $13 \pm k$, $25 \pm k$, $12n+1 \pm k$ ($n = 3, 4, \ldots,$) and negative sequence harmonics of orders $k-1$, $11 \pm k$, $23 \pm k$, $12n-1 \pm k$ ($n = 3, 4, \ldots$). These harmonic sequences are reflected back to the DC side as the kth harmonic and various high-order harmonics of $12 \pm k$, $24 \pm k$, $12n \pm k$ ($n = 3, 4, \ldots$). Among these harmonics, the most significant terms are the first-order kth harmonic on the DC side, and the positive sequence $k+1$ and negative sequence $k-1$ harmonics on the AC side. The higher harmonics are an order of magnitude smaller than the lower order harmonics. Therefore, for most analyses, particularly those with small distortion levels, it is reasonable to ignore the contribution from high-order harmonics.

With the high-order harmonics ignored, the presence of a second-harmonic distortion on the DC side will result in a positive sequence third harmonic and a negative sequence fundamental frequency component on the AC side.

The interaction mechanism described above can be extended to non-harmonic frequencies [12]. For instance, if there is a distortion near the fundamental frequency such as at 51 Hz on the DC side, the distortions on the AC side would be near the second harmonic (101 Hz) for

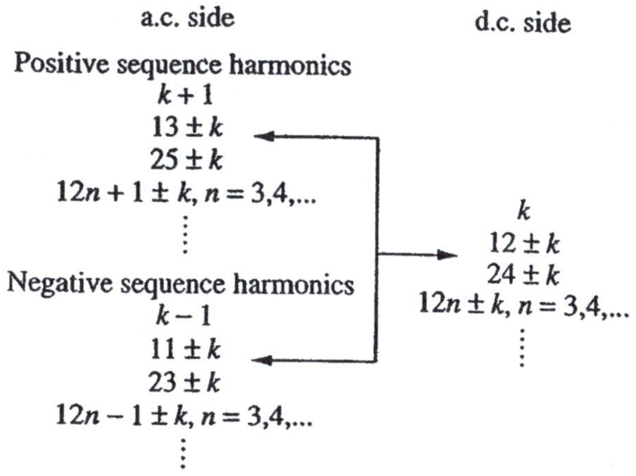

Figure 3.22 Associated three-port harmonic orders

the positive sequence component and near DC (1 Hz) for the negative sequence component. As the frequency of the DC side distortion approaches fundamental frequency, the lower corresponding frequency component on the AC side, which is in the negative sequence format, will be approaching 0 Hz, i.e. approaching DC.

If the DC side distortion is exactly at the fundamental frequency, the negative sequence component on the AC side is true DC, but with different levels in the three phases, which is customarily regarded as 'unbalanced DC' generated by the converter. However, the sum of the DC distortions in the three phases will be zero, and these distortions can be written mathematically in a negative sequence format as follows:

$$\begin{aligned} I_a &= |I| \cos(0.t + \delta + 0^\circ) \\ I_b &= |I| \cos(0.t + \delta + 120^\circ) \\ I_c &= |I| \cos(0.t + \delta + 240^\circ) \end{aligned} \tag{3.56}$$

This form of distortion has been referred to as 'negative sequence DC' [13], a useful term in the analysis of converter transformer core saturation instability because the distortion on the converter DC side, related to this instability, is close to the fundamental frequency. Therefore, the significant harmonic distortions on the converter AC side concerning this instability are the positive sequence second harmonic and the negative sequence DC. The DC will tend to saturate the transformer core which may ultimately lead to an instability.

3.12 Summary

This chapter has described the structure and analysed the voltage and current waveforms of the basic three-phase static converter and its extension to the double bridge configuration that has become standard in HVDC transmission. In early designs the harmonic contribution from the static power converter was approximately assessed, under the assumption of specified terminal AC voltage and DC conditions. Such conditions, however, can change as a result of the interactions that exists between the converter and the AC and DC systems. Thus, special consideration has been given to the effect of these interactions on the converter waveforms.

Extensive pre-design AC system measurements, computer simulation studies, and even a more rigorous approach including the effective contribution of the converter to the resonant frequencies, are now carried out to provide better designs and minimise the risk of harmonic instabilities. Therefore, after four decades of continuously improved designs, the present LCC-based HVDC converter has become a very robust and reliable transmission system component.

Considerations of energy efficiency and operational reliability still make thyristor-based LCC, despite being the least flexible solution, the preferred option for large-power and long-distance DC transmission. The extent to which alternative self-commutating converters, both of the VSC and CSC types, is expected to match the proven performance and reliability of the LCC conventional solution constitutes the subject of the following chapters.

References

1. Adamson, C. and Hingorani, N.G. (1960), *High Voltage Direct Current Power Transmission*, Garraway, London, Chap. 3.

2. Arrillaga, J. and Smith, B.C. (1998), *AC-DC Power System Analysis*, Power and Energy Series 27, IEE, London.
3. Arrillaga, J. and Watson, N.R. (2003), *Power System Harmonics*, Second Edition, John Wiley & Sons, Ltd, Chichester.
4. Paice, D.A. (1996), *Power electronic converter harmonics: Multipulse methods for clean power*, IEEE Press, New York.
5. Baird, J.F. and Arrillaga, J. (1980), 'Harmonic reduction in dc ripple reinjection', *Proceedings of the IEE*, C, **127**, 294–303.
6. Arrillaga, J. and Villablanca, M. (1990), 'A modified parallel hvdc convertor for 24-pulse operation', *IEEE Transactions*, paper 90SM281-6 PWRD.
7. Ainsworth, J.D. (1968), 'The phase-locked oscillator: a new control system for controlled static convertors', *Transactions of the IEEE*, **PAS-87**(3), 859–865.
8. Giesner, D.B. and Arrillaga, J. (1972), 'Behaviour of h.v.d.c. links under unbalanced a.c. fault conditions', *Proceedings of the IEE*, **119**, 209–215.
9. CIGRE WG 14-03 (1989), 'AC harmonic filters and reactive compensation with particular reference to non-characteristic harmonics', Complement to ELECTRA 63 (1979).
10. Ainsworth, J.D. (1981), 'Harmonic instabilities', Paper presented at the Conference on Harmonics in Power Systems, UMIST (Manchester).
11. Kimbark, E.W. (1971), *Direct Current Transmission*, Vol. I, John Wiley & Sons, Inc., New York.
12. Wood, A.R. (1993), 'An analysis of non-linear hvdc converter behaviour in the frequency domain and a new control proposal', PhD thesis, University of Canterbury, New Zealand.
13. Chen, S., Wood, A.R. and Arrillaga, J. (1996), 'HVDC converter transformer core saturation instability: a frequency domain analysis', IEE Proceedings, Generation, Transmission and Distribution, **143**(1), 75–81.

4

Self-Commutating Conversion

4.1 Introduction

The main shortcomings of LCC, discussed in Chapter 3, are large reactive power requirements (both during rectification and inversion), injection of low-order harmonic currents, risk of inverter commutation failures and their dependence on reasonably strong AC systems to provide the commutating voltages. These problems can be eliminated in self-commutated conversion by the use of more advanced switching devices with turn-on and turn-off capability.

The present self-commutating HVDC technology favours the use of IGBT-based VSC, combined with high-frequency sub-cycle switching carried out by PWM (the subject of Chapter 5).

Although the IGBT switch can also be used in CSC, a diode is needed in series with the IGBT in this case to provide sufficient reverse voltage withstand capability and the extra diode increases the converter losses.

Multi-level conversion (the subject of Chapters 6), unlike PWM, uses fundamental frequency switching and can, therefore, be designed with thyristor-type switching devices (such as the GTO and IGCT). In particular, the IGCT is an ideal switch for HVDC application due to its higher voltage and current ratings, high reverse voltage blocking capability (without the need for the series diode) and low snubber requirements.

Most of this chapter is concerned with the basic structure and operating principles of self-commutating VSC. A brief description is also made of self-commutating CSC and of the concept of self-commutating reinjection conversion, a subject considered in greater detail in Chapter 7.

4.2 Voltage Source Conversion

4.2.1 VSC Operating Principles

To simplify the description of the VSC operating principles, the converter valves are assumed lossless, the DC capacitors ripple negligible and the converter transformer lossless. The

Flexible Power Transmission: The HVDC Options J. Arrillaga, Y.H. Liu and N.R. Watson

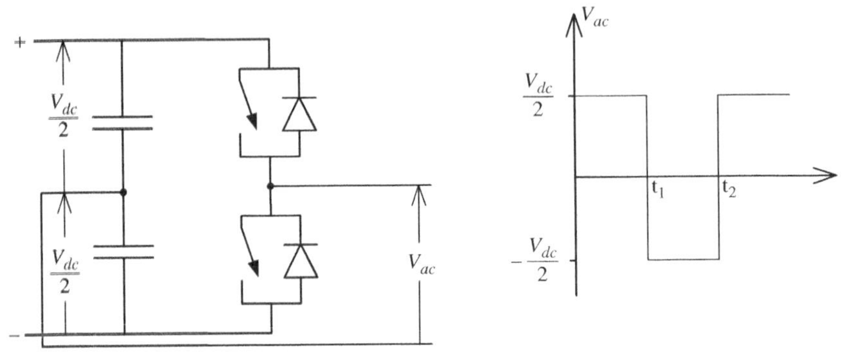

Figure 4.1 Two-level single phase VSC (Reproduced by permission of CIGRE.)

description is made here with reference to the two-level converter, where the AC terminals are switched in a bipolar way between two discrete voltage levels ($+V_d$ and $-V_d$); the switching circuit for one phase and its corresponding voltage output are shown in Figure 4.1.

It should be noted that the valve conducting periods are now 180°, whereas in the CSC case (described in Chapter 3) the width of the pulse was 120° and, therefore, the output phase (the current in that case) was $+I_d$, 0 or $-I_d$, i.e. the bridge AC phase current had three levels.

In practice the high harmonic content of the two-level output waveform needs to be reduced by PWM. Alternatively, instead of two levels, VSC can use the variety of multi-level topologies described in Chapters 6 and 7.

Since the conduction in solid-state switches is unidirectional, anti-parallel diodes are connected across them, to ensure that the bridge voltage has only one polarity, while the current can flow in both directions. In Figure 4.1 the midpoint of the capacitors is used as a reference for the AC output voltage.

There are four possible current paths in a single phase two-level converter [1], as shown in Figure 4.2. When the upper switch is ON, the output voltage is $+V_d/2$ and the current flows through the upper free-wheeling diode if the current is negative (Figure 4.2(a)) or through the upper switch if the current is positive (Figure 4.2(b)). When the lower switch is ON, the output voltage is $-V_d/2$ and the current flows through the lower switch (Figure 4.2(c)) if the current is negative or through the lower free-wheeling diode (Figure 4.2(d)) if the current is positive.

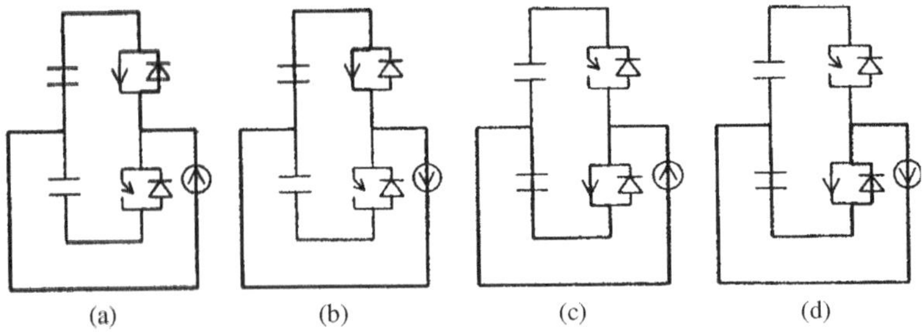

Figure 4.2 Current paths in the single phase two-level VSC (CIGRE copyright)

When the two main switches are blocked, the anti-parallel diodes form an uncontrolled rectifier. In this condition the application of an external AC voltage charges the DC capacitors, via the uncontrolled rectifiers, to the peak value of the AC voltage across the upper and lower DC capacitor. Once the DC capacitor is charged and the external source connected, VSC can commence.

The main switches can be turned on and off in any desired pattern; however, immediately before one switch is turned on, the opposite switch must be turned off, as their simultaneous conduction would create a short circuit of the DC capacitors. This causes a small blanking period (of a few microseconds) where none of the switches are ON and the current path is via the free-wheeling diodes. To explain this condition let us start at a point, after $t = 0$ in Figure 4.1, where the upper switch is ON and, thus, the AC terminal is connected to the plus terminal of the DC capacitors; this results in current flowing through the upper switch (as shown in Figure 4.2(b)). When the upper switch is turned off (at instant t_1 in Figure 4.1), the reactance of the AC circuit maintains the present current, thus the diode in parallel with the lower switch turns on (creating the conducting path shown in Figure 4.2(d)). As a result, the output voltage changes from plus to minus $V_d/2$, i.e. the polarity reversal has been initiated by turning off the upper switch.

Thus there is no need to turn on the lower switch immediately following the turn-off of the upper one. When, after the blanking time the lower switch is turned on, this switch takes over the current when the current reverses its polarity (via the conducting path of Figure 4.2(c)); there is a delay dependent on the reactance of the transformer and phase reactor as well as the AC system voltage.

A similar procedure, via the opposite switches and diodes, can be used to explain the transfer from the negative to positive output voltage waveform. This process applies to all VSC configurations irrespective of the fundamental power flow operating condition (i.e. in the four quadrants). Therefore at the fundamental frequency the VSC can be represented as a voltage phasor, with the magnitude and phase angle of the AC output voltage determined by the DC voltage and the firing pulse patterns. The power implications are discussed in the following section.

Power transfer control

If the VSC is connected to a passive network on the AC side, the power flows unidirectionally from the DC input side to the passive AC load. If the VSI is connected to an active AC system, the power can be made to flow in either direction by making the phase angle of the converter AC output voltage positive or negative with respect to that of the AC system voltage.

The switches must block a unidirectional voltage but need to be able to conduct current in either direction if bidirectional power flow is required.

Figure 4.3(a) shows the fundamental frequency phasor representation of the VSC operating as an inverter and supplying active and reactive power to the AC system. In this operating condition the diagram shows that the VSC output voltage (V_2) has a larger amplitude and is phase advanced with respect to the AC system voltage (V_1).

As described in the introductory chapter, the active and reactive power exchanges (P and Q) between two active sources are expressed as

$$P = (V_2 \sin(\delta)/X)V_1 \tag{4.1}$$

$$Q = (V_2 \cos(\delta) - V_1)/X)V_1 \tag{4.2}$$

where δ is the phase angle and X the reactance between V_1 and V_2.

If the converter is used only as a reactive power compensator, then there is no need for an active DC system, and the converter is terminated by a DC capacitor. However, the capacitor size should be sufficient to permit the charge and discharge operations caused by the switching sequence of the converter valves without exceeding the specified ripple.

Figure 4.3(b) illustrates the case when the VSC operates purely as a reactive power compensator, i.e. the converter (V_2) and supply system (V_1) voltages are in phase with each other. Thus, when they both have the same magnitude there is no exchange of reactive power between the converter and the system (i.e. the current is zero): when V_2 is larger than V_1 the current leads the voltage by 90°, i.e. the converter behaves as a capacitor and, thus, generates reactive power; when V_2 is smaller than V_1 the current lags the voltage by 90°, i.e. the converter behaves like an inductor and, thus, absorbs reactive power.

In this mode of operation the VSC is analogous to an ideal rotating synchronous compensator (and is usually referred to as a STATCOM) [2]; however, the VSC has no inertia and its response is practically instantaneous. Thus the VSC acts as an AC voltage source, controlled to operate at the same frequency as the AC system to which it is connected.

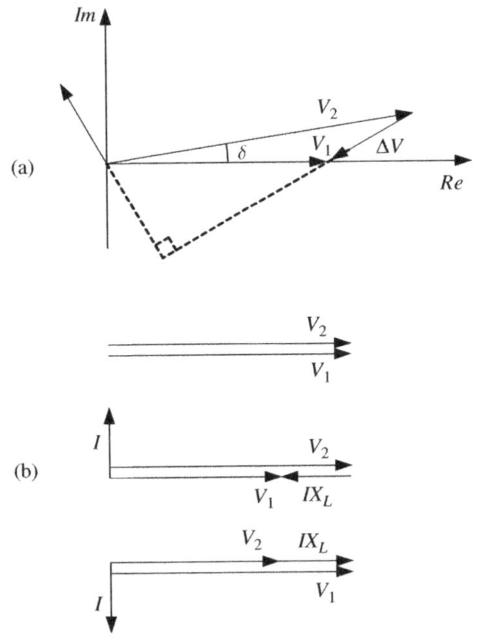

Figure 4.3 Operating modes of the VSC: (a) phasor diagram when providing active and reactive power; (b) phasor diagram when operating as a reactive power controller

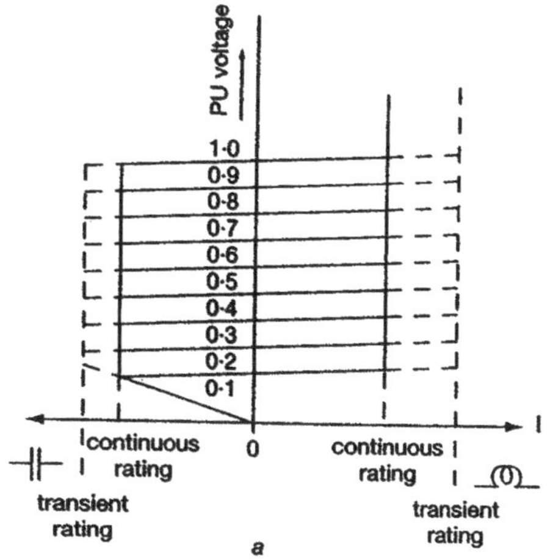

Figure 4.4 Voltage/current characteristics of a VSC operating as a STATCOM

The voltage/current characteristics of the VSC operating as a STATCOM, shown in Figure 4.4, illustrate that this type of converter can provide full capacitive support for voltages down to approximately 0.15 per unit. This is a considerable advantage over conventional static VAR compensators, such as the thyristor-based static VAR compensator (SVC) (in which the maximum capacitive current capability at low voltage is very limited).

If the converter is connected to an active DC system (e.g. to another converter as in the case of Figure 4.5) the AC current and voltage can have any phase relationship, and the converter can act as a rectifier or an inverter, and with leading or lagging reactive power (i.e. four-quadrant operation is possible as illustrated in Figure 4.6). Thus the VSC can be controlled to operate at any point within the circle of Figure 4.6, the radius of which represents the converter MVA rating.

There are, of course, active and reactive power limits determined by the maximum allowable valve current and maximum allowable DC voltage respectively on the storage capacitor. For a given AC system voltage, the DC voltage rating is determined by the maximum AC output voltage that the converter must generate to provide the maximum required reactive power.

The *PQ* characteristics depend on the AC converter voltage, and, therefore, the interface transformer ratio can be used to optimise them. Moreover, with the addition of an on-load tap-changer (OLTC), the transformer ratio can be continuously optimised to maximise the steady-state power capability of the converter. Also the OLTC can be used to minimise the power losses of the VSC transmission system.

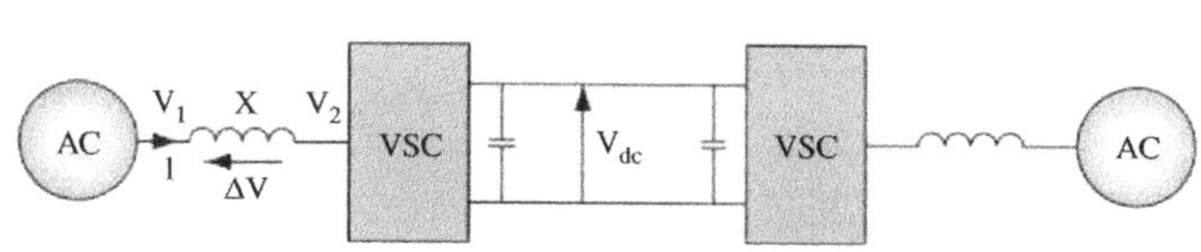

Figure 4.5 Basic VSC interconnection

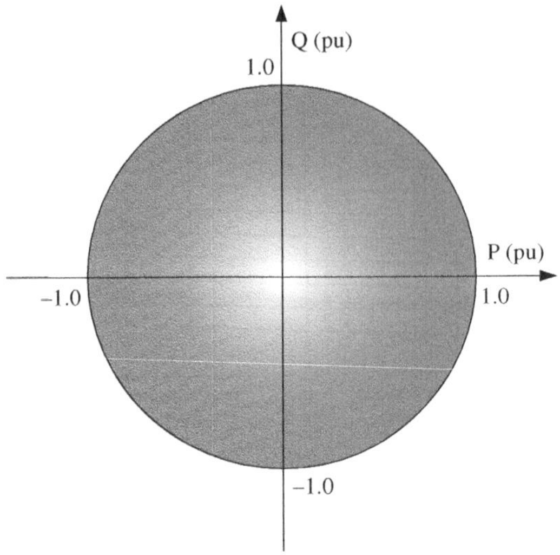

Figure 4.6 *PQ* diagram of VSC power transfer

4.2.2 Converter Components

As the name indicates, VSC power conversion implies the presence of a voltage source on the DC side. A voltage source maintains a prescribed voltage across its terminals irrespective of the magnitude or polarity of the current flowing through the source.

The circuit diagram of Figure 4.7 [1] shows the basic components of a VSC substation, which consist of one or more converter units including the high-voltage valve chains, DC capacitors, reactors, transformers and filters, as well as (not shown) control, monitoring, protection and auxiliary equipment.

DC capacitor

In a VSC the DC side is strongly capacitive, and thus voltage stiff. In this respect, the DC capacitor serves the purpose of stabilising the DC voltage. Consequently the voltages are well defined and are normally considered independent of the converter operation.

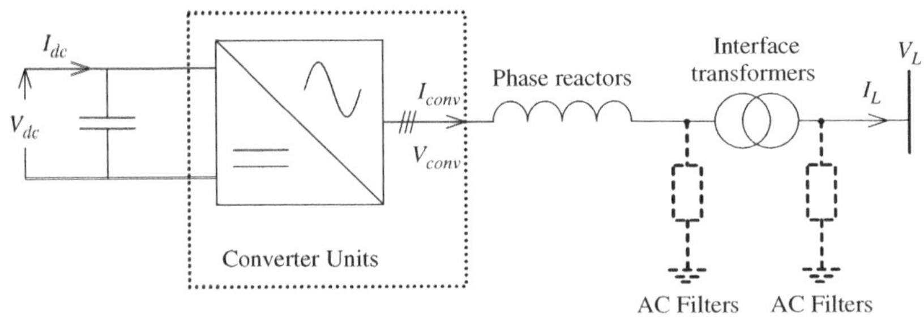

Figure 4.7 Basic diagram of a VSC substation (Reproduced by permission of CIGRE.)

However, in practice, VSC operation produces some harmonic currents in the DC circuit. These currents, due to the presence of DC circuit impedance, give rise to corresponding harmonic voltages, and therefore DC voltage ripple. They also affect the ripple on the capacitors of other stations connected to the same DC system. The DC voltage ripple is mainly influenced by the following factors:

- the VSC valve switching strategy;
- the capacitance of the DC capacitor;
- AC system imbalance and distortion.

The size of the DC capacitor reduces with increasing switching frequencies, because every time the valves are switched the DC current in the capacitor changes direction, thus reducing the effective ripple.

Coupling reactance

The VSC is not directly connected to the AC supply when the latter is a strong system (i.e. has a large short-circuit capacity compared with the converter rating). In such a case, some coupling reactance is needed between the system busbar and the converter terminals. As well as reducing the fault current, this coupling reactance stabilises the AC current, helps to reduce the harmonic current content and enables the control of active and reactive power from the VSC. This component includes the inductance of the interface transformers and any extra phase reactor possibly added for these purposes. The interface reactors and transformers must withstand any high-frequency voltage stresses imposed by VSC operation. Thus, depending on the converter configuration, it may be economical to place a high-frequency blocking filter between the converter and the interface transformer.

The high-voltage IGBT valve

The switches constituting the valves of self-commutating converters are either turned off (i.e. non-conducting) or saturated (i.e. completely turned on, with only a small voltage drop across the switch). To withstand the very high voltage ratings of HVDC transmission, the converter valves comprise many series-connected IGBTs. For instance, the valves used in the original 150 kV HVDC Light schemes had up to 300 series-connected IGBTs. The IGBT valve will in principle experience the same requirements as the thyristor valve of the conventional line-commutated HVDC converter (described in Section 8.2).

A typical IGBT building block consists of a single switching device, snubber components and the gate drive. The gate board is equipped with overvoltage, undervoltage and overcurrent protection. High reliability is provided by mounting the IGBT on a separate heat sink, thermally decoupled from the snubber resistor. For diagnostic purposes the switches are connected by fibre optics to a pulse distribution board.

Good voltage sharing is critical to ensure that all the devices in series experience similar operating conditions. The use of snubber circuits reduces this problem but adds to the complexity of the valve and increases the power losses. It is possible to design the gate control of the IGBT valve to maintain acceptable voltage sharing without snubbers, provided that the spread in the devices' characteristic switching times, switching transient behaviour and blocked state leakage currents are sufficiently small.

Static voltage sharing is achieved by means of a parallel-connected resistor per IGBT module. The function of the resistor is to provide a leakage current which is dominant over the IGBT, thus forcing the voltage sharing by potential divider action. The resistor current is typically 10 times the IGBT leakage current, in order to swamp all the variations of leakage currents under different operating and temperature conditions.

All IGBTs experience a variation of turn-on delay time with temperature, the nominal value varying from device to device. The rise time also causes significant variations in total turn-on time between individual devices in the series assembly. The slower the device performs at turn-on, the greater is its share of total valve voltage.

The temperature, commutated current and fall time cause variations in turn-off delay time between the series-connected devices; the slower the device turn-off, the greater its share of the total voltage.

For series-connected IGBTs, the use of a snubber capacitor improves the dynamic voltage sharing (i.e. the spread in turn-off delay), while a *di/dt* inductor helps the IGBTs to share the voltage (due to spread in turn-on delay time) and diode voltage sharing under reverse recovery condition.

The addition of a small *di/dt* limiting inductor helps to reduce the turn-on losses. Although the contribution of the turn-off snubber capacitor to loss reduction is less significant, it is still justifiable. The snubber shifts the switching losses from the silicon to a low-cost passive component where it can be dissipated away from the IGBT or recovered into the supply or load.

The number of series-connected IGBTs required to sustain the converter voltage rating must be increased to provide the VSC valve with redundancy to continue operating following the failure of individual devices. Also, for the valve to continue operating, a faulty IGBT must not create an open circuit, i.e. the faulty device must establish a short circuit and be capable of conducting current indefinitely, until the next scheduled maintenance period. This is achieved by means of specially developed packaging of IGBTs for use in VSC transmission, because the standard IGBT modules normally use bond wires that create an open circuit following failure.

In contrast with the thyristor valves of LCC, the individual switches (GTOs, IGCTs or IGBTs) in VSC transmission schemes have to withstand high voltage for a substantial part of its operation time and, therefore, the probability of an incident particle initiating a destructive current avalanche during the blocking stage is greatly increased.

In contrast with CSC transmission converters, which normally operate in stable conditions (i.e. the power changes occur at moderate rates), the higher controllability of VSCs puts heavier demands on their use as adaptive system controllers.

As well as the rated current, the valves have to withstand peak current due to ripple and transient overcurrent levels and need to be provided with an additional protection margin; they also need to be capable of turning off the current following a short circuit close to the valve itself. The gate unit must maintain safe device operation during short-circuit conditions. The IGBT is better behaved in this respect.

The anti-parallel diodes

The anti-parallel diodes in the VSC bridge constitute an uncontrolled bridge rectifier and, thus, unlike the IGBTs (which can be switched off in a few microseconds during transients), the diodes have to be designed to withstand the fault-created stresses.

A DC short-circuit fault, as shown in Figure 4.8, creates a current path through the diodes and the fault current is limited only by the AC system and converter impedances. This fault current needs to be cleared by a circuit breaker on the AC side, which will require a few cycles to operate.

The diode will also experience inrush current (as well as an overvoltage on the DC bus) if the VSC is energised by the circuit breaker under zero voltage on the DC side. This overvoltage must be limited by external components (such as pre-injection breaker resistors). Alternatively, the DC capacitor can be charged independently prior to converter energisation. In this case a DC blocking voltage will appear after the initial current surge.

4.2.3 The Three-Phase VSC

The basic configuration of a six-pulse VSC, shown in Figure 4.9, consists of a three-phase bridge connecting the AC source to a predominantly capacitive DC system.

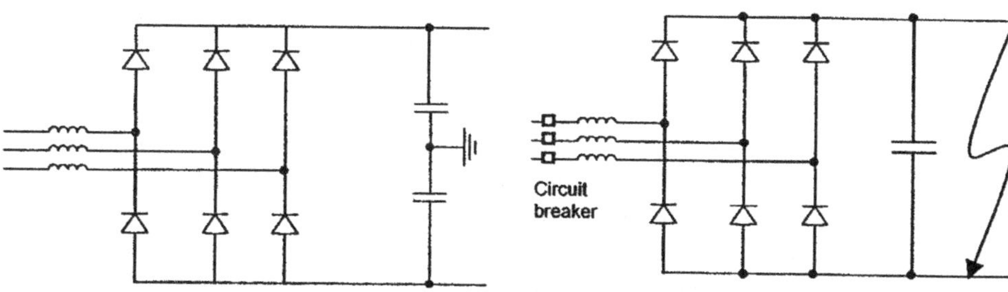

Figure 4.8 A two-level VSC bridge with the IGBTs turned off

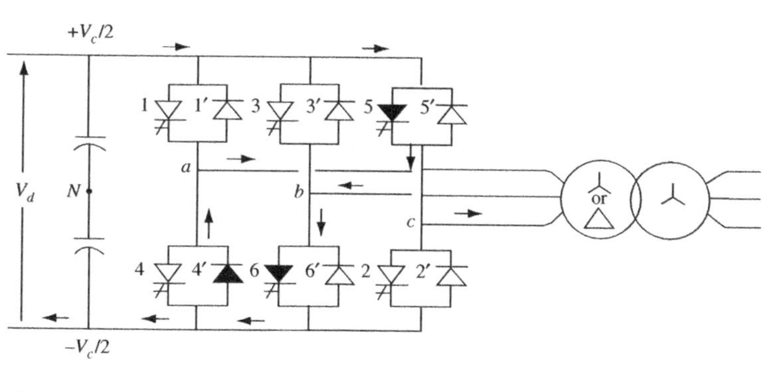

Figure 4.9 Three-phase full-wave bridge VSC (Reproduced by permission of CIGRE.)

The bridge valve unit is an asymmetric turn-off device (shown in Figure 4.9 as a GTO) with a reverse-connected diode in parallel. The turn-off device carries out the instantaneous inverter function, while the diode is needed to provide a path for the transfer of energy from the AC to the DC side (i.e. rectifier operation) to charge the capacitor. During each cycle there will be periods of rectification and inversion determined by the phase angle, and, therefore, the average current will determine the net power flow direction (i.e. rectification or inversion). If the converter operates as a rectifier with unity power factor, only the diodes conduct the current, whereas during inverter operation with unity power factor only the turn-off devices conduct.

By appropriate control of the turn-on and turn-off switchings, a three-phase AC waveform is produced at the AC output.

For fast transients the DC capacitor can be regarded as a perfect voltage source. Thus in the short time sequence in which the switching devices are controlled to interconnect the DC and AC terminals, the DC voltage remains practically constant and the VSC produces a quasi-square-wave AC voltage.

The introductory material used in Chapter 3 to describe the operation of the line commutated three-phase CSC applies equally to the VSC process, which can be considered its dual. There are, however, two important differences between CSC and VSC. One is the duration of the valve conducting period, which in the case of the VSC is 180° instead of the 120° generally adopted for the CSC configuration. The 180° conduction is needed in the VSC case to avoid the condition in which both arms of the bridge leg are in the off-state; this condition would occur if a 120° period were used, which would cause uncertainty in the output voltage. The other important difference is the absence of commutation overlap, which makes VSC operation more predictable and easier to analyse.

As for the line-commutated configuration, the valves in the circuit of Figure 4.9 are controlled to commutate only at the power system frequency. The time sequence of valve conduction is as described for the line-commutated CSC and displayed in Figure 3.2(a). However, instead of producing the rectified voltage waveforms $R_1Y_3B_5$ (on the common cathode side) and $B_2R_4Y_6$ (on the common anode side), the VSC configuration converts half of the DC voltage into 180° rectangular waveforms, both on the common cathode and common anode sides of the converter bridge.

Figure 4.10 shows the following waveforms:

- The phase voltages v_{aN}, v_{bN}, v_{cN} with respect to the capacitor midpoint (not with respect to the transformer neutral, as these two points are not directly connected).

- The line voltages on the converter side (i.e. v_{ab}, v_{bc}, v_{ca}). Again, as in the line-commutated case, the AC side has only three-phase conductors (i.e. the transformer neutral is floating) and, therefore, these voltage waveforms are of 120° duration.

- The voltage of the floating neutral in the secondary of the star-connected transformer v_n, with respect to the midpoint N. This potential is the sum of the instantaneous potentials of the three phases (v_{aN}, v_{bN}, v_{cN}) and is a square AC waveform varying at three times the fundamental frequency, with a magnitude equal to one-sixth of the DC voltage.

- The phase voltage v_{an} across the star-connected transformer secondary.

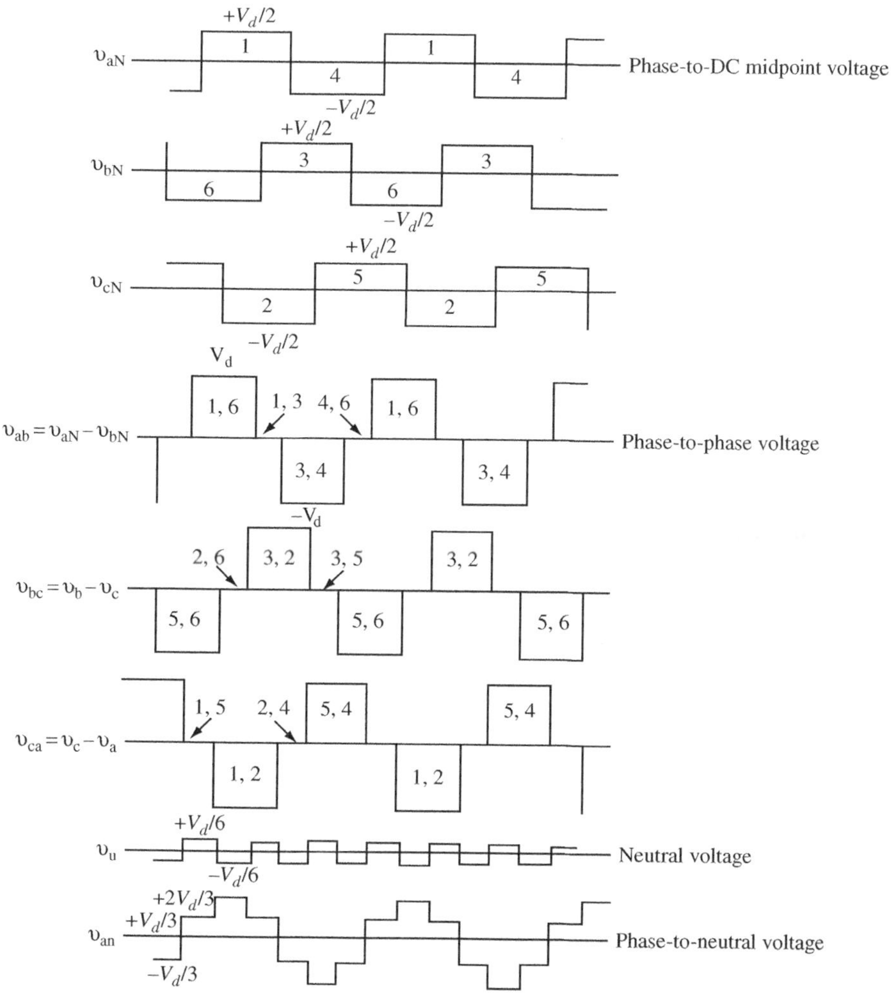

Figure 4.10 AC voltage waveforms of the VSC in Figure 4.9 (Reproduced by permission of CIGRE.)

Analysis of the fundamental and harmonic voltage components

Fourier analysis of the phase voltage waveforms v_{aN}, v_{bN}, v_{cN} which, as shown in Figure 4.10, are rectangular waves of amplitude $V_d/2$ (V_d being the DC side voltage) and 180° width, provides the following time domain expressions for the three phases [2]:

$$v_{aN} = (4/\pi)(V_d/2)[\cos(\omega t) - (1/3)\cos(3\omega t) + (1/5)\cos(5\omega t) - (1/7)\cos(\omega 7t)\ldots] \tag{4.3a}$$

$$v_{bN} = (4/\pi)(V_d/2)[\cos(\omega t - 2\pi/3) - (1/3)\cos 3(\omega t - 2\pi/3) + (1/5)\cos 5(\omega t - 2\pi/3) - (1/7)\cos 7(\omega t - 2\pi/3)\ldots] \tag{4.3b}$$

$$v_{cN} = (4/\pi)(V_d/2)[\cos(\omega t + 2\pi/3) - (1/3)\cos 3(\omega t + 2\pi/3) + (1/5)\cos 5(\omega t + 2\pi/3) - (1/7)\cos 7(\omega t + 2\pi/3) \ldots] \quad (4.3c)$$

These three equations indicate the following sequences for the fundamental and harmonic frequency components:

- The fundamental component is a symmetrical set of positive sequence.
- The triplen harmonics are all of zero sequence (i.e. their values are the same in the three phases at all instants).
- The fifth harmonic is of positive sequence.
- The seventh harmonic is of negative sequence etc.

Subtracting the neutral voltage v_n from v_{aN} (refer to Figure 4.10) gives the phase-to-neutral voltage (v_{an}), i.e.

$$v_{an} = (2V_d/\pi)[\cos(\omega t) + (1/5)\cos(5\omega t) - (1/7)\cos(7\omega t) - (1/11)\cos(11\omega t) \ldots] \quad (4.4)$$

which is a three-level voltage of magnitudes 0, $V_d/3$ and $2V_d/3$ and is free from triplen harmonics. It should be noted that v_{aN} and v_{an} are in phase. Voltages v_{bn} and v_{cn} are the same but phase shifted by 120° and 240° respectively from v_{an}. The phase-to-phase voltage (v_{ab}) consists of two levels 0 and V_d; its fundamental component is phase shifted by 30° with respect to v_{an} and its amplitude is $\sqrt{3}$ times the v_{an} amplitude.

The time domain expression of the 120° phase-to-phase voltage is

$$V_{ab} = (2\sqrt{3}/\pi)V_d[\cos(\omega t) - (1/5)\cos(5\omega t) + (1/7)\cos(7\omega t) \ldots] \quad (4.5)$$

which shows that the triplen harmonics have now been eliminated.

The total rms value of V_{ab} is $0.816V_d$, and the fundamental component (V_c) of the converter voltage is

$$V_c = (\sqrt{6}/\pi)V_d = 0.78V_d \quad (4.6)$$

Thus in the stiff VSC the AC voltage output is a function of the DC voltage only and, therefore, to alter the former requires a corresponding change in the latter. This can be achieved by charging or discharging the DC capacitor either from a separate source or from the AC system. It will be shown in the following chapter that it is possible to vary the AC voltage without changing the DC voltage using PWM.

The harmonic voltages are related to the fundamental frequency voltage by

$$V_n = V_c/n \quad (4.7)$$

Thus the harmonic analysis of the VSC process is extremely simple as compared with that of LCC due to the practically instantaneous turn-off provided by the advanced switching devices.

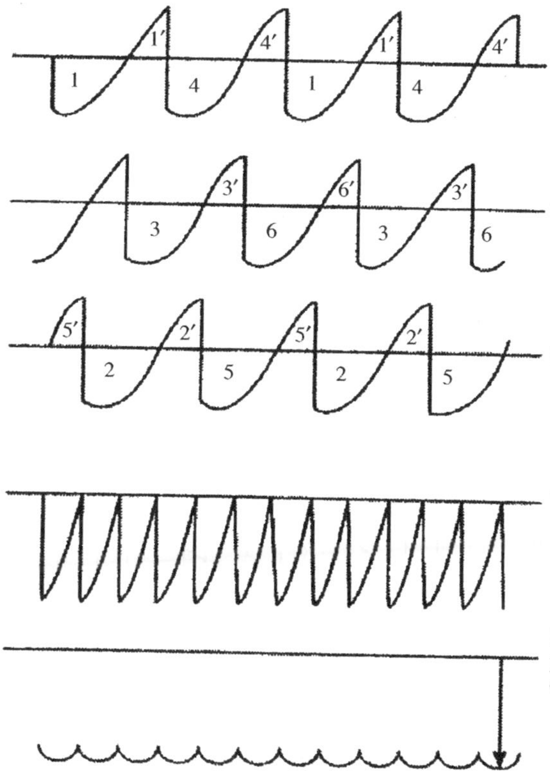

Figure 4.11 DC waveforms of the VSC in Figure 4.9 (Reproduced by permission of CIGRE.)

Figure 4.11 shows that the DC current waveforms of the VSC of Figure 4.9 are the same as the voltages of the six-pulse LCC if, in the latter case, the commutation overlap is ignored.

Thus, following the same reasoning as in Section 3.4, the current waveform contains a direct component and a series of harmonics of orders $n = 6k$. The direct component is equal to

$$I_d = \frac{3\sqrt{2}}{\pi} I \cos(\theta) \tag{4.8}$$

where I is the AC phase current rms and θ the firing angle, which in self-commutation is also the power factor angle (thus replacing both the terms α and ϕ used in LCC in Chapter 3). Current I_d changes from $+1.35I$ to $-1.35I$ and vice versa as the firing angle changes from full rectification to full inversion.

The ratio of the nth harmonic to the peak value of the instantaneous DC current (refer to Equations 3.29–3.30) will have a minimum value of $\sqrt{2}/(n^2 - 1)$, which occurs when the power factor is unity, and a maximum of $\sqrt{2}n/(n^2 - 1)$, which occurs when the power factor is zero.

All these equations show that the higher the value of n (the harmonic order), the lower the harmonic amplitude, which explains the importance of eliminating the low orders. However, as explained in Chapter 3, during unbalanced conditions the second-harmonic and other low-frequency orders reappear.

In the above development it is assumed that the AC is perfectly sinusoidal. In practice the phase currents will contain small contents of other harmonics that may be present in the AC system and converter voltage waveforms.

The 12-pulse VSC configuration

Similar to CSC LCC, the combination of two bridges with their AC voltage waveforms phase shifted by means of the star and delta interface transformer connections constitutes a 12-pulse converter. The difference with respect to the CSC configuration is that in the VSC case the primaries of the two interface transformers must be connected in series, so that the individual bridge voltages are added at the output. The 12-pulse structure and voltage waveforms are illustrated in Figure 4.12.

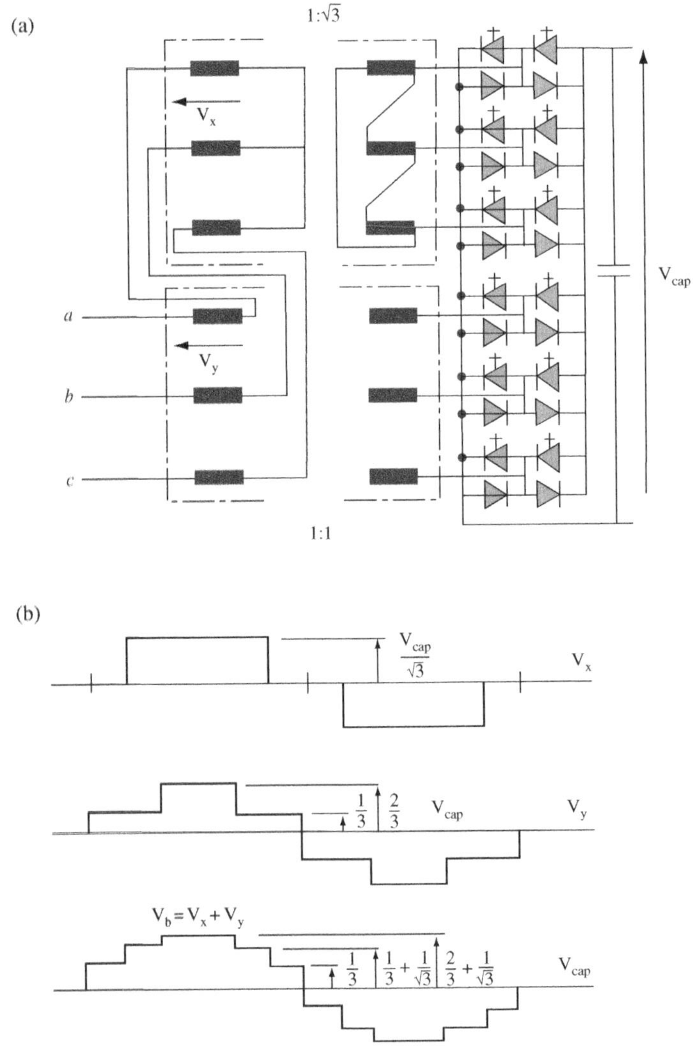

Figure 4.12 The 12-pulse VSC: (a) circuit structure; (b) voltage waveforms

Multi-bridge conversion

A possible option for HVDC application is the connection of several phase-shifted bridges in series on the DC side, to share the voltage ratings as well as reduce the harmonic content. This alternative to PWM only uses fundamental frequency switching and, thus, reduces the harmonic content with lower switching losses. For example, the use of four transformers in series with relative phase displacements of 15° giving rise to 24-pulse operation; similarly eight transformers having 7.5° phase shift will produce 48-pulse operation, etc. However, this will require the use of complicated zigzag connections. Alternatively, the multi-stepped waveform can be created by phase shifting the switching instants of the series-connected bridges with respect to each other.

The magnetic-transformer-coupled multi-pulse VSC has been implemented in 18-pulse and 48-pulse converters for battery energy storage and static condenser applications respectively [3, 4]. It must be understood that the nominal pulse number of the multi-bridge configuration can only be achieved under ideal conditions. In practice the individual harmonics cannot be completely cancelled, due to imperfections in the system components and asymmetry and distortion in the AC system voltage. However, their residual magnitudes will normally be acceptably small.

The main disadvantage of this configuration is the multiplicity of transformers, each producing extra losses and requiring different levels of isolation from earth. This, added to the need for a balanced impedance between phases and windings, limits the practical number of windings to four.

Harmonics penetration

Since the VSC acts as a voltage harmonic source, some harmonic voltage will appear at the PCC (Point of Common Coupling) with other consumers. The voltage harmonic transfer at each frequency will be proportional to the ratio of the system impedance at that frequency to that of the VSC impedance at the same frequency; the VSC impedance will include the interface transformer, phase reactors and filters.

The system impedance at some frequencies may be capacitive (due to the charging capacitance of transmission lines and cables, the presence of shunt capacitors, etc.), and may create series resonances. Thus comprehensive harmonic studies are required to determine the frequency variation of the AC system impedance under different operating conditions. This procedure is no different from that used for LCC design, described in detail in [5], and should include the presence of already existing system harmonic content.

When these studies indicate that the VSC contribution at the PCC exceeds the permissible level, harmonic filters will be needed following the recommendations of [6]. As the low-order harmonics will normally be suppressed by PWM or multi-level topologies, these filters will be tuned to higher frequencies and will, therefore, be cheaper and more compact. Moreover, the absence of low-order harmonic filters also eliminates the need for extra switchgear required to subdivide the filter branches following variations in the (LCC) reactive power requirements.

The ability of the converter to compensate for any variations in the exchange of reactive power avoids the need for filter branches' subdivisions and related switchgear.

Current and power components

The AC current results from the difference between the AC system and converter voltages divided by the transformer and any extra interface impedance; to simplify the explanation it is assumed that this current is perfectly sinusoidal. By altering the turn-on and turn-off instants of the converter valves, the converter voltage can be made to lead or lag the system voltage and thus the current can be placed in any of the four quadrants. This implies four-quadrant power controllability, i.e. it permits rectification and inversion with absorption or generation of reactive power. The response is practically instantaneous as the VSC has no inertia. Of course bidirectional exchange of active power would require stored energy on the DC side, e.g. in the form of a large battery or superconducting magnet. If real power exchange is not required, the VSC behaves as an ideal synchronous compensator. In this case the VSC exchanges reactive power with the AC system by maintaining the rectangular output voltage waveform in phase with the power system voltage and varying its magnitude. In this mode of operation, as there is no real power exchange the active power on the AC side of the converter is equal to the DC side power [7], i.e.

$$V_d I_d = V_{ter} V_c \sin(\delta)/(\omega L) = B_t (V_{ter} V_c \sin(\delta)) \tag{4.9}$$

where

V_{ter} = the AC system voltage at the converter terminals (i.e. on the primary side of the converter transformer)
V_c = the converter AC side voltage
δ = the phase angle difference between V_{ter} and V_c (which can be altered by varying the position of the turn-on and turn-off switching instants)
ωL = the transformer reactance (with the winding resistance ignored)
B_t = the value of the susceptance across the transformer.

Substituting Equation (4.6) in (4.9), the DC becomes

$$I_d = (\sqrt{6}/\pi)\ V_{ter} \sin(\delta)/(\omega L) = B_t \left(\frac{\sqrt{6}}{\pi} V_{ter} \sin(\delta) \right) \tag{4.10}$$

and since the capacitor voltage is proportional to the charge (or current) going into the capacitor, by making δ positive (V_{ter} leading V_c) the capacitor voltage will increase. Similarly, when δ is negative (V_{ter} lagging V_c) this will reduce V_d. Moreover, in the steady state δ will need to be slightly positive to supply the internal power losses in the converter.

Power losses

There are three contributors to the IGBT losses. First the on-state or conduction loss, which depends on the conducting resistance, the rms current through the device and the power factor. The other two contributors are the turn-on and turn-off losses, which depend on the current and voltage at each switching event, the switching frequency and the characteristics of the device during turn-on and turn-off. These three factors also apply to the diode part of the valve, although the latter experiences much lower turn-on loss, because it switches on as soon as a forward voltage appears.

Some extra losses are produced in the gates, snubber circuits and leakage currents during the off-state. The power factor is normally different for rectifier and inverter operation; the highest conduction loss tends to occur during inverter operation for the IGBT and in rectifier operation for the diode.

Because the IGBT switching losses are due to the simultaneous presence of high voltage and high current during much of the switching region, the switching processes should be completed as soon as possible, i.e. under the largest possible dv/dt and di/dt conditions. However, the series connection of IGBTs requires moderate values of dv/dt per device. Moreover, due to the presence of stray inductance in the commutating circuit a high di/dt will produce an extra voltage ($L\ di/dt$) across the IGBT; it is therefore important to design the converter layout to keep the stray inductance as low as possible.

A comparison of the power losses in two-level (operated at 1650 Hz) and three-level VSC topologies with the conventional LCC alternative, to maintain similar orders of waveform distortion, gives values of 3.5%, 2% and 0.8% respectively for the two-level, three-level and LCC solutions.

Electromagnetic interference

The VSC output waveform contains energy in the RF range and is, therefore, a source of electromagnetic interference (EMI). Although the radiated component is normally attenuated by the metal enclosure of the VSC housing, some of the EMI energy will be transmitted by conduction into the power lines and the noise level may be orders of magnitude higher than the radiated noise. This effect is illustrated in Figure 4.13 and contains a differential mode voltage (creating a current I_{dm} flowing between the lines) and a common mode voltage (creating a current I_{cm} flowing from lines to ground). These two modes will have to be taken into account in the design of the high-frequency filter. The problem can be reduced by means of properly designed snubbers at the expense of some extra losses.

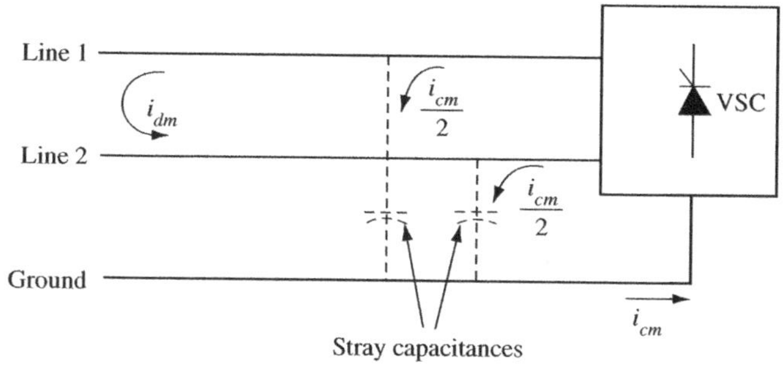

Figure 4.13 Conducted interference of a VSC

4.3 Comparison of LCC and VSC

The main differences between the conventional LCC and self-commutating VSC configurations result from the fact that the latter do not require an AC voltage source for the commutations, and that these are practically instantaneous, whereas the duration of the commutations in LCC depends on the voltage source level and the commutation reactance (which is dominated by the converter transformer leakage reactance). Thus, even when the firing angle is 0°, LCC consumes reactive power, the amount of which can be as much as 60 % of the active power in an HVDC converter. Moreover, the reactive power requirement in LCC varies with the amount of active power conversion and, therefore, the converter filters and VAR compensation structures need to be adjusted as the load varies. VSC, on the other hand, can provide any reactive power (sink or source) independently of the active power transfer.

In LCC, as the active power transfer varies, the corresponding changes in the reactive power produce AC system voltage fluctuations. The AC voltage drops cause additional reactive power consumption and further voltage reduction, which may lead to voltage instability. To prevent the instability a minimum short-circuit ratio (SCR), defined as the ratio between the short-circuit power and converter power ratings, is required, which can be typically 2. The VSC has no minimum SCR requirement, but the active power transfer is limited by the impedance of the AC system viewed from the converter AC terminal.

Since each arm of the converter bridge switches only once per cycle, LCC needs filters for the 12-pulse related harmonics, the filter capacity being typically 20 to 30% of the converter rating. With the use of intercycle PWM switching, or multi-level configurations in the VSC schemes, the filter requirements are drastically reduced or even eliminated altogether.

A short circuit on the inverter AC system side will cause commutation failures in the case of LCC, with temporary interruptions in the power transfer. VSC does not suffer from commutation failures and can continue to transfer active power under such condition, the amount of power being limited only by the reduction of the AC voltage during the fault.

An advantage of LCC is the response to DC system faults, due to the combination of smoothing reactors (which limit the current peak) and fast converter control (which quickly reduces the DC current to zero); a restart is possible within 100 to 300 ms. With VSC the free-wheeling diode component of the valve devices will permit the fault current to continue indefinitely even when the IGBTs are blocked. Therefore the fault can only be cleared by circuit breaker action and this will involve large delays in restoring normal system operation.

Another important advantage of LCC is a substantial reduction in power losses as compared with the VSC configurations, particularly those using PWM.

4.4 Current Source Conversion

CSC requires a large inductance on the DC side to make the DC current well defined and slow to change. The AC side voltage is then the variable directly controlled by the conversion process. Since the AC system has significant line or load inductance, line-to-line capacitors must be placed on the AC side of the converter. The switches must block voltages of both polarities, but they are only required to conduct current in one direction. This naturally suits symmetrical devices of the thyristor type and, therefore, CSC constitutes the basis of the LCC process discussed in Chapter 3.

As the switches in CSC only carry the current unidirectionally, there is no need for an anti-parallel diode. In the CSC configuration shown in Figure 4.14(a), the roles of the DC capacitor and AC inductance are interchanged (with respect to the VSC alternative). In VSC the DC capacitor facilitates the rapid transfer of current from the outgoing valve to its opposite in the same phase leg, irrespective of the direction of the AC current; the capacitor must therefore be large enough to handle alternate charging and discharging with little change in DC voltage. In the self-commutating CSC configurations, although the valves can also be turned off at will, they also require an alternate path, otherwise the current turn-off will have to dissipate a large amount of energy in an inductive circuit. Therefore AC capacitors need to be connected between the phases to facilitate the rapid transfer of current.

The CSC injects AC current into the system or load with the necessary AC voltage behind to force the current injection; therefore, the DC source must be capable of driving such a current.

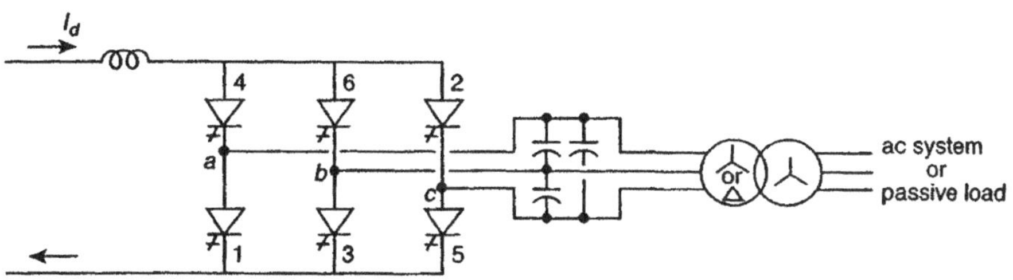

(a)

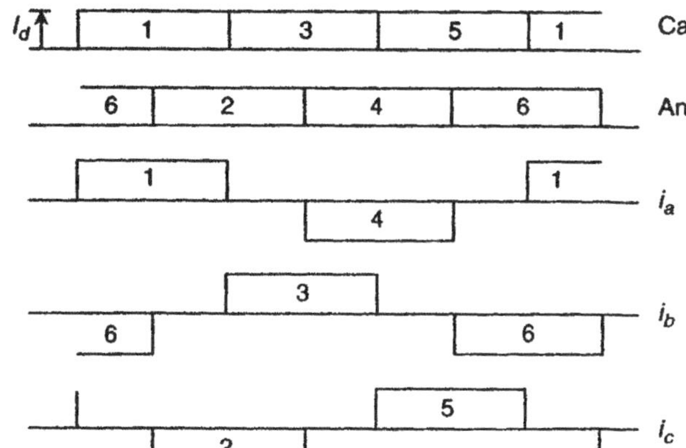

(b)

Figure 4.14 CSC: (a) circuit diagram; (b) current waveforms

Similar to the VSC, the CSC is capable of operating with leading power factor. Also the self-commutation of the valves permits the converter to operate as an inverter into a passive, as well as an active, load or system.

The introduction of self-commutation eliminates practically the main disadvantages of LCC described in Section 4.3 and makes the robust thyristor-based conversion an alternative worth considering for HVDC application.

4.4.1 Analysis of the CSC Waveforms [2]

The valve conduction sequence is the same as for the thyristor-based LCC process shown in Figures 3.6(c), (d) and (e), but without the commutation overlap. The three-phase injected currents are as shown in Figure 4.14(b).

In the time domain the phase current is expressed by

$$I = (1/\pi)2\sqrt{3}I_d[\cos(\omega t) - (1/5)\cos(5\omega t) + (1/7)\cos(7\omega t)\ldots] \tag{4.11}$$

The harmonic currents are of orders $6k \pm 1$ (for $k = 1, 2, \ldots$) and their rms values are obtained from the expression

$$I_n = (1/n)(\sqrt{6}/\pi)I_d \tag{4.12}$$

Thus the rms value of the fundamental component is

$$I_1 = (\sqrt{6}/\pi)I_d = 0.78I_d \tag{4.13}$$

The harmonic voltage content on the DC side is obtained from the expressions of the LCC (Section 3.7) but in the absence of commutation overlap, i.e. with $u = 0$.

4.5 The Reinjection Concept with Self-Commutation

4.5.1 Application to VSC

The reinjection principle has been described in Section 3.8.2 with reference to the line-commutated CSC (both for the single and double bridge configurations). When applied to VSC, the individual bridges need to be supplied, via the common or neutral point, by an appropriate voltage waveform derived from the DC voltage source and varying at six times the fundamental frequency. In the ideal case, the addition of the reinjection and main bridge voltages should produce a waveform that, when added to that of the second bridge, will achieve complete cancellation of the harmonic content at the converter system output terminals.

The ideal reinjection waveform

In Figure 4.15, the S_{Y1} arm of the star-connected bridge is switched on at $\omega t = \ldots, 0, 2\pi, 4\pi, \ldots$ and arm S_{Y4} at $\omega t = \ldots \pi, 3\pi, 5\pi, \ldots$. Similarly arm $S_{\Delta 1}$ of the

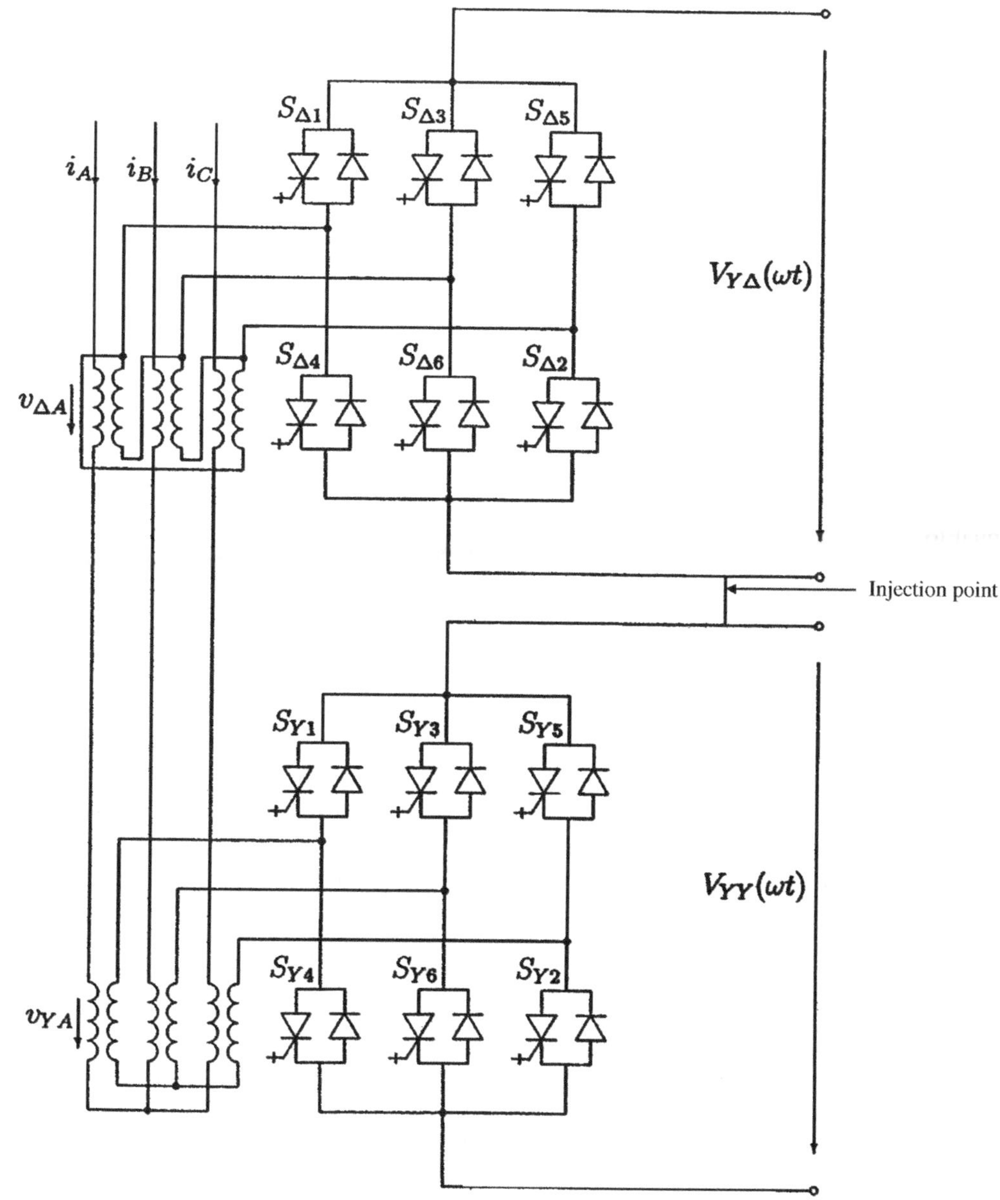

Figure 4.15 The two bridges of the 12-pulse VSC configuration

delta-connected bridge is switched on at $\omega t = \ldots, \pi/6, 13\pi/6, 25\pi/6, \ldots$ and arm $S_{\Delta 4}$ at $\omega t = \ldots, 7\pi/6, 19\pi/6, 31\pi/6, \ldots$.

The secondary winding voltages of the two interface transformers on the bridge side are given by

$$V_{Ya}(\omega t) = \begin{cases} V_{YY}(\omega t)/3 & 0 < \omega t < \pi/3 \\ 2V_{YY}(\omega t)/3 & \pi/3 < \omega t < 2\pi/3 \\ V_{YY}(\omega t)/3 & 2\pi/3 < \omega t < \pi \\ -V_{YY}(\omega t)/3 & \pi < \omega t < 4\pi/3 \\ -2V_{YY}(\omega t)/3 & 4\pi/3 < \omega t < 5\pi/3 \\ -V_{YY}(\omega t)/3 & 5\pi/3 < \omega t < 2\pi \end{cases} \tag{4.14}$$

$$V_{\Delta a}(\omega t) = \begin{cases} 0 & 0 < \omega t < \pi/6 \\ V_{Y\Delta}(\omega t) & \pi/6 < \omega t < 5\pi/6 \\ 0 & 5\pi/6 < \omega t < 7\pi/6 \\ V_{Y\Delta}(\omega t) & 7\pi/6 < \omega t < 11\pi/6 \\ 0 & 11\pi/6 < \omega t < 2\pi \end{cases} \tag{4.15}$$

If the voltage across the bridges connected to the Y/Y and Y/Δ transformers are made equal to

$$V_{YY}(\omega t) = V_{DC} + \sum_{k=1}^{\infty} A_{Yk} \cos(6k\omega t) \tag{4.16}$$

$$V_{Y\Delta}(\omega t) = V_{DC} + \sum_{k=1}^{\infty} A_{\Delta k} \cos(6k\omega t) \tag{4.17}$$

and considering that the harmonics of orders $n = 6(2l - 1) \pm 1 (l = 1, 2, \ldots)$ cancel in a balanced 12-pulse converter, it can be shown [8] that the addition of the winding voltages on the primary side will produce a perfectly sinusoidal waveform if the following relation exists between the DC and AC components of the bridge voltages:

$$\sum_{k=1}^{\infty} \frac{A_{Yk}}{(12l \pm 1)^2 - 36k^2} = \frac{V_{DC}}{(12l \pm 1)^2} \qquad \ell = 1, 2, \ldots \tag{4.18}$$

which consists of a set of linear algebraic equations and variables.

In practice the harmonics of high orders, i.e. $(12m + 1)$ and above, can be ignored and the number of variables and equations reduces to $2m$. For a sufficiently large value of the parameter m the values of V_k can be approximated by the expressions

$$A_k = \frac{A_{Yk}}{M_{DC}} = \frac{2[2(-1)^k - \sqrt{3}]}{(2 - \sqrt{3})(36k^2 - 1)} \approx \frac{14.9282(-1)^k - 12.9282}{36k^2 - 1} \qquad k = 1, 2, \ldots \tag{4.19}$$

$$\frac{A_{\Delta k}}{M_{DC}} = (-1)^k \frac{A_{Yk}}{M_{DC}} \approx \frac{14.9282 - 12.9282(-1)^k}{36k^2 - 1} \qquad k = 1, 2, \ldots \tag{4.20}$$

Using the results derived from Equations (4.19) and (4.20), the normalised reinjection voltage waveforms to be applied to the star- and delta-connected bridges become

$$V_Y(x) = 1 + \sum_{k=1}^{\infty} A_k \cos(6kx) \tag{4.21}$$

$$V_\Delta(x) = 1 + \sum_{k=1}^{\infty} (-1)^k \, A_k \cos(6kx) \tag{4.22}$$

These are shown in Figures 4.16(a) and (b) respectively, which always add to a DC level modulated by a small ripple (shown in Figure 4.16(c)). Figures 4.16(a) and (b) show that zero values appear at the points where the switches are turned on and off, which offers the possibility of achieving ZVS (Zero-Voltage Switching). The derivatives of the waveforms are limited, particularly around the zero values, and this ensures operation at low *dv/dt*. Figure 4.16(c) indicates that the two waveforms add to a DC level modulated by a small ripple. In practice, however, it is impractical to generate such ripple and some approximations to avoid it are considered next.

Symmetrical reinjection approximations

The reinjection waveforms in Figures 4.16(a) and (b) are symmetrical about the vertical axis but not about any of the crossing points of the waveform with its DC average (that is why their sum is not a constant DC).

To avoid the need to provide a DC power source with controllable ripple, the reinjection waveform must be made fully symmetrical. There are two possible solutions: (i) using a waveform that minimises the integration of the error square and the error derivative square (ESEDS) and (ii) using a linearly rising and linearly falling waveform (of constant derivative).

The ESEDS symmetrical waveform $V_{Ys}(x)$ is obtained by solving the expression

$$\min\left\{\int_0^{\pi/6}\left[[V_Y(x) - V_{Ys}(x)]^2 + \left(\frac{d[V_Y(x) - V_{Ys}(x)]}{dx}\right)^2\right]dx + \int_0^{\pi/6}\left[[V_\Delta(x) - V_{\Delta s}(x)]^2 + \left(\frac{d[V_\Delta(x) - V_{\Delta s}(x)]}{dx}\right)^2\right]dx\right\} \tag{4.23}$$

under the conditions of symmetry and area equality between the two groups of curves, i.e.

$$V_{Ys}(x) + V_{\Delta s}(x) = 2 \qquad \text{for } 0 < x < \pi/6$$

and

$$\begin{aligned}
&\int_0^{\pi/12} V_{Ys}(x)dx = \int_0^{\pi/12} V_Y(x)dx \qquad \int_{\pi/12}^{\pi/6} V_{Ys}(x)dx = \int_{\pi/12}^{\pi/6} V_Y(x)dx \\
&\int_0^{\pi/12} V_{\Delta s}(x)dx = \int_0^{\pi/12} V_\Delta(x)dx \qquad \int_{\pi/12}^{\pi/6} V_{\Delta s}(x)dx = \int_{\pi/12}^{\pi/6} V_\Delta(x)dx
\end{aligned} \tag{4.24}$$

Based on the numerical results of the symmetrical waveforms V_{Ys} and $V_{\Delta s}$, the Fourier components of the ESEDS waveform $V_{Ys}(x)$ are approximately given by

$$A_k = \frac{(7+4\sqrt{3})[1-(-1)^k]}{(36k^2-1)} \approx \frac{13.9282[1-(-1)^k]}{36k^2-1} \qquad k = 1, 2, \ldots \tag{4.25}$$

Similarly $V_{\Delta s}$ can be obtained by the application of a 30° phase displacement between them.

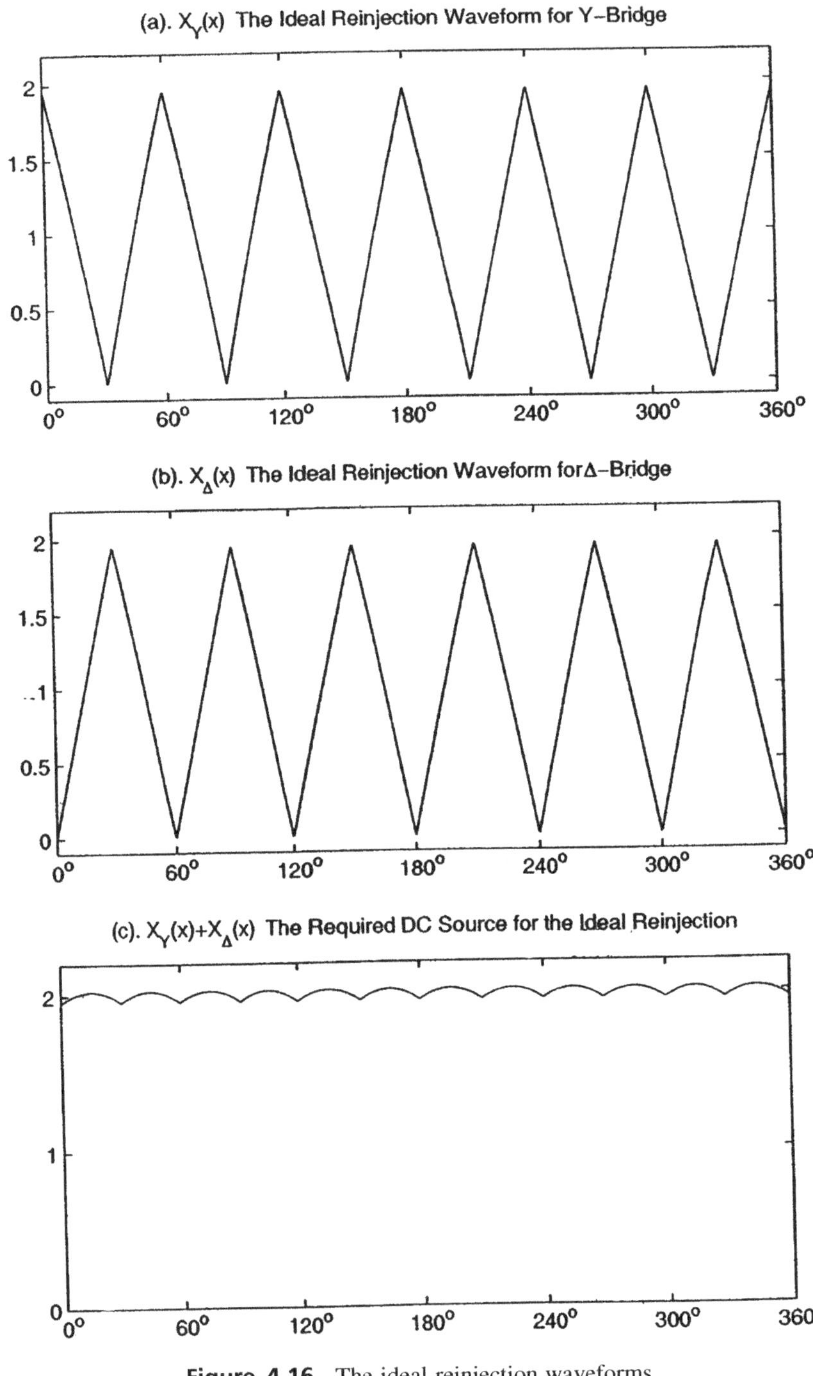

Figure 4.16 The ideal reinjection waveforms

The spectrum of the linear symmetrical waveform is

$$A_k = \frac{4[1-(-1)^k]}{k^2\pi^2} \approx \frac{0.4053[1-(-1)^k]}{k^2} \qquad k = 1, 2, \ldots \tag{4.26}$$

The maximum difference between the ESEDS and linear symmetrical waveforms is below 2.5 %. Therefore, considering its simplicity of implementation, the linear waveform is the preferred approximation.

The voltage waveforms of the linear approximation are shown in Figure 4.17, and the total harmonic distortion of the output voltage is

$$\begin{aligned}\mathrm{THD}_V &= \sqrt{\sum_{k=1}^{\infty}\frac{1}{(12k-1)^4}+\sum_{k=1}^{\infty}\frac{1}{(12k+1)^4}} \\ &= \sqrt{\frac{\pi^4(40+23\sqrt{3})}{8\times 12\times 81}-1} \approx 1.055\ 32\ \%\end{aligned} \tag{4.27}$$

The use of the ESEDS or linear symmetrical reinjection waveforms produces output waveforms with under 1 % THD, without the need to modify the DC power source. However, to derive these waveforms from the DC voltage is still impractical. A practical approximation in the form of a multi-level reinjection waveform [8] is described in Chapter 7.

4.5.2 Application to CSC

CSC can be viewed as the dual of VSC: that is, current waveforms in CSC are dual to voltage waveforms in VSC. CSC requires switching devices of different characteristics from those of VSC. The latter requires asymmetrical switches (with unidirectional voltage blocking and bidirectional current capability), while CSC requires symmetrical switches (with bidirectional voltage blocking and unidirectional current capability). Thus the IGBT cannot by itself be used in CSC; a diode connected in series with the IGBT can solve the problem, but at the expense of considerable extra power losses. The symmetrical GTO and IGCT types are more appropriate switching devices, due to the relatively low switching frequency of the CSC.

With reference to the 12-pulse CSC configuration, shown in Figure 4.18, the winding currents on the bridge side of the interface transformers are given by

$$I_{\Delta a}(\omega t) = \begin{cases} I_{B\Delta}(\omega t)/3 & 0<\omega t<\pi/3 \\ 2I_{B\Delta}(\omega t)/3 & \pi/3<\omega t<2\pi/3 \\ I_{B\Delta}(\omega t)/3 & 2\pi/3<\omega t<\pi \\ -I_{B\Delta}(\omega t)/3 & \pi<\omega t<4\pi/3 \\ -2I_{B\Delta}(\omega t)/3 & 4\pi/3<\omega t<5\pi/3 \\ -I_{B\Delta}(\omega t)/3 & 5\pi/3<\omega t<2\pi \end{cases} \tag{4.28}$$

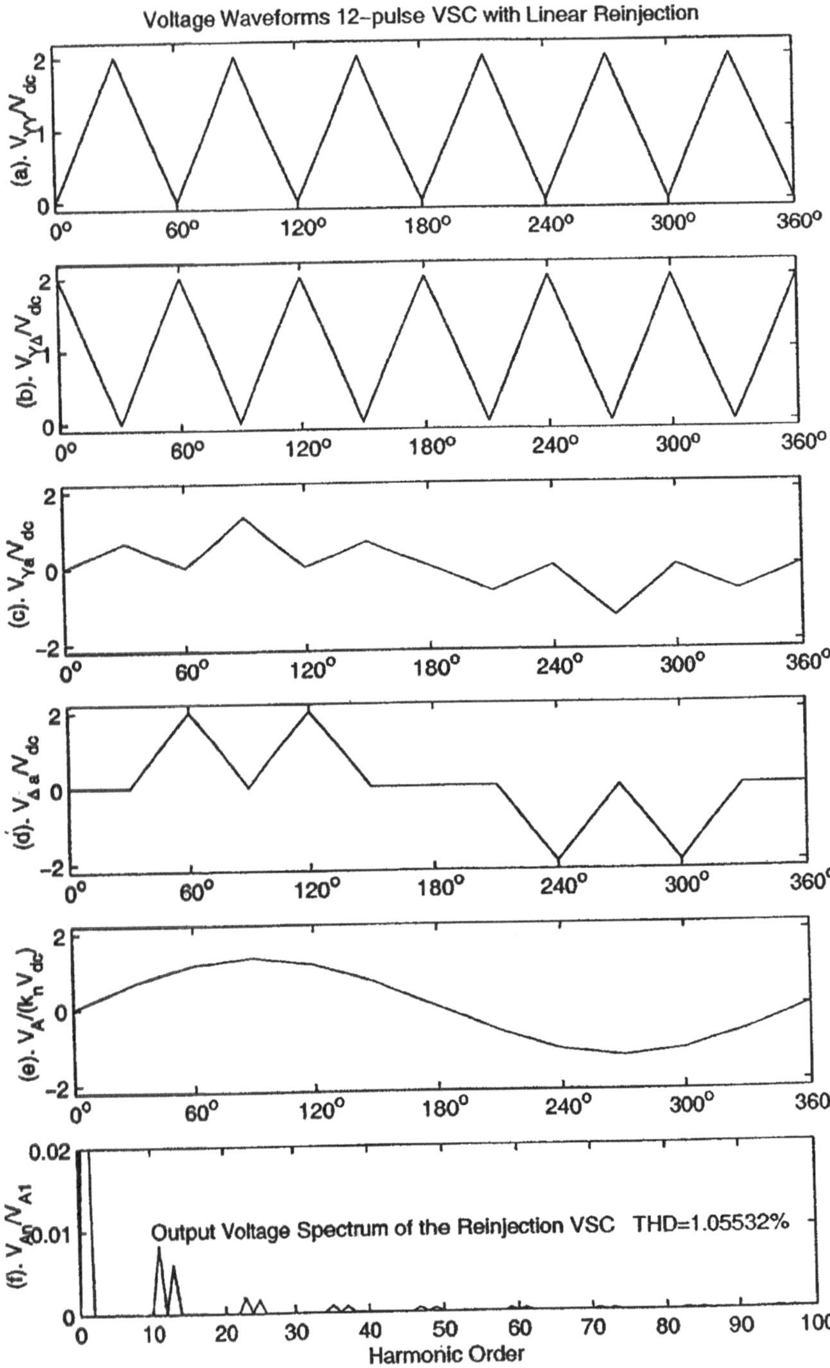

Figure 4.17 Voltage waveforms of a 12-pulse VSC with multi-level linear reinjection

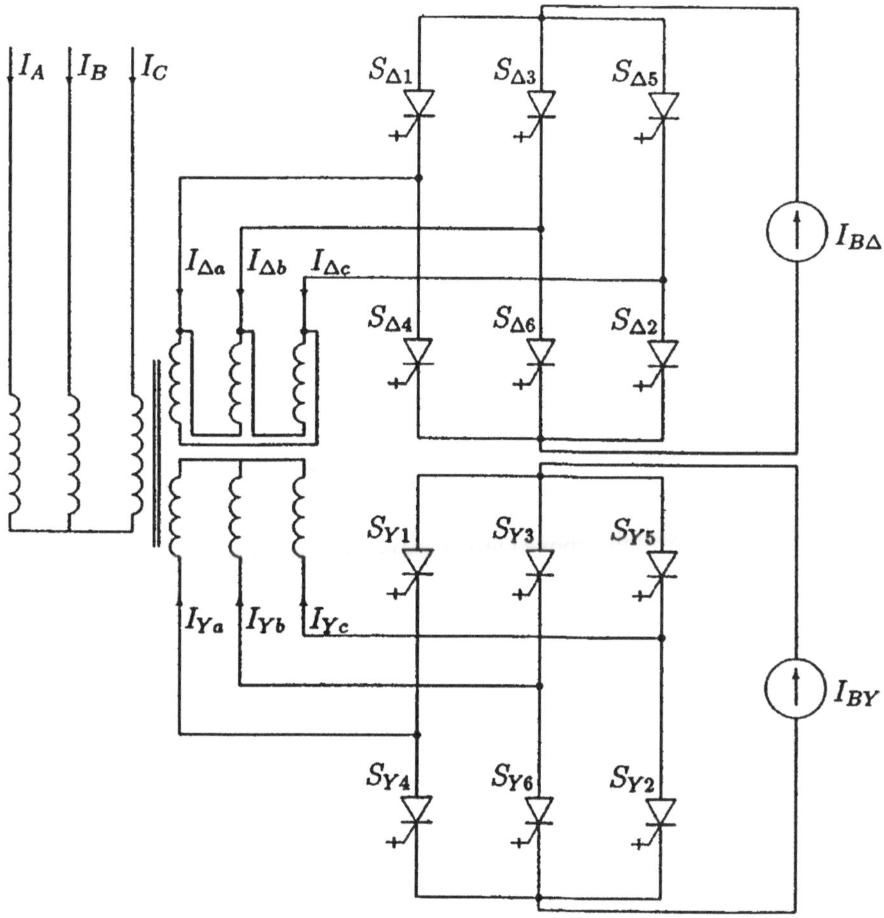

Figure 4.18 The two bridges of the 12-pulse CSC configuration

$$I_{Ya}(\omega t) = \begin{cases} 0 & 0 < \omega t < \pi/6 \\ I_{BY}(\omega t) & \pi/6 < \omega t < 5\pi/6 \\ 0 & 5\pi/6 < \omega t < 7\pi/6 \\ -I_{BY}(\omega t) & 7\pi/6 < \omega t < 11\pi/6 \\ 0 & 11\pi/6 < \omega t < 2\pi \end{cases} \tag{4.29}$$

By comparing Equations (4.14) and (4.15) with (4.28) and (4.29), and using the duality relation between the VSC and CSC configurations, the results obtained for the VSC can be easily extended to CSC, i.e. the currents to be supplied to the star- and delta-connected bridges are

$$I_{BY}(\omega t) = I_{DC} + \sum_{k=1}^{\infty} A_{Yk} \cos(6k\omega t) \tag{4.30}$$

$$I_{B\Delta}(\omega t) = I_{DC} + \sum_{k=1}^{\infty} A_{\Delta k} \cos(6k\omega t) \tag{4.31}$$

The corresponding currents in the secondary windings I_{Y_a} (ωt) and I_{Δ_a} (ωt) only include harmonics of orders $n = 6(2l - 1) \pm 1$ $(l = 1, 2, \ldots)$. Thus, as explained for the VSC configuration, the following relations must exist between the DC and AC components of the bridge currents for complete harmonic cancellation:

$$\sum_{k=1}^{\infty} \frac{(-1)^k A_{Yk}}{(12l \pm 1)^2 - 36k^2} = \frac{I_{DC}}{(12l \pm 1)^2} \qquad l = 1, 2, \ldots \tag{4.32}$$

$$\sum_{k=1}^{\infty} \frac{A_{\Delta k}}{(12l \pm 1)^2 - 36k^2} = \frac{I_{DC}}{(12l \pm 1)^2} \qquad l = 1, 2, \ldots \tag{4.33}$$

The resulting reinjection waveforms are also those shown in Figure 4.16, by changing the voltage for current terminology.

The multi-level current reinjection implementation [9], discussed in Chapter 7, provides a practical approximation to the above current waveforms.

4.6 Discussion

The description and analysis carried out in this chapter illustrate the capability of self-commutating conversion to provide four-quadrant power transfer at the converter terminals. The basic output waveform, however, is not acceptable for direct connection to the AC power system. In the line-commutated conversion described in Chapter 3 the most economic solution to this problem is the provision of passive filters, which is, to a large extent, justified by the need to provide reactive power compensation. This is not the case with self-commutation and thus the use of passive filters is not cost effective. Instead, the harmonic content is reduced by high-frequency switching (a solution discussed in Chapter 5) or by the use of complex multi-level converter configurations (discussed in Chapters 6 and 7).

VSC, currently the preferred option for HVDC application, has the following advantages:

- The converter is internally protected from line voltage transients by the DC capacitor.

- The voltage-source-type termination requirements of the converter are satisfied at the DC terminals by the presence of the DC capacitor; therefore no additional termination components, such as filters, are normally needed.

- The converter generates voltage harmonics. Therefore the harmonic currents drawn from the AC system are a function of the transformer leakage inductance, which increases rapidly for higher frequencies, thus greatly reducing the higher order harmonic currents.

The main disadvantages of the basic VSC solution are:

- The three-phase bridge produces a rectangular voltage waveform. The combination of appropriately shifted individual bridges' waveforms by means of multi-winding transformers is not practical here, because of the high circulating harmonic currents in those windings. Therefore, a multi-pulse configuration requires the use of separate small transformers or of a magnetic structure (at low voltage) with a separate high-voltage interface transformer. This disadvantage does not apply to multi-level VSC configurations.

- The converter must have fast output current limitation to avoid exceeding the capability of the semiconductor devices following system faults.
- It is difficult to protect the converter against internal faults (which may even require the use of fuses).

The CSC alternative also has some advantages, such as the following:

- Better current control. The short-term overcurrent protection is inherent by the presence of the DC side inductor, while the long-term protection is achieved by the current control loop. Internal faults, however, may still require fast interruption.
- The large power inductor, providing the DC bus energy storage, is simpler and more reliable than the large capacitor of the VSC.
- It is better suited to higher power devices, such as the IGCT, which can block voltage in either direction but conduct current only in the forward direction.
- Soft switching is easier to provide.

On the negative side the CSC configurations have some disadvantages, such as the following:

- Since the converter DC terminals are shunted by the smoothing reactor, the naturally current-source-type termination (which consists of system and transformer impedance) of the AC terminals must be compensated by an additional voltage-source-type termination, such as a capacitive AC filter across the AC terminals.
- The losses of the smoothing reactor are high, compared with those of a VSC DC capacitor.
- The switch voltages are poorly defined; most semiconductor switches tolerate transient overcurrent better than transient overvoltages.

References

1. CIGRE (2004), 'VSC transmission', Document B4-WG 37, Paris.
2. Hingorani, N.G. and Gyugyi, L. (2000), *Understanding FACTS: Concepts and technology of Flexible ac Transmission Systems*, IEEE Press, New York.
3. Walker, L. (1990), '10 MW GTO converter for battery peaking service', *IEEE Transactions on Industry Applications*, **26**(1), 63–72.
4. Schauder, C. *et al.* (1994), 'Development of a ±100MVAR static condenser for voltage control of transmission systems', IEEE paper 94SM479.6PWRD.
5. Arrillaga, J. and Watson, N.R. (2003), *Power System Harmonics*, Second Edition John Wiley & Sons, Ltd, Chichester.

6. CIGRE (1999), 'Guide for the specification and design evaluation of ac filters for hvdc systems', Document WG-14.30, Paris.
7. Moore, P. and Ashmole, P. (1998), 'Flexible ac transmission systems - Part 4', *IEE Power Engineering Journal*, 95–100.
8. Liu, Y.H., Arrillaga, J. and Watson, N.R. (2004), 'A new STATCOM configuration using multi-level dc voltage reinjection for high power application', *IEEE Transactions on Power Delivery*, **19**(4), 1828–1834.
9. Perera, L.B., Liu, Y.H., Watson, N.R. and Arrillaga, J. (2005), 'Multi-level current reinjection in double bridge self-commutated current source conversion', *IEEE Transactions on Power Delivery*, **20**(2), 984–991.

5

Pulse Width Modulation

5.1 Introduction

A very flexible way of controlling the output voltage waveform of a VSC is the use of high-frequency switching of the essentially constant DC voltage [1–4]. The output waveform is then averaged (filtered) to produce a specified fundamental component of controllable magnitude as well as eliminate the low-frequency harmonics. This process is called pulse width modulation (PWM), since the desired fundamental and harmonic voltages are controlled by modulating the width of the voltage pulses.

To achieve an acceptable attenuation of the high-frequency components, the switching frequency should be many times the fundamental frequency. This requirement, however, conflicts with the need for efficiency because of the extra energy loss imposed by the frequent switching.

PWM was originally designed around the needs of inverter-fed AC drives, as these require fast and continuous control of the frequency and voltage magnitude. Although such high level of controllability is not required in power transmission, where the frequency and voltage magnitude are only permitted to vary within narrow limits, the PWM concept has so far been adopted as the basis of a more flexible HVDC technology. An additional benefit of the absence of passive filters for the low-order harmonics is the prevention of the converter/system harmonic interaction experienced by the line-commutated CSCs.

This chapter describes the main PWM strategies available for the control of the fundamental component and minimisation of the harmonic content of the converter output waveform.

5.2 PWM Operating Principles

PWM provides an effective solution to the simultaneous control of the voltage (or current) and harmonic content of the power conversion process. With reference to VSC this is

Flexible Power Transmission: The HVDC Options J. Arrillaga, Y.H. Liu and N.R. Watson

achieved by high-frequency chopping of the square wave voltage output waveform produced by the basic conversion process.

However, to be of practical use in power transmission the following two conditions must be met:

1. The output waveform should be free from even harmonics, i.e. either the phase waveforms possess half-cycle symmetry or any prevailing asymmetry is eliminated in the phase-to-phase voltages. Under this condition the phase-to-phase voltage will contain only odd-harmonic orders other than those multiples of three.

2. The three-phase system should be symmetrical, i.e. the potentials of the three phases have the same waveform but displaced by 120° from each other.

A critical factor in the choice of the modulation principle is the ratio of the modulation frequency to the output frequency

$$p = f_p / f$$

This factor determines the harmonic spectrum for a certain degree of control and for a given modulation pattern. A sufficiently large value of this ratio will reduce all the low-order harmonics to be within the specified limits. However, a high-frequency ratio will cause high switching losses. Moreover, the achievable voltage–time area will be reduced and with it the fundamental component at full output voltage.

Two other commonly used factors are:

1. The control ratio γ, defined as the ratio of the fundamental components of the modulated (V_1) and unmodulated ($V_{1(OM)}$) waveforms

$$\gamma = \frac{V_1}{V_{1(OM)}} \tag{5.1}$$

2. The utilisation ratio, which is the measure of how well the modulation principle utilises the maximum available voltage–time area at full output voltage:

$$k = \frac{V_{1(MAX)}}{V_{1(OM)}} \tag{5.2}$$

5.3 Selective Harmonic Cancellation

To control the fundamental and harmonic voltages simultaneously, the phase potentials must be reversed a number of times during each half-cycle at predetermined angles of the square wave.

Generally, at any fundamental switching frequency, each chop (or reversal) provides one degree of freedom, which permits either cancelling a harmonic component or controlling the magnitude of the fundamental voltage [2]. Thus for m chops per half-cycle one chop must be utilised to control the fundamental amplitude and so $m-1$ degrees of freedom remain. The

$m-1$ degrees of freedom may be utilised to eliminate completely $m-1$ specified low-order harmonics or to minimise the total harmonic distortion.

By way of example, Figure 5.1 shows three waveforms with a different number of voltage reversals. The notches are placed symmetrically about the centre line of the half-cycle. With the reversal angles shown $(\alpha_1, \alpha_2, \alpha_3, \ldots, \alpha_p)$ the following rms value of the nth harmonic voltage results:

$$V_n = \frac{1}{\sqrt{2}}\frac{4}{\pi}\frac{V_d}{2}\left[-\int_0^{\alpha_1}\sin(n\alpha)d\alpha + \int_{\alpha_1}^{\alpha_2}\sin(n\alpha)d\alpha - \int_{\alpha_2}^{\alpha_3}\sin(n\alpha)d\alpha \ldots + \int_{\alpha_p}^{90}\sin(n\alpha)d\alpha\right]$$

$$= \frac{\sqrt{2}}{\pi n}V_d\left\{[\cos(n\alpha)]_0^{\alpha_1} - [\cos(n\alpha)]_{\alpha_1}^{\alpha_2} + [\cos(n\alpha)]_{\alpha_2}^{\alpha_3} \ldots - [\cos(n\alpha]_{\alpha_p}^{90}\right\}$$

$$= \frac{\sqrt{2}}{\pi n}V_d\{2[\cos(n\alpha_1) - \cos(n\alpha_2) + \cos(n\alpha_3) + \cdots] - 1\} \tag{5.3}$$

For instance, to control the fundamental component and eliminate the fifth and seventh harmonics will need three values of α, which can be determined from the equations

$$\frac{\pi V_1}{\sqrt{2}V_d} + 1 = 2[\cos(\alpha_1) - \cos(\alpha_2) + \cos(\alpha_3)] \tag{5.4a}$$

$$1 = 2[\cos(5\alpha_1) - \cos(5\alpha_2) + \cos(5\alpha_3)] \tag{5.4b}$$

$$1 = 2[\cos(7\alpha_1) - \cos(7\alpha_2) + \cos(7\alpha_3)] \tag{5.4c}$$

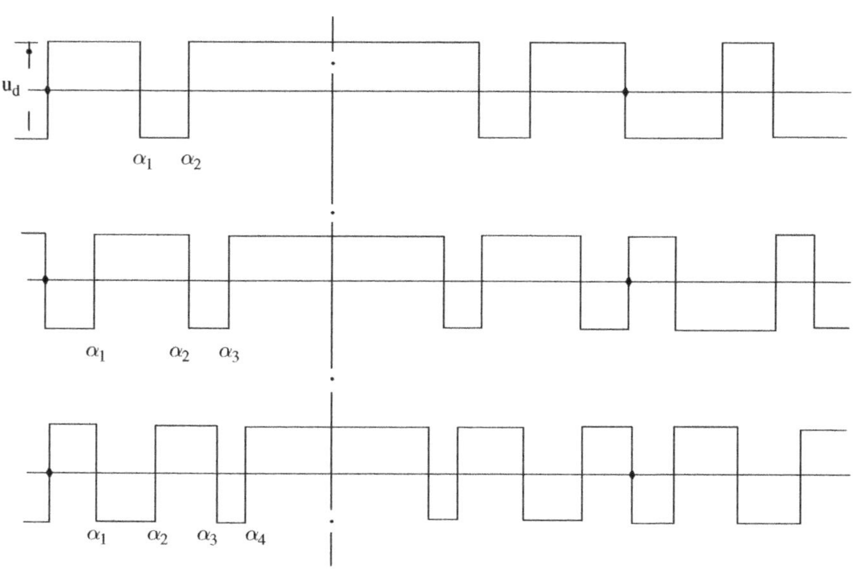

Figure 5.1 Voltage reversals

The remaining harmonic orders, i.e. 11th, 13th, 17th, etc., can then be calculated from

$$V_n = \frac{\sqrt{2}}{\pi n} V_d \{2[\cos(n\alpha_1) - \cos(n\alpha_2) + \cos(n\alpha_3)]\} \tag{5.5}$$

where α_1, α_2, α_3 are the values obtained from Equations (5.4).

The results of this analysis are plotted in Figure 5.2.

At any fundamental frequency, elimination of the lower order harmonics from the phase waveforms will cause the portion of the rms which was provided by the eliminated harmonics to be spread over the remaining harmonic magnitudes. However, the integrating filter characteristic of the system impedance is more effective in reducing the current harmonics at higher orders. The complete elimination of selected low-order harmonics will increase the magnitudes of the first uncancelled harmonics.

An alternative solution to selective harmonic cancellation is the minimisation of the total harmonic content, or of a range of selected harmonics [5–6]. These techniques, however, do not respond quickly to transient conditions, because the pulses do not occur at fixed intervals and closed-loop control is implemented cycle by cycle. More advanced closed-loop controls [7–9] are found in the literature.

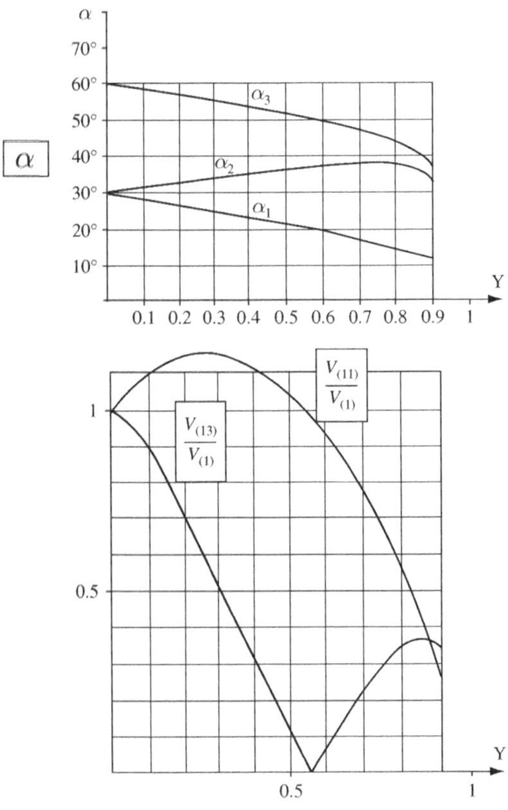

Figure 5.2 Elimination of the 5th and 7th harmonics by means of voltage reversals

Another possible solution is the use of a hysteresis band modulator, which calculates the error between the specified output and the measured output; the state of the switches is then changed in response to the magnitude of the error. This closed-loop technique provides a faster transient response. However, the variable nature of the switching period produces a continuous spread of the output spectra which includes sub-harmonics and is thus impractical for low switching frequencies.

5.4 Sinusoidal (Carrier-Based) PWM

In this case the reversal instants, instead of being fixed, are determined by the intersections of a sinusoidal reference voltage of fundamental frequency and an amplitude (or modulation index) M, with a sawtooth modulating waveform of a (carrier) frequency pf (where p is the frequency ratio) [10]. The dominant harmonic of this alternative is that of order p.

For low-frequency ratios, p should be an integer, to ensure that the two intersecting waves are synchronous, thus avoiding discontinuities and fluctuations. The basic method, illustrated in Figure 5.3, controls the line-to-line voltage from zero to full voltage by increasing the magnitude of the sawtooth or the sine-wave signal, with little regard to the harmonics generated; for instance, the output voltage waveform of Figure 5.3 contains even harmonics.

Preferably, the carrier frequency should be an odd multiple of three ($n = 3, 9, 15, \ldots$). This alternative, shown in Figure 5.4(a), provides half- and quarter-wave symmetries, which eliminates the even harmonics from the carrier spectrum and allows symmetrical three-phase voltages to be generated from a three-phase sine wave set and one sawtooth waveform.

For high values of p (i.e. over 20) the value of p can be increased continuously as the output frequency decreases. This means that the fixed triangular wave will be compared with a sinusoid of variable frequency and amplitude.

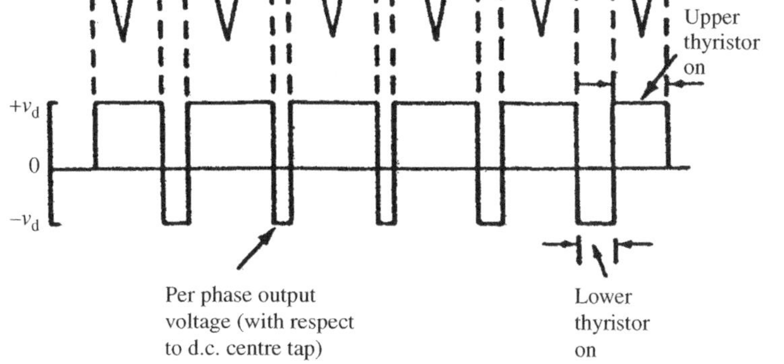

Figure 5.3 Principle of PWM

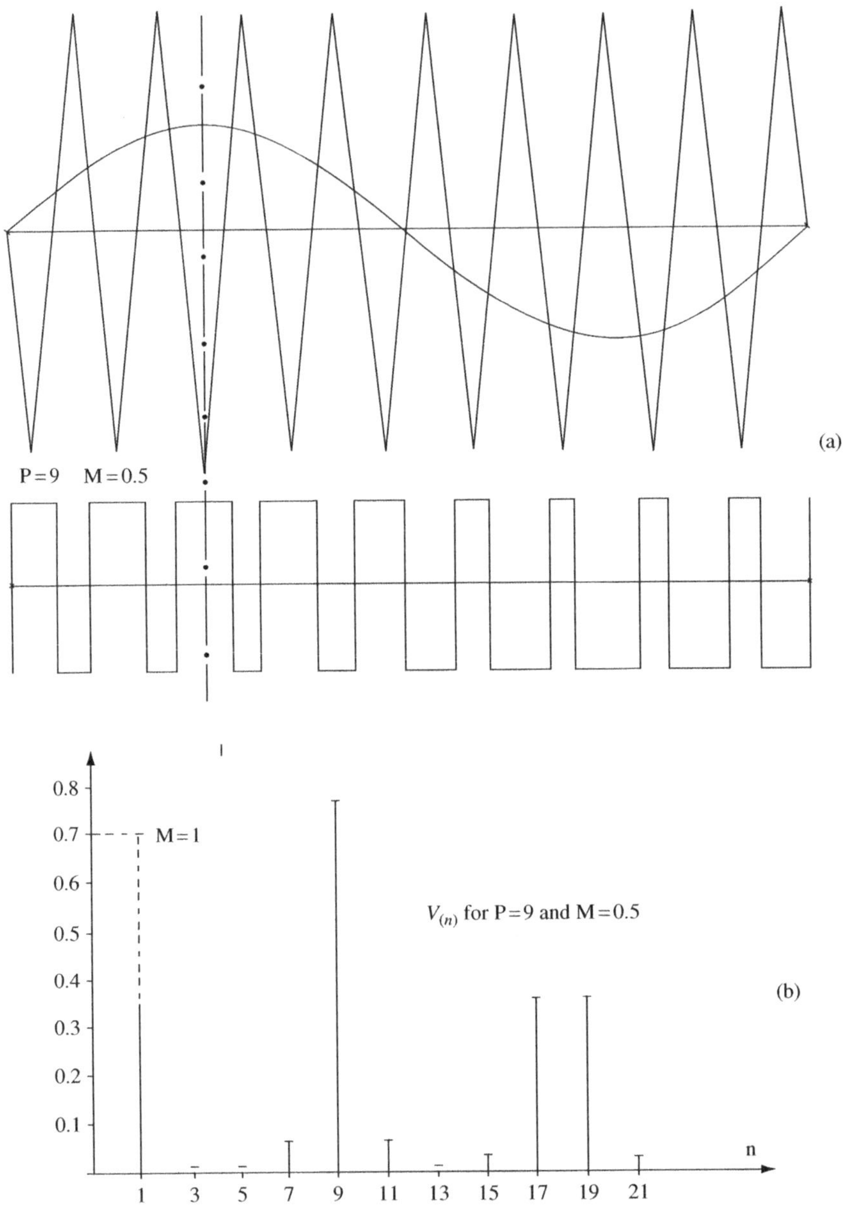

Figure 5.4 Sinusoidal PWM

A computer simulation of the harmonic content will first find the values of the intersecting points $(\alpha_1, \alpha_2, \alpha_3, \ldots)$ from the equations

$$M \sin(\alpha_i) = (-1)^i \left(\frac{2p}{\pi}\alpha_i - 2i\right) \qquad \text{for } 1 \leq i \leq \frac{p-1}{2} \tag{5.6}$$

With those values the rms content of the harmonics can be calculated from Equation (5.3).

The following comments apply when $M < 1$ and p is an odd integer greater than 6:

- Even harmonics, odd harmonics of orders $n < p - 2$ and some higher harmonics are eliminated.
- Harmonics of an order equal to p or in its neighbourhood, as well as the multiples of p, are amplified (Figure 5.4(b) shows these effects).
- The ideal utilisation ratio is

$$k_i = \frac{\pi}{4}$$

5.5 PWM Carrier-Based Implementation

PWM, as described in the previous section, is amenable to natural sampling by analogue techniques. The triangular carrier for instance can be generated by integrating a square wave carrier clock. In this case naturally sampled PWM reacts instantly to changes in the input modulating commands and produces no distortion of the synthesised waveform.

Carrier-based modulators can also be implemented digitally. In the digital implementation the required pulse width value is placed directly as a digital quantity into a counter, which is then clocked to generate the edge. An alternative approach is to make a digital comparison of the pulse width stored in a register with a free-running timer. In both of these alternatives the response can be substantially delayed when using low carrier frequencies.

In carrier-based PWM the switching instants are decided cycle by cycle. The aim of the modulation is to ensure that the average value of each output pulse is proportional to the averaged value of the input signal over the switching period. This condition provides a linear, distortion-free transfer function.

In the case of a constant input signal, the above condition is met by the intersection of a triangular carrier waveform with the input value. This is shown in Figure 5.5 for single trailing and double edge modulation.

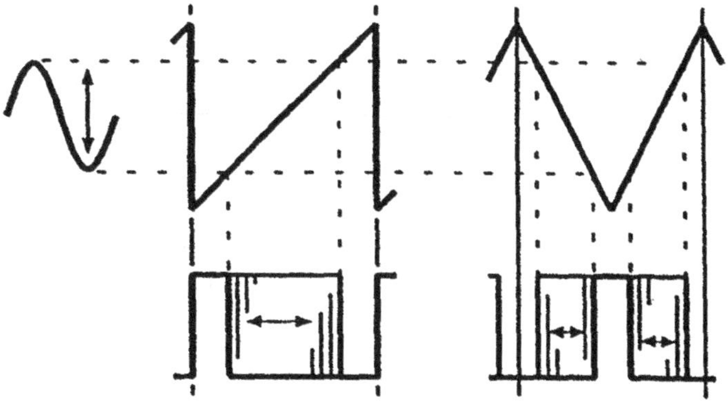

Figure 5.5 One switch cycle of carrier-based PWM with single trailing-edge modulation (on the left) and double edge modulation (on the right)

When the input signal is not constant during the switched period, and since the signal may not be known ahead of time, it will not be possible to meet the modulation condition. In this case a distortion-free transfer function is only possible if a delay is acceptable.

5.5.1 Naturally Sampled PWM

As already explained, in naturally sampled PWM the switching instants are generated by the intersection of the carrier and input waveforms and the sampling instants occur exactly at the same time as the output edges.

For a constant modulation signal, as shown in Figure 5.6(a), a Fourier decomposition of the naturally sampled PWM voltage (V_{nat}) waveform [11] leads to the following expression:

$$\begin{aligned} V_{nat}(\omega_c t) &= k + (2/\pi) \sum_{m=1}^{m=\infty} \frac{\sin(km\pi)}{m} \cos(m\omega_c t) \\ &= \frac{1}{\pi} \sum_{m=-\infty}^{m=\infty} \frac{\sin(km\pi)}{m} \cos(m\omega_c t) \end{aligned} \tag{5.7}$$

where $0 \leq k \leq 1$ and ω_c is the carrier frequency (in radians).

In general, PWM may have some arbitrary phase displacement ϕ_c and Equation (5.7) is better expressed as a sum of complex exponentials:

$$\begin{aligned} V_{nat}(\omega_c t) &= k + (2/\pi) \sum_{m=1}^{m=\infty} \frac{\sin(km\pi)}{m} e^{jm(\omega_c t - \phi_{c_c})} \\ &= (1/\pi) \sum_{m=-\infty}^{m=\infty} \frac{\sin(km\pi)}{m} e^{-jm\phi_c} e^{jm\omega_c t} \end{aligned} \tag{5.8}$$

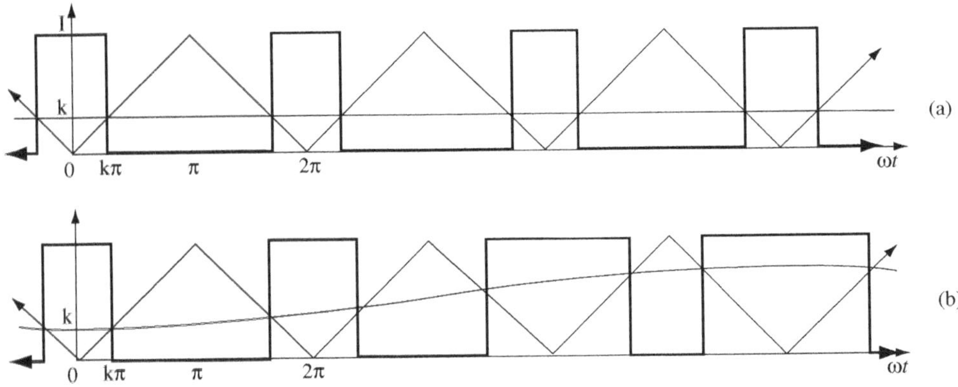

Figure 5.6 Natural sampled PWM of pulse or carrier frequency: (a) for constant DC modulation $k(0 \leq k \leq 1)$; (b) for fixed sinusoidal modulation, $k = 1/2 + (M/2)\cos(\omega_1 t)$

When the modulating waveform is a sinusoid (Figure 5.6(b)), i.e.

$$k = \frac{1}{2} + \frac{k_\lambda}{2}\cos(\omega_1 t)$$

where k_λ is the modulation depth (varying between 0 and 1) and ω_1 the input signal frequency (in radians), the timings of the edges of the PWM waveform are themselves functions of time. In this case the Fourier description of a double edge, naturally sampled waveform of height stepping between 0 and 1 is expressed in terms of Bessel functions of the first kind [11]

$$\begin{aligned}
V_{nat}(\omega_c t, \omega_1 t) &= \frac{1}{2} + \frac{k_\lambda}{2}\cos(\omega_1 t) \\
&- \frac{2}{\pi}\sum_{p=1}^{p=\infty}\cos(p\pi)\frac{J_0[k_\lambda(2p-1)\pi/2]}{(2p-1)}\cos[(2p-1)\omega_c t] \\
&- \frac{2}{\pi}\sum_{p=1}^{p=\infty}\sum_{q=1}^{q=\infty}\cos[(p+q)\pi]\frac{J_{2q}[k_\lambda(2p-1)\pi/2]}{(2p-1)}\cos[(2p-1)\omega_c t \pm 2q\omega_1 t] \\
&- \frac{2}{\pi}\sum_{p=1}^{p=\infty}\sum_{q=1}^{q=\infty}\cos[(p+q)\pi]\frac{J_{2q-1}(k_\lambda p\pi/2)}{2p}\cos[2p\omega_c t \pm (2q-1)\omega_1 t]
\end{aligned} \tag{5.9}$$

In this equation the first term is the ideal output term (of frequency ω_1), the second the harmonics of the carrier frequency ($m\omega_c$) and the last two the sidebands of the carrier and its harmonics ($m\omega_c \pm n\omega_1$).

Equation (5.9) can be simplified by combining terms and then generalised by expressing it as a sum of exponential terms, i.e.

$$V_{nat}(\omega_c t, \omega_1 t) = \frac{1}{2} + \sum_{m=-\infty}^{m=\infty}\sum_{n=-\infty}^{n=\infty}\sin[(m+n)\pi/2]\frac{J_n(k_\lambda m\pi/2)}{m\pi/2}e^{j(n\phi_1 - m\phi_c)}e^{j(n\omega_1 t + m\omega_c t)} \tag{5.10}$$

where $J_n(x)$ is the nth order Bessel function. The first five Bessel functions $J_0(x)\ldots J_4(x)$ are plotted against x in Figure 5.7(a).

In order to show more clearly the behaviour for small values of x (the usual situation for PWM) Figure 5.7(b) plots the results in logarithmic scale. As $x \to 0$, $J_0(x) \to 1$ and $J_n(x) \to 0$.

For $x > n$ the Bessel functions look like sine waves with their amplitudes decaying as $x^{-1/2}$ [12]. For $x < n$ they follow simple power laws, with an asymptotic form for $0 < x \ll n$,

$$\begin{aligned}
V_n(x) &\sim \frac{1}{\Gamma(n+1)}\left(\frac{1}{2}x\right)^n && \text{for } n \geq 0 \\
&\sim \frac{1}{n!}\left(\frac{1}{2}x\right)^n && \text{for } n \geq 0 \text{ and } n \text{ an integer}
\end{aligned} \tag{5.11}$$

These Bessel functions describe the amplitude of the carrier and sidebands ($m = 1, 2, 3, \ldots$). They also define the magnitude, and thus attenuation, of the modulating signal ($m = 0,\ n = 1$) and generation of harmonics of that signal ($m = 0,\ n = 3, 5, 7, \ldots$).

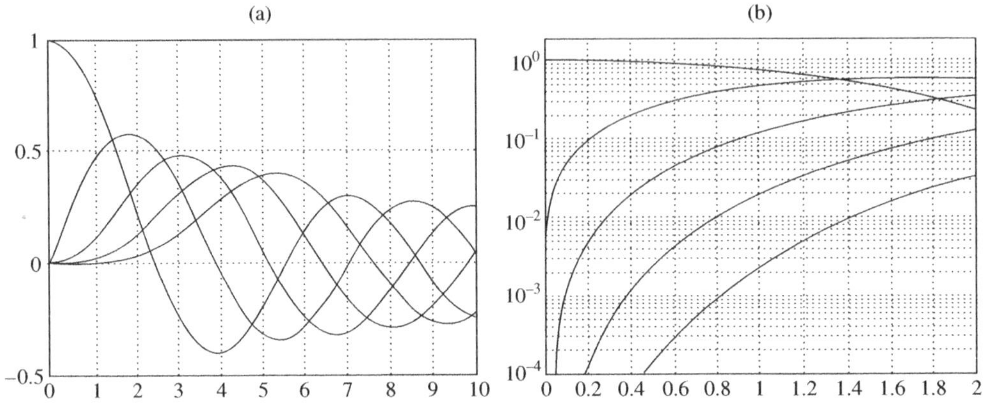

Figure 5.7 The first five Bessel functions of the first kind, $J_o(x)...J_4(x)$ plotted versus x: (a) in normal scale; (b) in logarithmic scale

With natural sampling the argument of the Bessel function is only a function of m. For $m = 0$, the Bessel function becomes $J_n(0)$, which is either one ($n = 0$) or zero ($n \neq 0$). This means that the original modulating signal appears in the PWM output spectra unattenuated, undistorted and without phase delay, regardless of the frequency of modulation depth of that signal. The gain of the modulator/converter cascade is equal to the peak amplitude of the output waveform divided by the peak amplitude of the triangular carrier.

In summary, naturally sampled PWM does not attenuate the synthesised fundamental component and does not produce distorting components that are multiples of the fundamental.

5.5.2 Uniformly Sampled

In uniform PWM the input signal is regularly sampled and held constant at the beginning of each switching cycle prior to being compared with the triangular carrier. This process is illustrated in Figure 5.8, which shows natural and uniform sampling; the latter introduces a delay of $T_{sam}/2$ (on average), which causes a 'sinc'-like frequency response

Using the same analysis as for the natural sampling, the spectra of the uniformly sampled, double edge PWM voltage (V_{uni}) is determined from the expression

$$V_{uni}(\omega_c t, \omega_1 t) = 2\sum_{m=0}^{m=\infty}\sum_{n=-\infty}^{n=\infty} \sin[(m+n)\pi/2]\frac{J_n[x(m+n\omega_1/\omega_c)\pi/2]}{(m+n\omega_1/\omega_c)\pi/2} \times e^{j[n\phi_1 - m\phi_c - (m+n\omega_1/\omega_c)\pi/2]}e^{j(n\omega_1 t + m\omega_c t)} \quad (5.12)$$

In this case, the extra term $(n\omega_1/\omega_c)$ makes the argument of the Bessel function non-zero even when $m = 0$. The transfer function of the modulating input signal rolls off as the ratio f_1/f_c increases; this is accompanied by a phase group delay of about one-half of the sampling period.

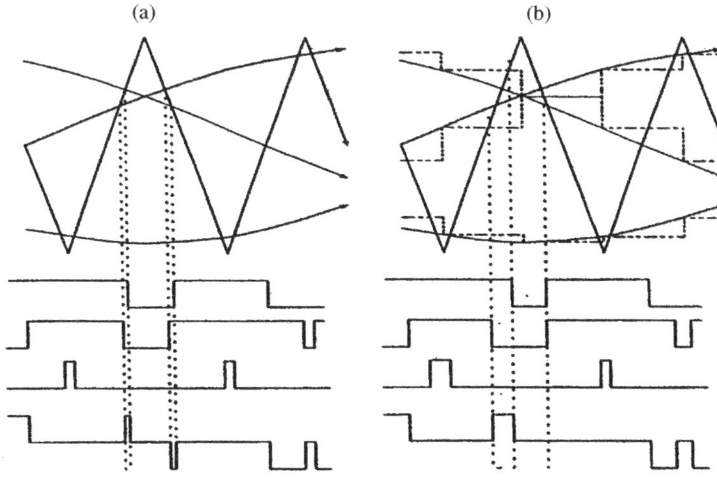

Figure 5.8 The process of natural (a) and uniform (b) sampling

The digitally generated uniform PWM can achieve good consistent cancellation of the carrier terms. It does, however, attenuate the input signal and generates input signal harmonics. Moreover, odd harmonics of the fundamental modulating signal (nf_1) are now present. This solution is thus only applicable to relatively low carrier frequencies.

5.6 Modulation in Multi-Bridge Converters [13]

Rather than increasing the frequency of the PWM pattern to reduce the harmonic content of the output voltage, a multi-bridge configuration can be subjected to a phase-shifted carrier. Considering n units, each carrier is shifted by T/n, where T is the period of the fundamental reference wave. Thus the voltage harmonic components of the individual units are shifted with respect to each other and can be designed to be cancelled when the outputs of the various bridges are added.

Each bridge of a transformer-connected multi-bridge converter is modulated independently and the overall output can be described by the same equations presented in the earlier sections. Figure 5.9 illustrates the operating principle with reference to the four-bridge configuration. In Figure 5.9(a), the three sine waves of frequency f_1 intersect with four triangular carriers of frequency f_c to produce 12 PWM waveforms. The four PWM control waveforms associated with phase A are shown in Figure 5.9(b) and the resultant five-level PWM waveform for this phase in Figure 5.9(c). Finally, when the latter is combined with the corresponding waveform in phase B (not shown), their difference produces a nine-level phase-to-phase waveform.

The phase of the spectral components of the natural and uniformly sampled PWM implementations are $n\phi_1 - m\phi_c$ and $n\phi_1 - m\phi_c - (m + n\omega_1/\omega_c)\pi/2$ respectively. In both cases, however, the phase of the modulator signal term f_1 is not affected by changes in the carrier phase ϕ_c (since $m = 0$); therefore, the input signal frequency component will add regardless of the phase relationship of the carriers used for the individual bridges.

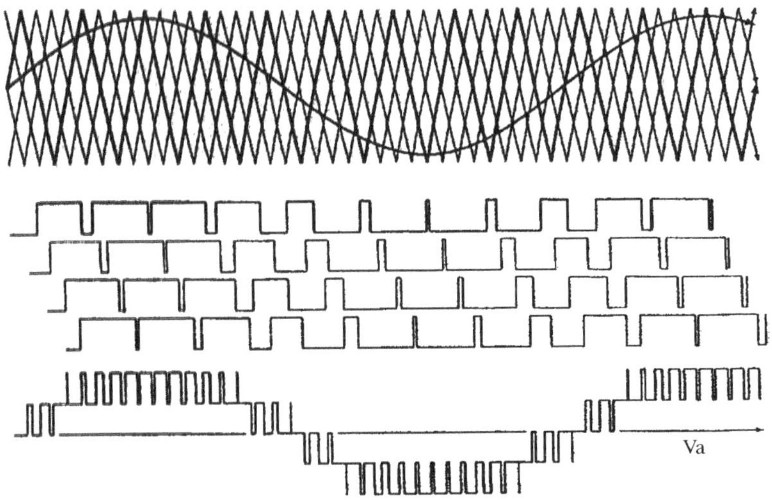

Figure 5.9 Five-level PWM waveforms: (a) triangular carriers and modulating wave; (b) individual PWM waveforms; (c) output voltage

Changing the phase of the carrier signal affects the phase of the carrier terms and their sidebands (since $m \neq 0$). However, choosing ϕ_{ci} for each modulator such that $\Sigma\ m\phi_{ci} = 0$ causes all carriers and sideband terms of order m to sum to zero. Consider as an example the case of a three-phase, three-bridge, four-level converter. The phases of the triangular carriers of the individual phases are $-2\pi/3$, 0 and $2\pi/3$, and thus the phasor sum of the harmonic terms will be zero for $m = 1, 2, 4, 5, 7, 8, \ldots$. Only the spectral components associated with $m = 0, 3, 6, \ldots$ will appear in the PWM output, i.e. for a switching frequency of 450 Hz, the lowest carrier and sidebands are centred around $3 \times 450 = 1350$ Hz.

The PWM control strategies can to a large extent be developed independently of the converter topology. Thus the results discussed in this section with reference to the transformer-summed multi-bridge converter are generally applicable to the multi-level configurations to be described in the next two chapters.

A collection of plots of the spectra of a naturally sampled multi-level modulator for different ratios of the modulating and carrier frequencies are shown in Figure 5.10. These show that the amplitude of the harmonic terms in the natural PWM spectrum is independent of the pulse number.

A reported high-power application uses 48 modules with 120 Hz carrier frequency to synthesise a 60 Hz sinusoid [14]. Despite the low pulse number ($N = 2$) used, the first uncancelled carrier appears at $48 \times 120 = 5760$ Hz.

5.7 Summary

From the operational viewpoint the use of PWM adds considerable flexibility to the VSC. The modulation process develops independent fundamental and harmonic frequency voltage control from a constant DC source; this in turn provides four-quadrant power controllability and avoids the use of low-order harmonic filters.

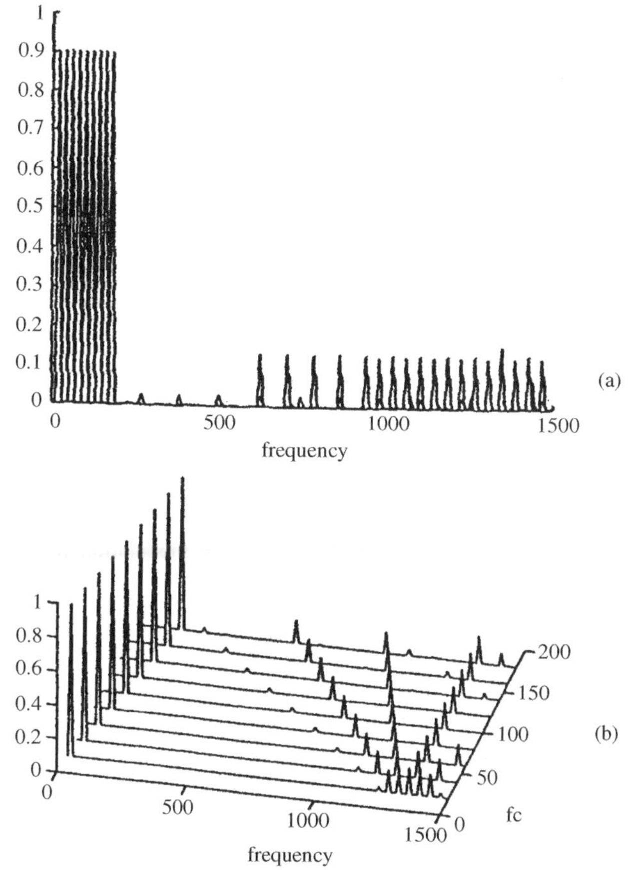

Figure 5.10 Spectra of a naturally sampled four-level converter for different input frequencies ($M = 0.9$, $f_c = 450\,\text{Hz}$): (a) waterfall plot; (b) front elevation

From the design viewpoint, however, the frequent and rapid switching of the AC bus voltage between two levels results in repetitive, high-frequency transient stresses, which affects the design of the VSC components and, unless appropriately filtered, will cause electromagnetic interference. Wound components are particularly vulnerable to repetitive, high-frequency transient stresses and the dielectrics of insulation materials cause premature ageing.

Among the advantages of increased switching frequency are: simpler filtering arrangements, smaller DC ripple and smaller size of VSC stations. An additional benefit of PWM is that the size of the DC capacitor can be reduced. The size of the capacitor is decided to ensure that the voltage ripple is between 2 and 10 %. Every time the valves are switched the current direction in the DC capacitor changes direction and this effect reduces the voltage ripple.

The main disadvantages are higher converter losses, reduced power capability for each device and increased voltage (dv/dt) stresses.

The ability to control the transfer of active power as well as the terminal voltage has made VSC–PWM conversion (based on IGBT switching) attractive for HVDC transmission. The question is whether such level of flexibility is necessary and affordable. These matters are critical to the prospective acceptance of alternative self-commutating DC transmission technologies, and will be discussed in later chapters.

References

1. Schonung, A. and Stemmler, H. (1973), 'Reglage d'un motor triphase reversible a l'aide d'un convertisseur statique de frequence commande suivant le procede de la sous-oscillation', *Revue Brown Boveri*, **51**, 557–576.
2. Patel, H.S. and Hoft, R.G. (1973), 'Generalised technique of harmonic elimination in voltage control in thyristor inverters. Part 1. Harmonic elimination'. *Transactions of the IEEE*, **IA-9**(3), 310–317.
3. Patel, H.S. and Hoft, R.G. (1974), 'Generalised technique of harmonic elimination and voltage control in thyristor-inverters. Part 2. Voltage control techniques'. *Transactions of the IEEE*, **IA-10**(5), 666–673.
4. Buja, E.S. and Indri, G.B. (1977), 'Optimal pulse width modulation for feeding a.c. motors', *Transactions of the IEEE*, **IA-13**(1), 38–44.
5. Goodarzi, G.A. and Hoft, R.G. (1987), 'GTO inverter optimal PWM waveform', IEEE Industry Applications Society Annual Meeting, 1: 312–16.
6. Enjeti, P.N., Ziogas, P.D. and Lindsay, J.F. (1990), 'Programmed PWM techniques to eliminate harmonics: a critical evaluation', *IEEE Transactions on Industry Applications*, **26**(2), 302–316.
7. Bowes, S. and Clark, P. (1995), 'Regular sampled harmonic elimination PWM control of inverter drives', *IEEE Transactions on Power Electronics*, **10**(5), 521–531.
8. Bowes, S. (1993), 'Advanced regular sampled PWM control techniques for drives and static power converters', IEEE IECON'93, 662–669.
9. Candy, J.C. and Gabor, C.T., editors, (1992), *Oversampling delta-sigma data converters: theory, design and simulation*, IEEE Press, Piscataway, NJ.
10. Holtz, J. and Beyer, B. (1992), 'Optimal PWM for ac servos and low cost industrial drives', IEEE Industry Application Meeting, 1: 1010–1017.
11. Wood, P. (1981), *Switching Power Converters*, Van Nostrand Reinhold, New York.
12. Press, W.H. *et al.* (1986), *Numerical recipes: the art of scientific computing*, Cambridge University Press, Cambridge.
13. Walker, G. and Ledwich, G. (1994), 'Bandwidth considerations for multilevel converters', AUPEC, Adelaide, 514–519.
14. Zhang, Z. and Ooi, B. (1993), 'Multimodular current source SPWM converters for a superconductive magnetic energy storage system', *IEEE Transactions on Power Electronics*, **8**(3), 250–256.

6

Multi-Level Conversion

6.1 Introduction

As explained in Chapter 2, the IGCT and press pack IGBT power switches are now suitable to form reliable high-voltage valves. The series-connected power switches can be fired synchronously or asynchronously. Synchronous control, the basis of the two-level schemes, causes static and dynamic voltage sharing problems as well as high dv/dt.

Possible alternatives to two-level conversion for high-voltage applications are the multi-pulse and multi-level topologies. Increasing the pulse number has been traditionally achieved in CSC by the series or parallel connection of bridges, their respective voltage waveforms being phase shifted with respect to each other.

Applying the multi-bridge concept to self-commutating VSC can also improve the converter output waveforms without the assistance of high-frequency switching; however, for high pulse conversion the increased number of converter transformers required makes this solution unattractive.

A more effective alternative for high-voltage application is the multi-level concept with asynchronous firing control; this improves the dynamic voltage balancing of the valves, while the steady-state voltage sharing is achieved by means of clamping devices.

As in the case of PWM, multi-level converters can vary the phase position of the converter fundamental frequency voltage with respect to the AC system voltage waveform; however, their effect on the magnitude of the fundamental frequency voltage is very different. In the PWM solution the magnitude of this voltage can be varied independently from the DC voltage, whereas in the multi-level alternative the voltage magnitude is fixed by the DC voltage. The main object of multi-level conversion is to generate a good high-voltage waveform by stepping through several intermediate voltage levels, i.e. the series-connected devices are switched sequentially producing an output waveform in steps. This eliminates the low-order harmonics and reduces the dv/dt rating of the valves by forcing them to switch against a fraction of the DC voltage.

In principle, the main valves of multi-level converters are switched at the fundamental frequency, which eliminates the higher frequency components typical of PWM. However,

Flexible Power Transmission: The HVDC Options J. Arrillaga, Y.H. Liu and N.R. Watson

when the level number is low, the latter is still needed (though at lower switching frequencies) to reduce the harmonic content of the output waveform to meet the specified standards. On the other hand, the level number has to be limited due to the increasing structural converter complexity for higher numbers.

The multi-level principles are discussed in this chapter with reference to VSC. Even though most of the HVDC projects use line-commutated CSC, the latter has so far not been considered cost effective in self-commutating HVDC conversion. However, VSC and CSC are dual topologies and thus the analysis carried out for the former can also be used in the latter by considering the converter output in terms of current instead of voltage.

The following multi-level VSC topologies are discussed in this chapter:

- diode-clamped circuit (DCC)
- flying capacitor circuit (FCC)
- cascaded H-bridge circuit (IHC).

A more recent multi-level alternative, based on the reinjection concept described in Sections 3.8.2 and 4.5, is the subject of Chapter 7.

6.2 Diode Clamping

6.2.1 Three-Level Neutral Point Clamped VSC

In the conventional two-level VSC, each phase leg of the three-phase converter consists of a switch pair (which for a high-voltage application requires a number of series-connected switches); the DC side capacitor is shared by the phase legs. The switch pairs are gated in a complementary (bipolar) way, such that the output is connected to either the positive or negative potential of the capacitor. The main drawbacks of this configuration are static and dynamic voltage sharing problems (requiring complex balancing techniques) and the high *dv/dt* generated by the synchronous commutation of all the switches.

These problems are substantially reduced in the three-level neutral point clamped (NPC) topology [1], where the converter unit AC terminals are switched between three discrete voltage levels. This scheme has the advantage of 'effectively' doubling the switching frequency as far as the output harmonics are concerned, compared with the bipolar voltage switching scheme. Also the output voltage steps are halved with respect to those of the two-level configuration. The three-phase unit, shown in Figure 6.1 for a GTO-based converter, has an extra terminal on the DC side connected to the centre point of the equally split DC source. The switching configuration has two sets of valves in series, with their intermediate point connected to the DC supply centre tap via extra diodes. For a given valve voltage rating, however, the total DC supply voltage is doubled so that the output voltage per valve remains the same as in the two-level configuration.

The output phase-to-neutral and the phase-to-phase voltage waveforms are shown in Figure 6.2 with respect to the midpoint of the DC capacitor. The phase-to-neutral voltage

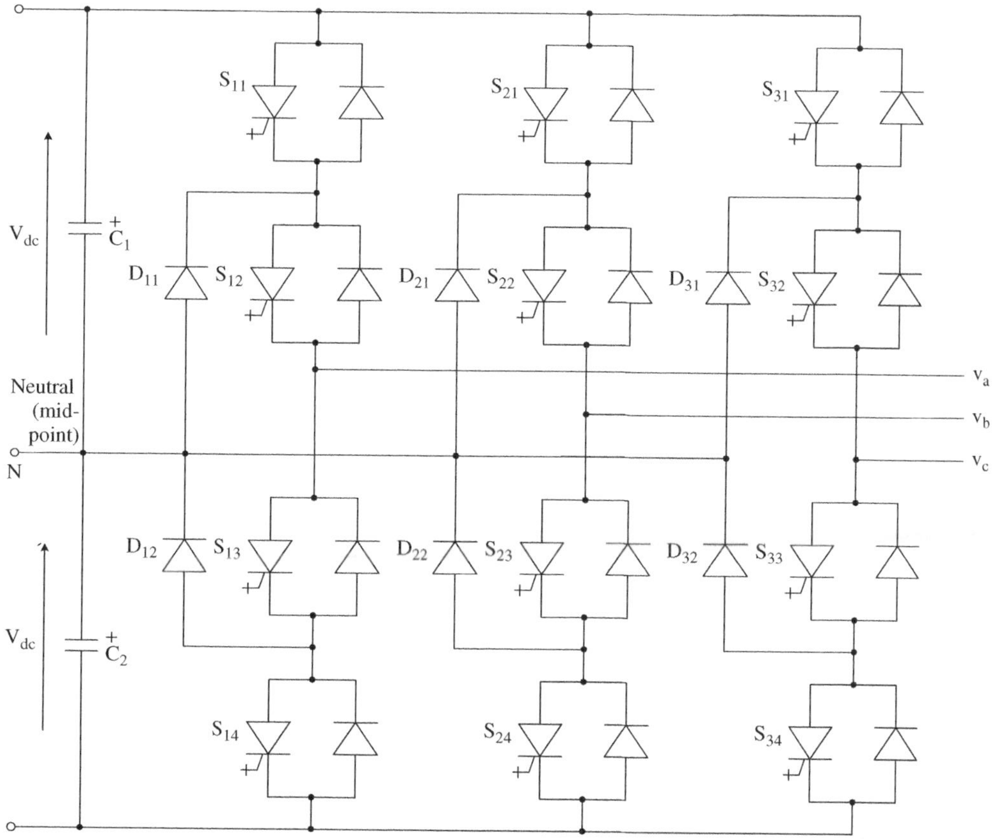

Figure 6.1 Diagram of a three-phase NPC converter

consists of three levels: positive, negative and zero. The positive level is produced by switching on the two series-connected upper valves of the phase unit. Similarly, the negative level is produced by switching the two series-connected lower valves. The zero level is produced by switching the upper and lower middle valves, thus connecting the centre tap of the DC supply to the output via the two extra diodes. In the zero-voltage output region the current continues to flow via the upper middle GTO device and upper centre tap diode (when positive), or the lower middle GTO and the lower centre tap diode (when negative). This process, however, increases the flow of triplen harmonics ripple current through the midpoint of the DC supply.

The frequency components of the phase-to-N (the capacitor neutral or midpoint) voltage are given by

$$v_{aN} = \frac{2V_d}{\pi}\left[\sin\left(\frac{\sigma}{2}\right)\sin\left(\omega t+\frac{\sigma}{2}\right)-\frac{1}{3}\sin\left(\frac{3\sigma}{2}\right)\sin\left(\omega t+\frac{\sigma}{2}\right)+\frac{1}{5}\sin\left(\frac{5\sigma}{2}\right)\sin\left(\omega t+\frac{\sigma}{2}\right)-\ldots\right] \tag{6.1}$$

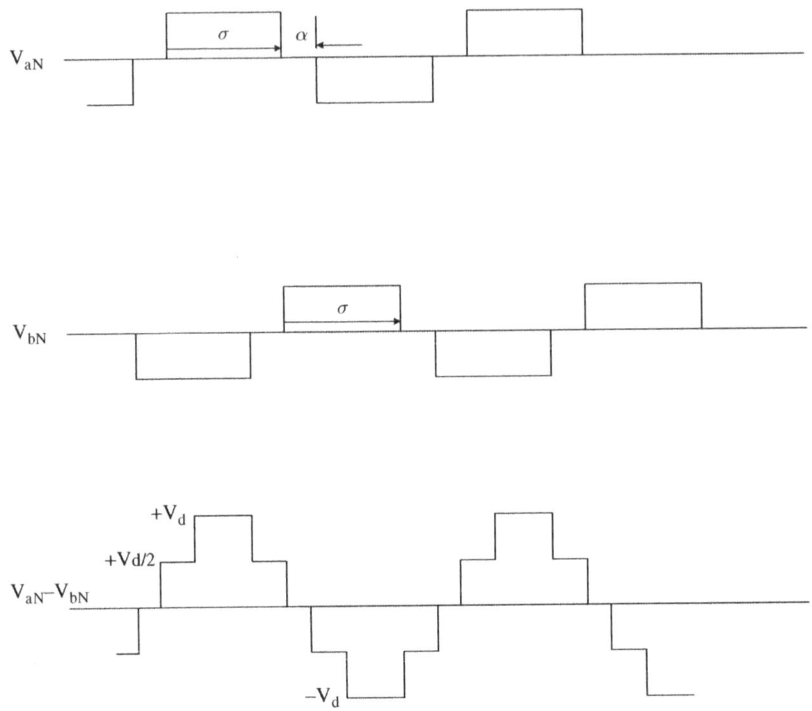

Figure 6.2 AC voltage waveforms for a three-level VSC

and the rms value of the voltage harmonic of order n is

$$V_n = \frac{\sqrt{2}V_d}{\pi n} \sin\left(\frac{n\sigma}{2}\right) \tag{6.2}$$

The relative durations of the positive, negative and zero regions are a function of the control angle σ, which defines the conduction interval of the top upper and the bottom lower valves. Therefore this parameter controls the magnitude of the fundamental rms component of the output voltage, which according to Equation (6.2) reaches a maximum of $\sqrt{2}V_d/\pi$ at $\sigma = 180°$ and becomes zero at $\sigma = 0°$. Thus an important advantage of the three-level configuration is its capability to control the magnitude of the output voltage without changing the number of valve switchings per cycle. Another advantage is that, with judicious choice of the zero-voltage time (α in Figure 6.2), selected harmonic components of the output waveform can be eliminated. For example, for α equal to 30° the positive and negative intervals are 120° and 60° respectively and that means that the waveform is free of triplen harmonics.

Figure 6.3 shows the variation of the fundamental and harmonic components of the output voltage (as a ratio of the maximum value) versus the pulse width.

The operating advantages of this scheme are realised at the expense of some increase in the circuit complexity and a more rigorous control of the current transfers between the four valves, with constrained voltage overshoot. Another requirement is the need to accommodate the increased triplen harmonic content flowing through the midpoint of the

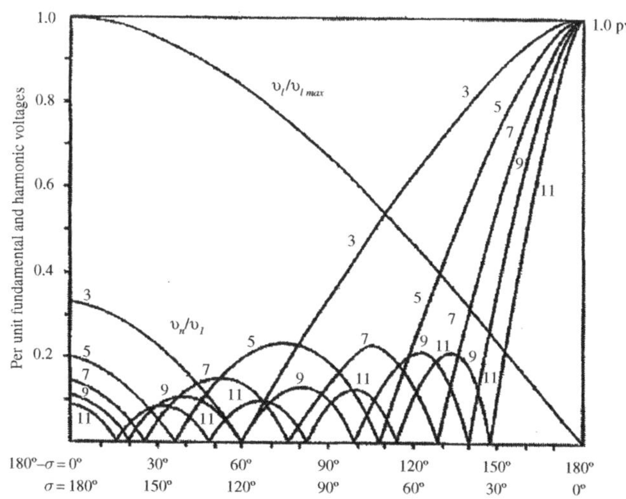

Figure 6.3 Fundamental and harmonic voltages for a three-level VSC (Reproduced by permission of CIGRE.)

DC supply; thus some extra DC storage capacitors will be needed to reduce the fluctuation of the midpoint voltage.

The three-level configuration reduces *dv/dt* across the valves during switchings, as well as the frequency requirements of the PWM control and, therefore, the switching losses. To illustrate this point, a comparison of converter losses (including both the valves and converter transformers), to achieve the same harmonic reduction, shows that the losses in the three-level converter are of the order of 2 %, those of a two-level converter 3.5 %, while the corresponding figure for the line-commutated converter is 0.8 %.

6.2.2 Five-Level Diode-Clamped VSC

Figure 6.4(a) shows a three-phase, five-level diode-clamped configuration, where the DC bus consists of four capacitors (C_1, C_2, C_3, C_4) and the voltage across each capacitor is nominally $V_d/4$. Thus the voltage stress across each switching device is limited to $V_d/4$ and this is achieved by the clamping diode.

The output voltage waveform of the five-level pole is shown in Figure 6.4(b), with the capacitor midpoint used as a reference.

Let us now consider one phase of the five-level configuration, shown in Figure 6.5 [2]. If the negative DC rail (instead of the capacitor midpoint) is used as a reference (i.e. zero voltage) the switching pattern shown in Table 6.1 (where the symbols '1' and '0' indicate on- and off-states respectively) is used to synthesise the five-level waveform across the phase 'a' leg:

- turning on all the upper switches $(S_{a1}\text{–}S_{a4})$ produces $V_{a0} = V_d$;
- turning on three upper switches $(S_{a2}\text{–}S_{a4})$ and one lower switch (S'_{a1}) produces $V_{a0} = 3V_d/4$;

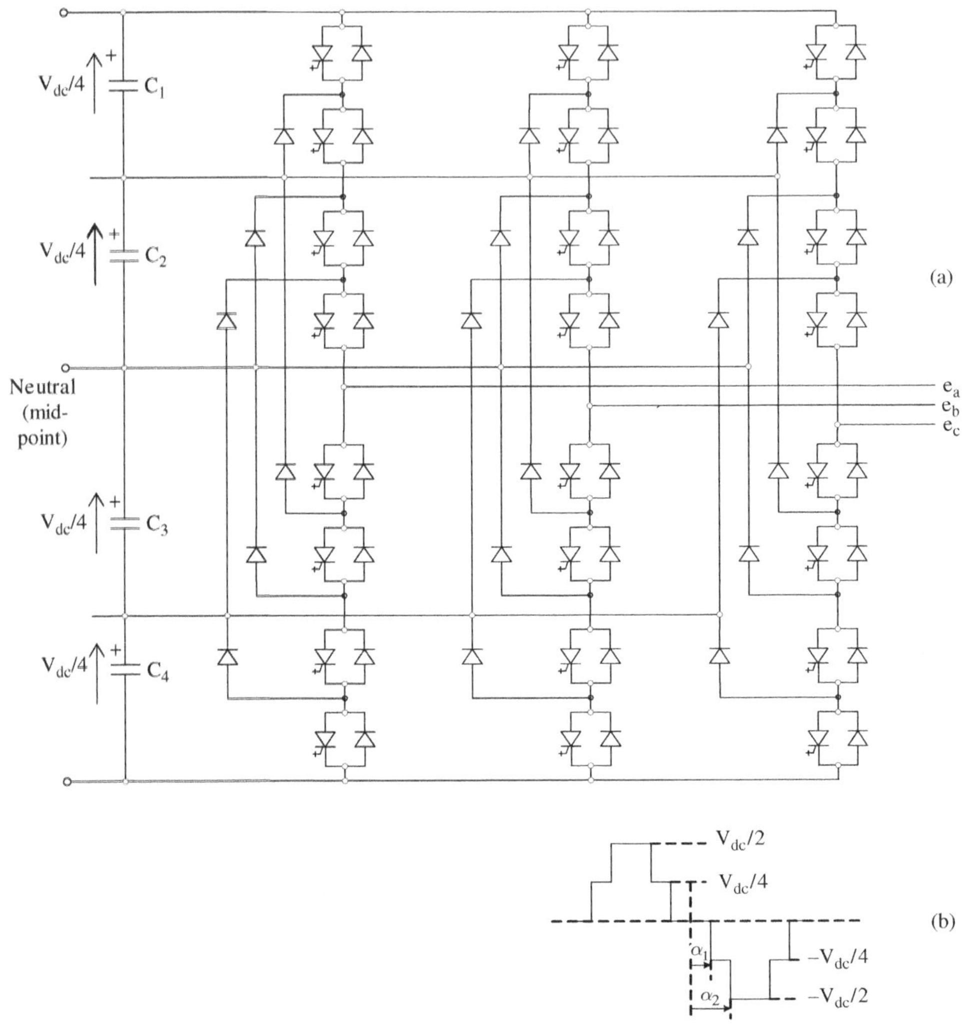

Figure 6.4 Diagram of a three-phase, five-level diode-clamped converter (a) and associated AC voltage (one-phase) waveform (b)

- turning on two upper switches (S_{a3}–S_{a4}) and two lower switches (S'_{a1}–S'_{a2}) produces $V_{a0} = V_d/2$;
- turning on one upper switch (S_{a4}) and three lower switches (S'_{a1}–S'_{a3}) produces $V_{a0} = V_d/4$;
- turning on all the lower switches (S'_{a1}–S'_{a4}) produces $V_{a0} = 0$.

The same switching pattern applies across the phase 'b' leg (if a is replaced by b) but phase shifted by 180° for the single phase configuration (in the three-phase configuration the shifts between the phases will be 120°). Thus the voltage between the 'a' and 'b' terminals

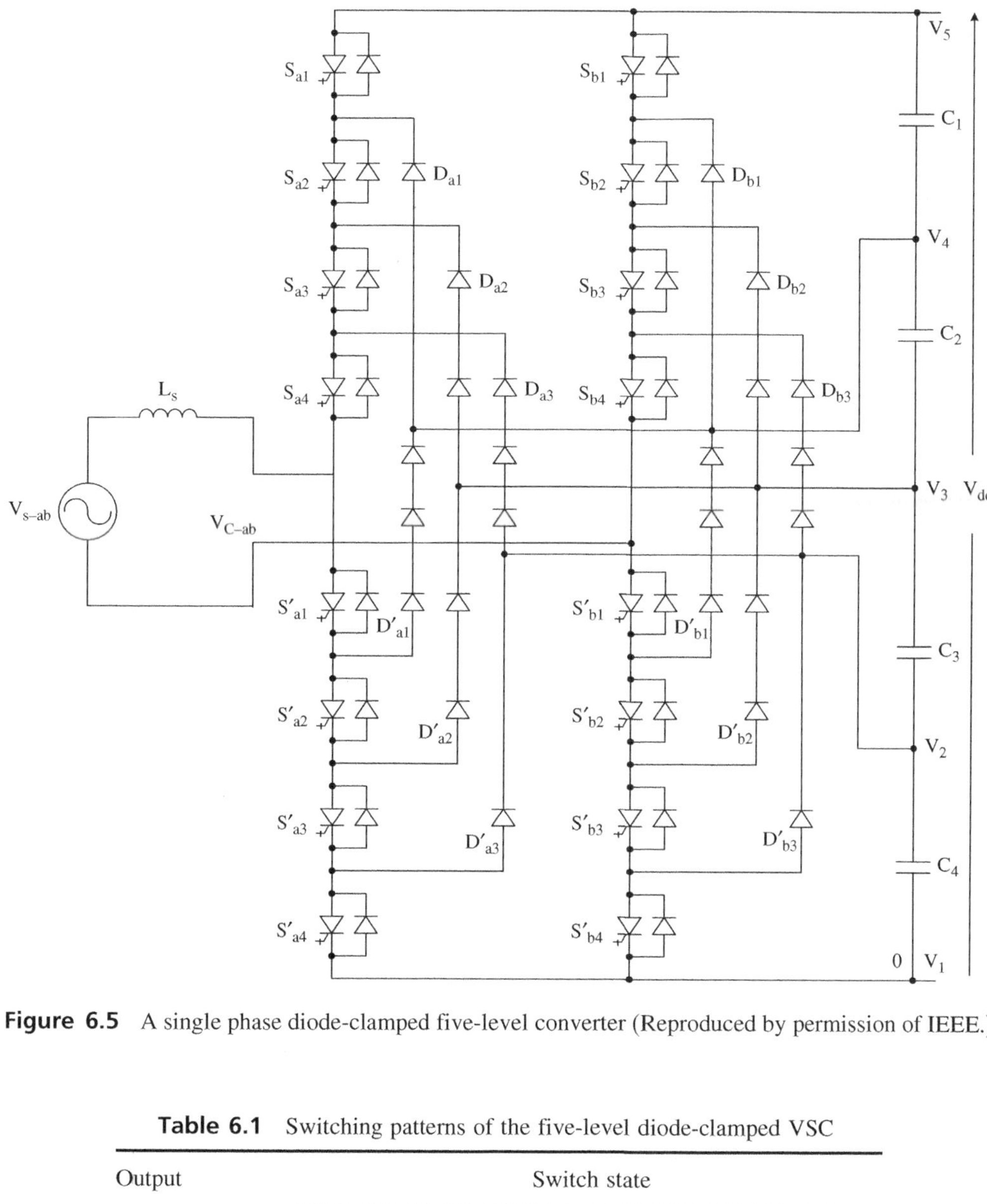

Figure 6.5 A single phase diode-clamped five-level converter (Reproduced by permission of IEEE.)

Table 6.1 Switching patterns of the five-level diode-clamped VSC

Output	Switch state							
V_{a0}	S_{a1}	S_{a2}	S_{a3}	S_{a4}	S'_{a1}	S'_{a2}	S'_{a3}	S'_{a4}
$V_5 = V_{DC}$	1	1	1	1	0	0	0	0
$V_4 = 0.75V_{DC}$	0	1	1	1	1	0	0	0
$V_3 = 0.50V_{DC}$	0	0	1	1	1	1	0	0
$V_2 = 0.25V_{DC}$	0	0	0	1	1	1	1	0
$V_1 = 0$	0	0	0	0	1	1	1	1

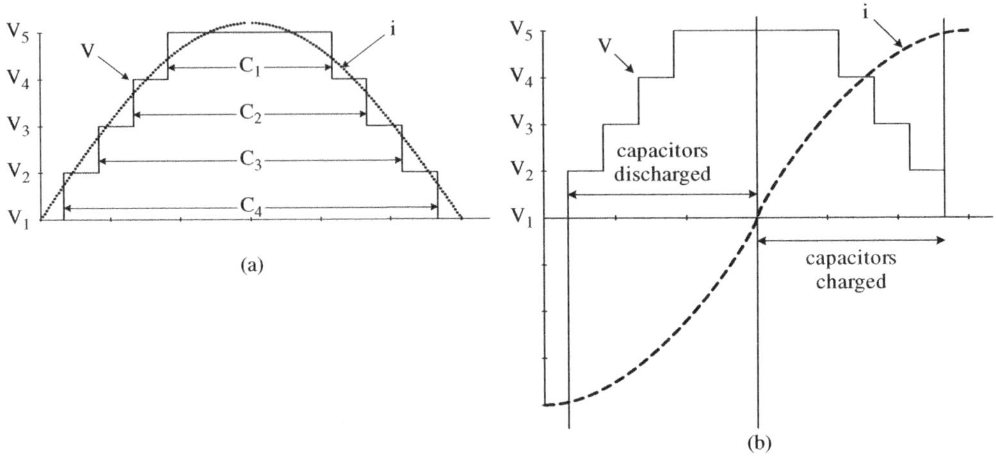

Figure 6.6 Capacitor charge profiles: (a) voltage and current in phase; (b) voltage and current out of phase by 90°

($V_{ab} = V_{a0} - V_{0b}$) is as shown in Figure 6.6. It is clear from the figure that switch S_{a1} conducts only for a small part of the half-cycle, while switch S_{a4} conducts almost for the entire cycle; thus, if all the switches have the same rating, the outer switches will be oversized and the inner switches undersized.

An important problem of this configuration is achieving capacitor voltage balancing, which is highly dependent on the displacement between the voltage and current fundamental components (or, ignoring waveform distortion, the power factor). When the converter operates at zero power factor (i.e. as a VAR compensator) the capacitor voltages are balanced by the equal charge and discharge in each half-cycle, as shown in Figure 6.6(b). At the other extreme, when operating at unity power factor, Figure 6.6(a) shows that the charging (for rectifier operation) and discharging (for inverter operation) times of the capacitors are different. Thus additional circuitry, such as an isolated supply for each of the levels, is needed to solve the capacitor balancing problems.

Also, apart from the two outer switches, i.e. S_1 and S_4', the rest are not directly clamped to the DC capacitors. Therefore, depending on the stray inductances of the structure, the indirectly clamped switches may be subjected to a larger proportion of the nominal blocking voltage during their off-state. Indirect clamping is also a problem in the diode structures, the blocking voltage of each diode depending on its position in the structure. This normally requires the use of more diodes in series; however, diversity in the diodes' switching characteristics as well as stray parameters call for large *RC* snubbers, which add to the cost and volume of the structure.

Moreover, the three inner DC rails carry bidirectionally controlled current, which prevents the use of polarised turn-on snubbers, as the latter will worsen the static overvoltage problem of the inner devices.

An alternative clamping diode structure, shown in Figure 6.7 [3], has been proposed to reduce the problem resulting from indirect clamping. In this alternative, as well as clamping the main switches, the clamping diodes also clamp themselves mutually.

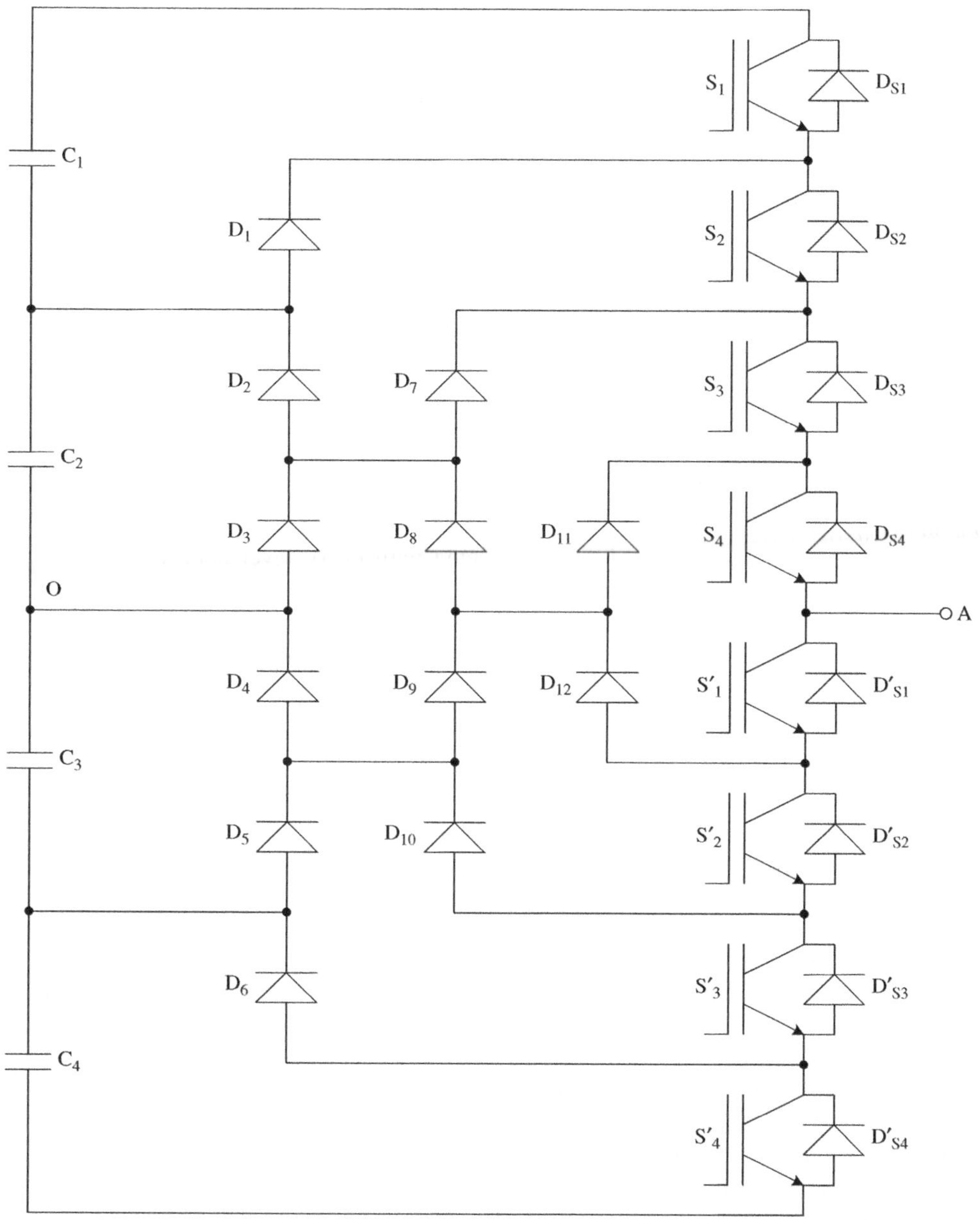

Figure 6.7 Structure of the alternative five-level diode-clamped converter (Reproduced by permission of IEEE.)

6.2.3 Diode Clamping Generalisation

In high-voltage source conversion the DC voltage has to be shared by a number of series-connected capacitors. Also the power converter valves require the use of switching devices in series to withstand the high voltage (which is much higher than the individual switch voltage rating). There are, therefore, multiple intermediate nodes available to implement the multi-level conversion.

An increase in the level number provides a corresponding reduction in the harmonic content. It also improves the steady-state and dynamic voltage balancing of the valves, because groups of switches are clamped to capacitors with equal voltage and, therefore, the output voltage waveform is controlled to change level by level. The multi-level diode-clamped (MLDC) configuration is a generalisation [4] of the three-level neutral-clamped converter concept [1] described in Section 6.2.1.

Figure 6.8 [5] shows one phase of the generalised (m-level) MLDC–VSC structure. In this case $(m-1)$ clamping units are needed with every clamping node, because the maximum blocking voltage across the clamping diodes is related to its position in the diode clamping network. In Figure 6.8 the clamping node with potential V_i is connected by a group of diodes to the node between valves G_{ui} and $G_{u(i+1)}$, and the maximum and minimum potentials are V_i and V_0. Thus the maximum voltage across the path from the node at potential V_i to the node between G_{ui} and $G_{u(i+1)}$ is $V_{m-1} - V_i = (m-1-i)V_L$ where $V_L = V_{dc}/(m-1)$; also, the maximum voltage across the path from the node at potential V_i and the node between G_{di} and $G_{d(i+1)}$ is $V_i - V_0 = iV_L$.

Thus the number of clamping diodes connected with every level node is $(m-1)$, and the clamping path from the node at potential V_i to the main upper switches is formed by $(m-1-i)$ clamping diodes; similarly, the clamping path from the node at potential V_i to the main down switches is formed by i clamping diodes. An m-level converter needs $(m-2)$ inner nodes to be clamped and, therefore, the total number of clamping diodes is given by the expression

$$N_{CD} = (m-1)\,(m-2)$$

The required number of main switches is easier to determine. Since the output voltage u_j in Figure 6.8 is varied from V_0 to $V_{(m-1)}$, for an m-level converter, the maximum voltage across the path from the output terminal to the lowest level node is $V_{(m-1)} - V_0 = (m-1)V_L$, and the maximum voltage across the path from the highest level node to the output terminal is also $V_{(m-1)} - V_0 = (m-1)V_L$. The number of main switches is, therefore, $2(m-1)$.

In principle, the use of MLDC–VSC may be justified in high-voltage reactive power controllers (such as the STATCOM) that require many series-connected switches; it needs a relatively small DC capacitance, has low switch blocking voltages and low switching losses. However the application of MLDC conversion to high-voltage active power control (as required in HVDC transmission) is less attractive because of the difficulty of maintaining the individual capacitor voltages balanced, particularly for high-level numbers.

Raising the level number beyond five is impractical because the number of clamping diodes required increases in proportion to the square of the level number, which makes the capacitor balancing system more complex, introduces unequal power ratings for the switching devices (causing design and maintenance difficulty), and makes it difficult to add redundant switches.

Analysis of the waveforms [6]

The MLDC waveforms are described here under ideal conditions, i.e. all the DC capacitances are of infinite size, the on-state switches' voltage drop is zero, their off-state impedance

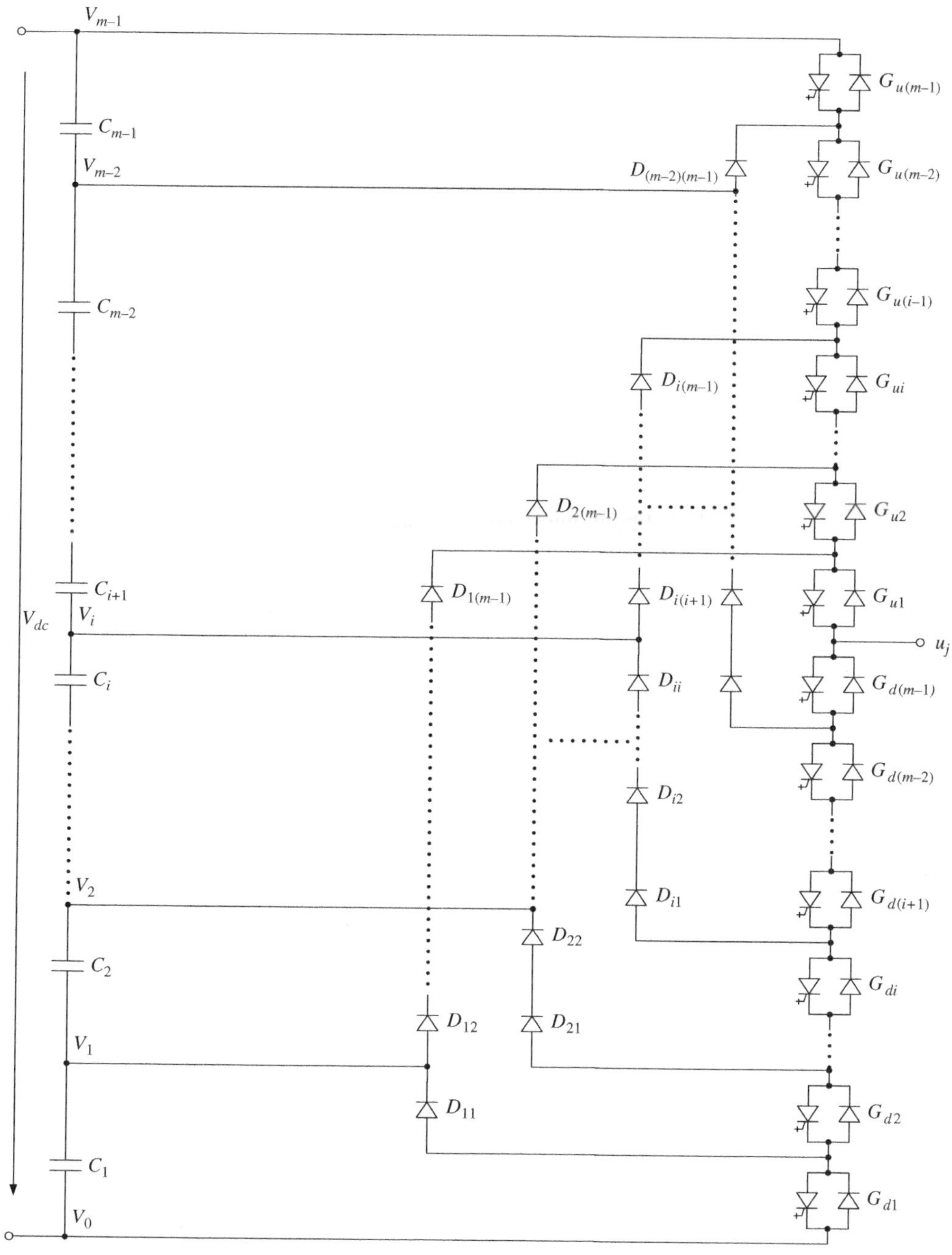

Figure 6.8 One phase of a generalised MLDC–VSC

infinite and the free-wheeling diodes are ideal. Also the voltages across the capacitors are constant and of the same value ($V_L = V_d/(m-1)$).

The Fourier components of the output voltage are given by the expression

$$V_n = \frac{4V_d}{n\pi(m-1)} \sum_{i=1}^{(m-1)/2} \cos(n\vartheta_i) \tag{6.3}$$

for odd-level numbers and

$$V_n = \frac{4V_d}{n\pi(m-1)} \left[\frac{1}{2} + \sum_{i=1}^{(m/2)-1} \cos(n\vartheta_i) \right] \tag{6.4}$$

for even-level numbers, where ϑ_i is the ith switching angle, m the number of levels and n the harmonic order (which in the simulation results presented here is limited to the 49th order).

The fundamental component and the individual harmonics of the phase voltage waveform, for a given number of levels, is calculated for the sets of switching angles which eliminate all possible low-order harmonics. There is no unique solution for the switching angles required to eliminate a given number of harmonics and, thus, the results shown relate to the switching set that gives minimum total harmonic distortion (THD).

Using a transformer reactance of 15 %, Figure 6.9 shows the variation of the AC line current THD with the number of levels. The results are given in normalised form for 1 per unit leading current operation, which is the worst case in terms of harmonic performance and device utilisation. The study shows that from three to six levels there is significant improvement, their respective THDs being 11 % and 2.5 %. For 14 levels this figure is reduced to about 0.6 % and there is no significant reduction in the THD above 12 levels.

The current waveform is also analysed assuming 1 per unit leading operation and quarter-wave symmetry (i.e. $0 < \vartheta < \pi/2$).

In an interval ($\vartheta_i - \vartheta_{i+1}$), where the converter phase voltage remains constant the following relationship applies:

$$L\frac{di}{dt} = \sqrt{2}V\sin(\omega t) - v_I \tag{6.5}$$

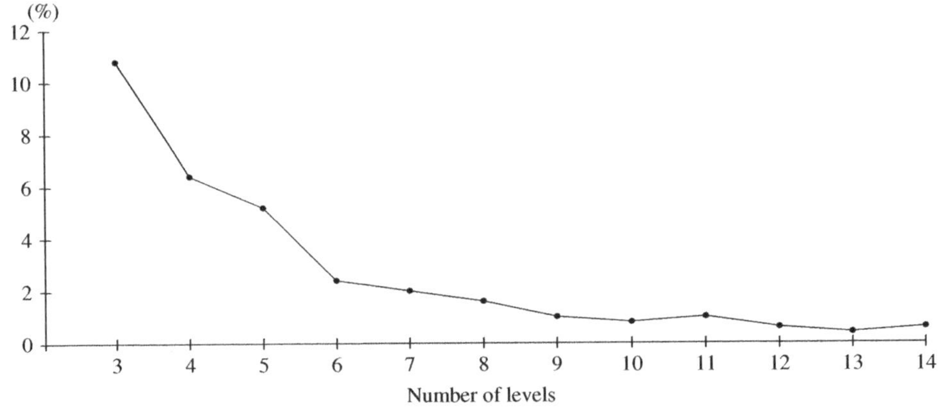

Figure 6.9 Total harmonic distortion in the AC line current

where
V = the rms value of the AC system voltage referred to the converter side
v_I = the instantaneous converter AC side voltage with respect to the AC neutral
L = the transformer leakage inductance (per phase).

Replacing the integration variable ωt by ϑ and using the terminal condition that the current is equal to zero at $\vartheta = \pi/2$, the current becomes

$$i(\vartheta) = -\frac{\sqrt{2}V\cos(\vartheta) + v_I\left(\vartheta - \frac{\pi}{2}\right)}{\omega L} \tag{6.6}$$

Figure 6.10 shows the nominal current waveform together with the corresponding voltage output for the five- and eight-level configurations.

Capacitor voltage balancing

The three-phase m-level converter has $(m-1)$ different DC side capacitors $(C_1 - C_{m-1})$, which are periodically charged and discharged to control the DC voltage and in turn the AC fundamental voltage component. Each capacitor charging current is controlled by the

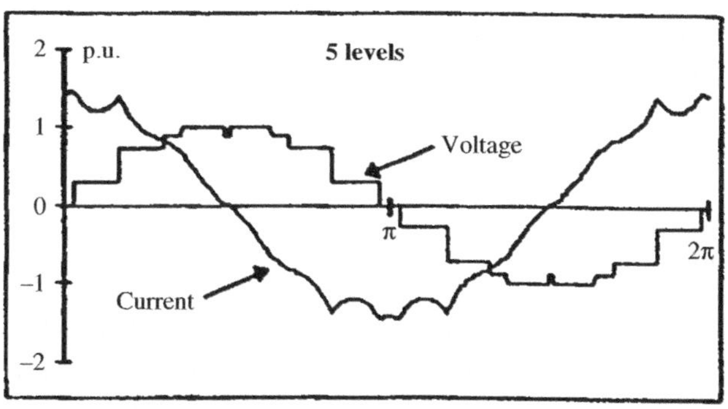

(a)

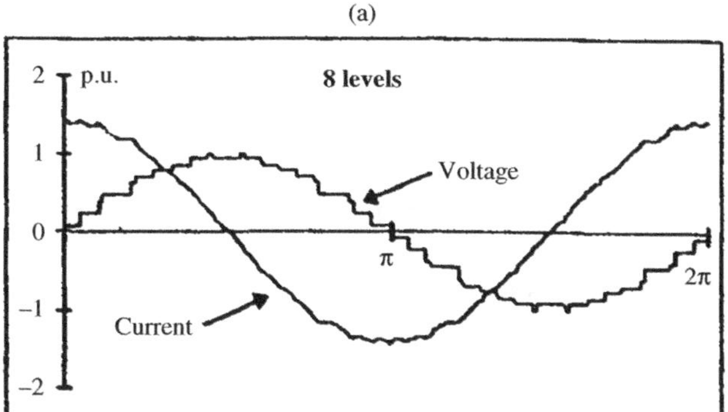

(b)

Figure 6.10 Leading AC and normalised converter phase voltage: (a) for five levels; (b) for eight levels

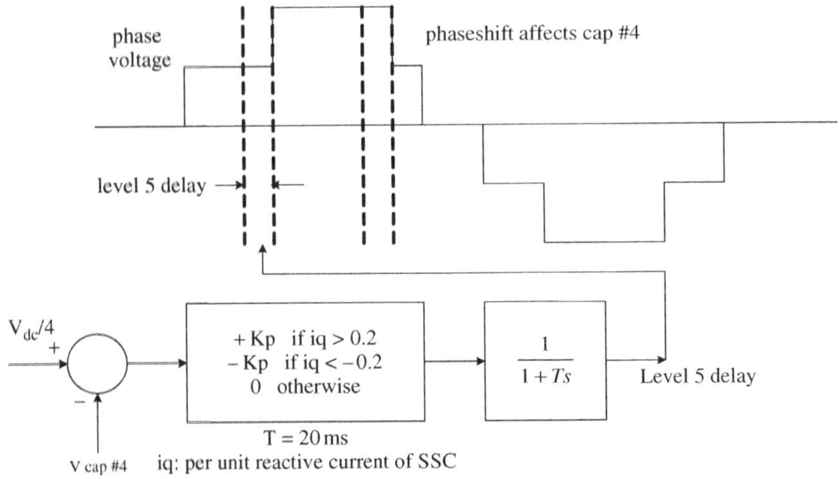

Figure 6.11 Balancing of the individual capacitor voltages [8] (Reproduced by permission of IEEE.)

switching strategy used and the AC side current. The voltage change in each capacitor results from the combined effect of the capacitor size and the charging current.

In normal operation, with balanced converter voltages and currents, the size of each capacitor can be decided to ensure that the voltages across the different capacitors change equally when the total DC side voltage is adjusted [7].

Even in the three-level configuration the capacitor voltages can become unequal in the presence of unbalanced currents. However, capacitor imbalance is more of a problem for higher level numbers and operating conditions requiring power angles different from 90°. The cause of the problem has already been explained in Section 6.2.2 with reference to the five-level configuration.

When the capacitors are not properly balanced, the voltage stress on the switching devices increases and the converter voltage waveform becomes more distorted. This condition can be avoided by using an adaptive control over the individual capacitor voltages, as shown in Figure 6.11 [8] for one of the capacitors. It can be achieved by generating an error signal when the individual capacitor voltage differs from its nominal share of the total DC voltage and this signal is used to phase-shift the related portion of the output voltage wave.

6.3 Flying Capacitor Configuration [9–11]

6.3.1 Three-Level Flying Capacitor

Figure 6.12 shows the flying capacitor circuit (also known as floating capacitor or imbricated cell conversion), as well as the corresponding output voltage waveform. The latter is identical to that of the neutral point clamped converter shown in Figure 6.1.

In the three-level flying capacitor solution, the additional voltage step is obtained by the use of a separate DC capacitor in each phase; this capacitor is pre-charged to one-half of the total DC voltage across the converter bridge.

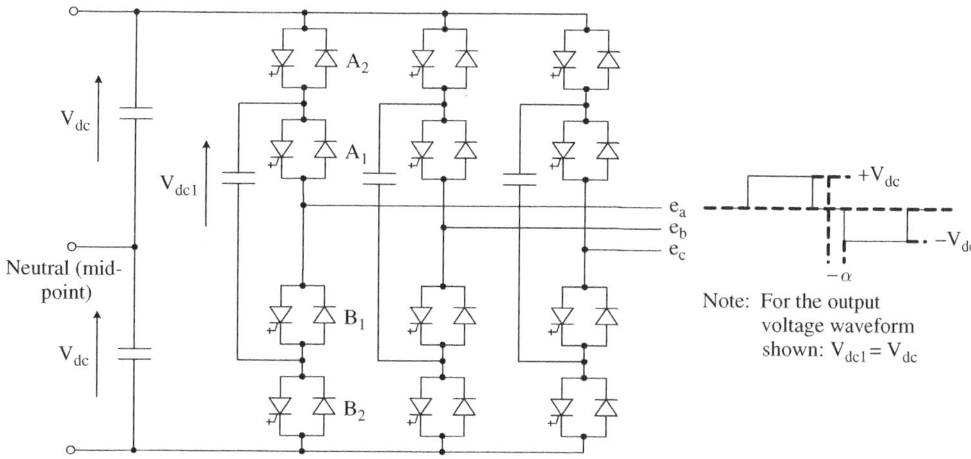

Figure 6.12 Diagram of a three-phase, three-level flying capacitor converter and associated AC voltage (one-phase) waveform

The converter valves switch the AC buses between the three voltage levels by directing the current path through the DC capacitors, adding to or subtracting from the voltages of the individual capacitors in any desired manner. The switches in each phase arm are arranged in two pairs, (A_1, B_1) and (A_2, B_2). Within each pair, the switches need to be always in complementary states. Thus to create the $-V_{DC}$ level of the phase output waveform switches A_1 and B_2 must be turned on; similarly switches A_2 and B_1 need to be turned on to produce the $+V_{DC}$ level. The intermediate DC capacitor is bypassed for part of the fundamental frequency cycle.

The total capacitor power rating is considerably larger for this topology than for the two-level or three-level neutral point clamped topologies. This is partly due to the need for extra separate capacitors in each phase and partly because the DC capacitors carry a significant ripple current.

6.3.2 Multi-Level Flying Capacitor

The three-level flying capacitor concept can be extended to provide any number of voltage levels, by adding further capacitors and subdividing the VSC valves. The size of the voltage increment between two capacitors defines the size of the voltage steps in the output voltage waveform. As an example, Figure 6.13 shows one arm of a five-level configuration, as well as the output voltage waveform which is exactly the same as that of the five-level diode-clamped alternative of Figure 6.4.

The valves switch the AC buses between the different voltage levels by directing the current path through the DC capacitors as explained for the three-level case. The flying capacitor has greater flexibility than the diode-clamped converter. It is possible to create the same output voltage using alternative connections of series capacitors allowing current to flow in the direction required for recharging. For instance, in Figure 6.13, the output voltage level V_{dc1} can be produced by:

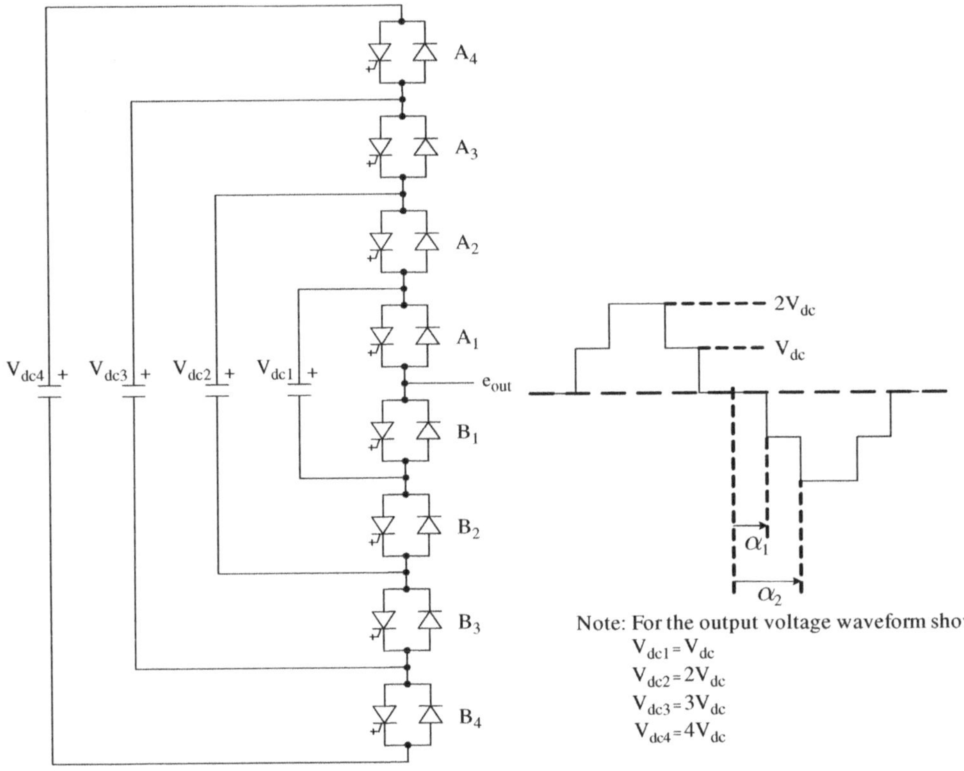

Figure 6.13 Five-level flying-capacitor-type pole

1. Turning on switches A_1, B_2, B_3, B_4.

2. Turning on switches A_2 and B_1 (thus connecting $V_{dc2} - V_{dc1}$ to the load); this reverses the current flow through the V_{dc1} capacitor.

3. Turning on switches A_3, B_1, B_2 (thus connecting $V_{dc3} - V_{dc2}$ to the load).

4. Turning on switches A_4, B_1, B_2, B_3 (thus connecting $V_{dc4} - V_{dc3}$ to the load).

The required number of flying capacitors for an m-level, three-phase converter, provided that the voltage rating of each capacitor used is the same as that of the main power switches is given by

$$3 \times 2[1+2+\ldots+(m-2)]+2(m-1)$$

which for the case of $m = 5$ results in 44 capacitor units.

Naturally, as the number of flying capacitors increases, so does the cost of equipment and the size of the converter station. However, compared with the three-level neutral point diode-clamped topology, the individual valves are switched approximately half the number of times in the five-level topology and the harmonic performance is still superior. Nevertheless, as indicated earlier, the control of a multi-level topology is complicated and higher frequency is required to balance each capacitor voltage [2].

Three-phase configuration

In the three-phase configuration shown in Figure 6.14 all capacitors have the same voltage rating and, thus, a series connection of capacitors is used to indicate the voltage level between the clamping points. The capacitors are directly connected to the output through the converter switches and the AC voltage waveform is very much the same as that produced by the diode clamping scheme. Thus, in general for an m-level converter, the phase voltage has m steps and the line voltage (for the single phase configuration), $(2m-1)$. For a three-phase unit, the main DC capacitors are shared by the three phases but the floating capacitors are not.

However, capacitor voltage control through the switching pattern becomes complex as the output is extended to three phases. Besides the difficulty of balancing the voltage in real power conversion, this configuration requires large numbers of storage capacitors, their size being largely proportional to the square of their nominal voltages and inversely proportional to the switching frequency. If the voltage rating of each capacitor is the same as that of a main power switch, an m-level converter will require a total of $(m-1)(m-2)/2$ auxiliary capacitors per phase leg in addition to $(m-1)$ main DC bus capacitors. To balance the capacitors' charge and discharge, it is possible to use two or more switch combinations for the middle voltage levels (i.e. $3V_d/4$, $V_d/2$ and $V_d/4$) in one or several fundamental cycles. Thus by appropriate selection of the switch combination, this configuration can be used for real power conversion; however, in this case the switch combination is rather complicated and the switching frequency needs to be higher than the fundamental frequency.

As the load current passes through the clamping capacitors, the current rating of these capacitors is higher.

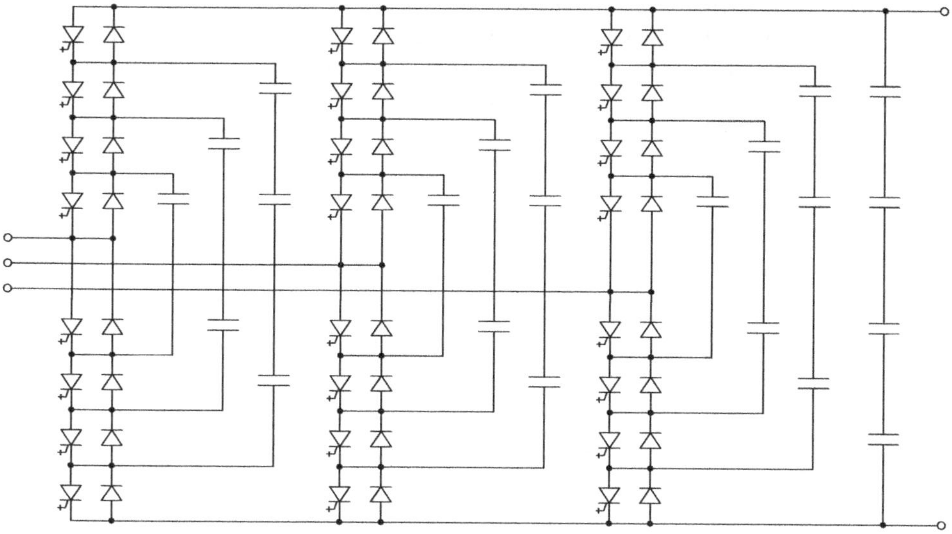

Figure 6.14 Three-phase flying capacitor multi-level configuration

Also, special complex pre-charging systems are required to maintain capacitor balance. On the other hand all the semiconductor switches have the same duty, which has a lower switch blocking voltage and offers the lowest (of the multi-level schemes) converter switching losses.

6.4 Cascaded H-Bridge Configuration

The basic building block of the H-bridge configuration, shown in Figure 6.15, is a single phase full bridge link consisting of four switches, S_1, S_2, S_3, S_4, connected to an isolated capacitor. The H-link can generate three voltage levels, i.e. $+V_{DC}$, 0, $-V_{DC}$. Turning on S_1 and S_4 gives $-V_{DC}$, turning on S_2 and S_3 gives $+V_{DC}$ and bypassing the cell (via S_1, S_2 or S_3, S_4) gives zero voltage.

Several links connected in series (cascaded) constitute a phase of the converter, and the phase output voltage waveform is the sum of the voltages of the activated links. For instance, Figure 6.16 shows the development of the phase voltage waveform for a chain of four links. The latter is synthesised by the sum of the individual link voltages ($v_{an} = v_1 + v_2 + v_3 + v_4$), with an appropriate selection of their conducting periods and contains nine steps [12]. Thus in general an m-level converter will require the use of $(m-1)/2$ H-links per phase. The complete structure of the three-phase configuration is shown in Figure 6.17.

The stepped waveform appears to be similar to that of the current waveform in a conventional multi-bridge CSC described in Chapter 3. However, while the latter consists of a series of pulses of varying height but equidistant in time, in the H-bridge VSC the pulses are of equal height and the intervals between switching instants vary.

The successive switching angles of the H-circuit may be controlled to minimise the total harmonic content or to eliminate several specific harmonics; in the latter case the H-chain can be controlled to eliminate N harmonics per phase, which for the three-phase configuration provides approximately $6N$-pulse operation.

When used for AC–DC or DC–AC power conversion the H-bridge configuration needs separate DC sources. This requirement suits some applications like the system connection from renewable energy sources such as fuel cells and photovoltaics.

The H-bridge configuration has considerable merit for applications where no real power transfer is involved, such as active filtering and reactive power compensation. In this case,

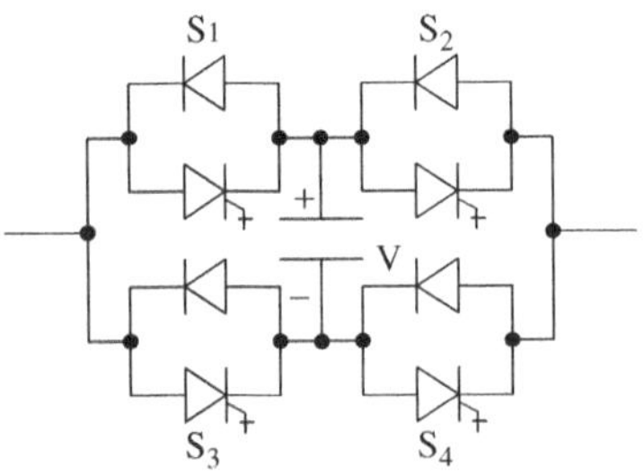

Figure 6.15 One link of the H-bridge configuration

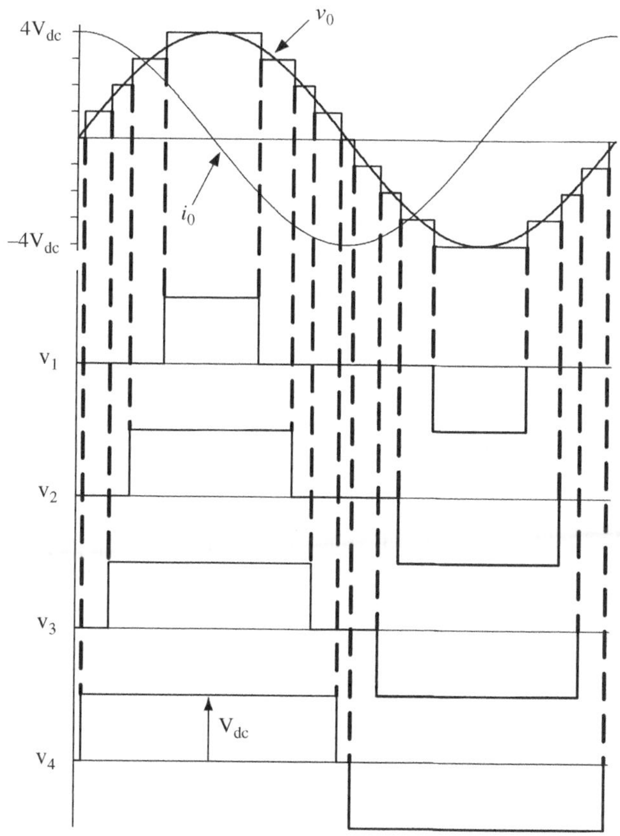

Figure 6.16 Waveforms of the five-level H-bridge converter (Reproduced by permission of IEEE.)

as the phase current I_{Ca} is leading or lagging the phase voltage v_{Can} by 90°, the average charge to each DC capacitor during each half-cycle is zero, i.e.

$$\int_{\theta_i}^{\pi-\theta_i} \sqrt{2}I\cos\theta d\theta = 0 \tag{6.7}$$

where $i = 1, 2, 3, 4$; $[\theta_i, (\pi - \theta_i)]$ represents the time interval during which the DC capacitor connects to the AC side and I is the rms value of the line current. As a result of the symmetrical charge flow, the voltages in all the DC capacitors remain balanced.

The H-configuration provides a well-defined operating environment for the power semiconductors, within a substantially isolated single phase bridge circuit. Its single phase converter structure can provide single phase compensation, an important requirement in applications like arc furnace reactive power compensation.

If a module fails (as a result of a short circuit) or is removed from service the converter can continue to operate at a correspondingly reduced voltage rating.

Under various system conditions the DC capacitor voltages can become unbalanced. This problem is solved by adaptive control action. Provision is made to exchange energy between

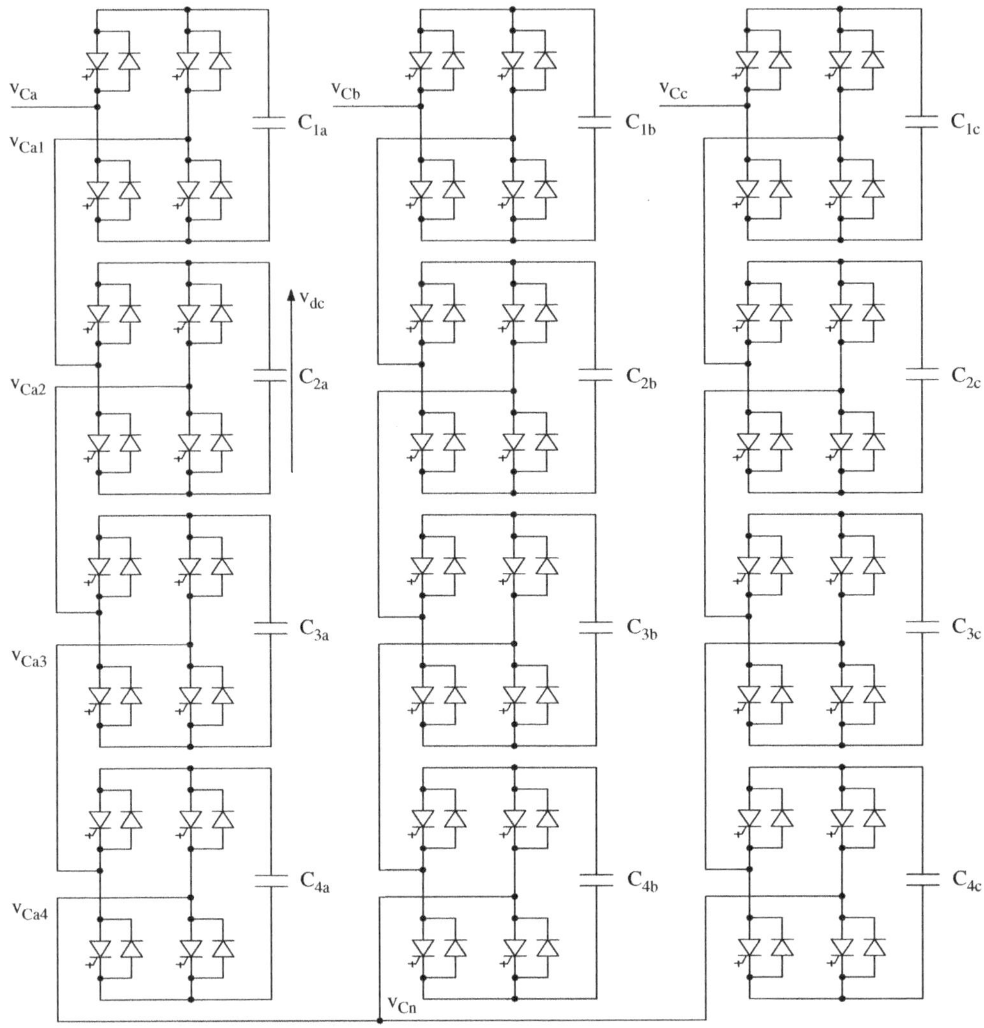

Figure 6.17 A three-phase star-connected H-bridge converter

the capacitors by means of dual IGBT inverters at each link [13]; these are connected to an auxiliary coupling bus via a power isolating transformer (APIT). The inverter/AIPT combination operates at a frequency of 7.5 kHz.

During startup, shutdown and system disturbances the links require an independent supply of energy. A ground-level power supply (GLPS) is used for this purpose. The GLPS also provides energy for pre-charging the DC capacitors.

In its present form, however, cascaded H-bridge conversion is not suitable for HVDC application, because the H-bridge high-voltage conversion process requires the use of many isolated DC sources in series. An interesting recent proposal is the use of a cascaded H-bridge as the basis for multi-level voltage reinjection, a subject discussed in Chapter 7.

6.5 Combined PWM/Multi-Level Conversion

Given the complexity of the multi-level configurations as the level number increases, the use of high-level numbers is unlikely to be accepted. However, the exclusive use of three levels is not sufficient to satisfy harmonic standards and needs to be complemented by PWM control.

The modulation strategies of multi-level converters can be developed independently of the type of topology used. Thus the multi-level PWM waveform results from the summation of individual two-level PWM waveforms. By appropriate phase shifting of the individual two-level waveforms, the amplitude of the voltage steps and hence the amplitude of the harmonic switching frequency terms are reduced in proportion to the number of switch pairs per phase, i.e. to the number of two-level waveforms. Moreover, the output PWM waveform also switches at a frequency which is the sum of the individual two-level switch frequencies. Thus the corresponding multi-level spectra only contain terms at multiples of the aggregate switch frequency.

From the above reasoning, the techniques described in Chapter 5 (i.e. selective harmonic elimination, hysteretic and carrier-based PWM), and their extension to multi-bridge converters (Section 5.7), are equally applicable to the multi-level topologies.

With reference to Figure 6.1, the outer switches of the bridge, i.e. $S_{11}, S_{14}, S_{21}, S_{24}, S_{31}, S_{34}$, can be subjected to PWM control, while the inner switches, i.e. $S_{12}, S_{13}, S_{22}, S_{23}, S_{32}, S_{33}$, are used to clamp the output terminal potentials in the off-periods of the PWM control to the neutral point potential via the D_{11}–D_{32} diodes. Figure 6.18 shows the output voltage waveform required to eliminate the fifth and seventh harmonics by the NPC-PWM converter. As this wave has quarter-wave symmetry, the Fourier components contain only the cosine terms, i.e.

$$\begin{aligned} C_n &= \frac{4V_d}{n\pi}\left[\int_0^{\alpha_1} \cos(n\theta)d\theta + \int_{\alpha_2}^{\alpha_3} \cos(n\theta)d\theta\right] \\ &= \frac{4V_d}{n\pi}[\sin(n\alpha_1) - \sin(n\alpha_2) + \sin(n\alpha_3)] \end{aligned} \tag{6.8}$$

If the fundamental component is expressed as $v_1 = V_m \cos(\omega t)$, the following simultaneous equations need to be solved to find the values of $\alpha_1, \alpha_2, \alpha_3$ that eliminate the fifth and seventh harmonics:

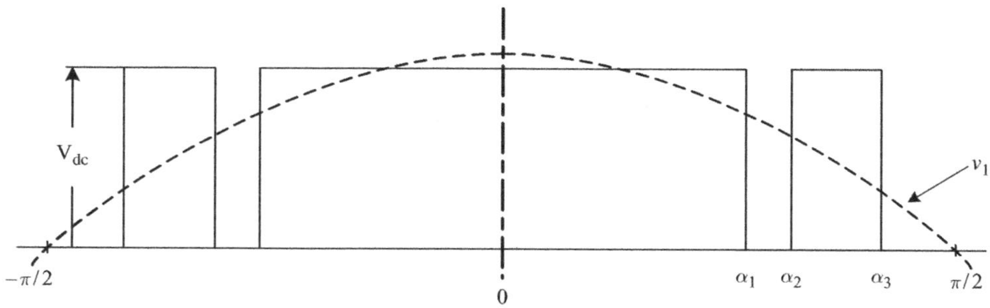

Figure 6.18 Elimination of the fifth and seventh harmonics from the phase voltage

$$\frac{4V_d}{\pi}[\sin(\alpha_1) - \sin(\alpha_2) + \sin(\alpha_3)] = V_m$$
$$\sin(5\alpha_1) - \sin(5\alpha_2) + \sin(5\alpha_3) = 0 \quad (6.9)$$
$$\sin(7\alpha_1) - \sin(7\alpha_2) + \sin(7\alpha_3) = 0$$

The maximum value of V_m that satisfies these equations is $1.19V_d$.

The harmonics resulting from the NPC-PWM (Figure 6.19) show that the 11th and 13th harmonics are considerably reduced.

The use of three different PWM combinations of the multi-level carrier signal is analyzed and compared in [14]. These are APO (shown in Figure 6.20(a)), where all carriers are alternatively in opposition to each other; PO (shown in Figure 6.20(b)), where the carriers above the zero reference are in phase with each other, but in opposition to those below the reference line; PH (shown in Figure 6.20(c)), where all the carriers are in phase with each other.

The parameters of the modulation process used in [14] are:

$p = f_c/f_m$, the ratio of carrier to modulation frequency
$M = A_m/(N'A_c)$, the modulation index, where A_m and A_c are the amplitudes of the modulation and carrier signals respectively
$N =$ the level number of the output phase voltage and $N' = (N-1)/2$
$\phi =$ the phase angle displacement between the sinusoidal reference and the first positive triangular carrier signal.

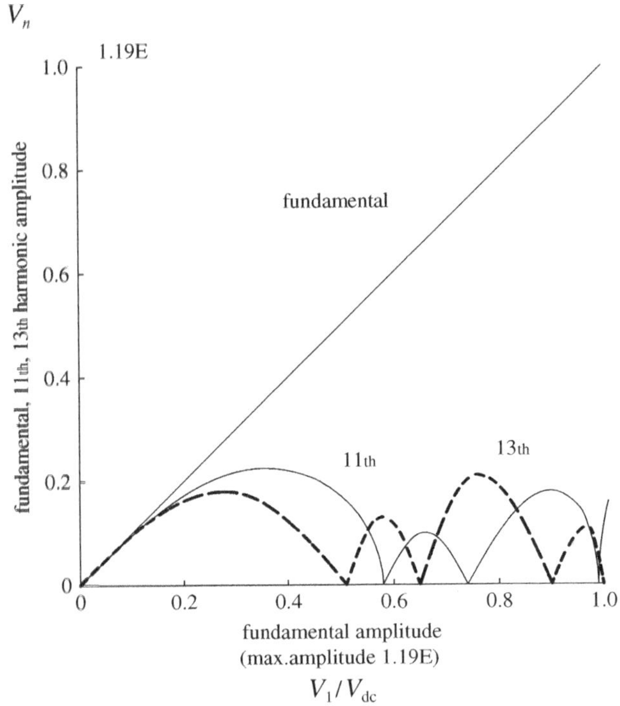

Figure 6.19 Amplitudes of the fundamental, 11th and 13th harmonics in the NPC-PWM inverter (with constant V/f control)

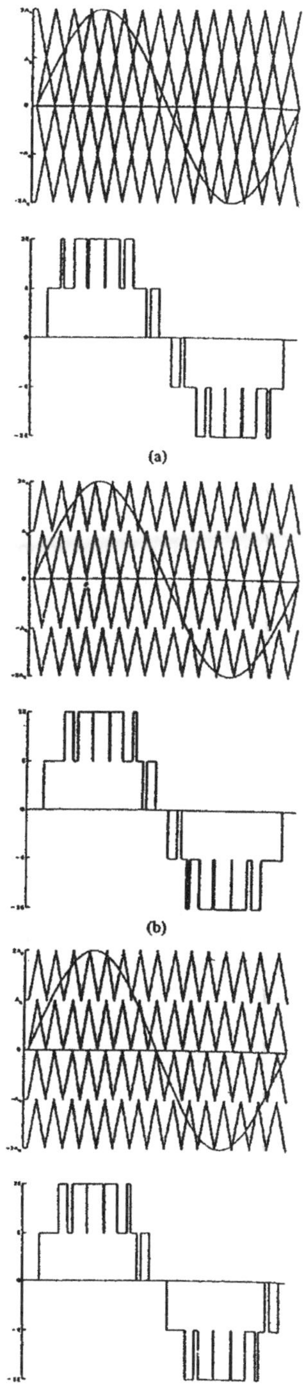
(a)

(b)

(c)

Figure 6.20 Three carrier combinations: (a) APO; (b) PO; (c) PH (for $M = 1$, $P = 15$ and $\phi = 0$)

The comparison shows that when p is high there is practically no difference between the three alternatives, because all the harmonics are shifted to the high-frequency range, where they are easily filtered. For low values of p, the PH combination results in a large value of the component of order p, but reduces the values of the remaining orders. However, in a three-phase system the p-order harmonic component can be made to be of zero sequence and thus be eliminated from the output; under these conditions the PH combination is generally the most satisfactory solution for large-power applications, where the carrier frequency needs to be kept low to reduce the switching losses.

6.6 Relative Merits of the Multi-Level Alternatives

Multi-level conversion involves considerably more hardware than the two- and three-level topologies and increases the station size. On the other hand, the valve design is simpler and the stresses lower, because for the same switching frequency each valve (in the nth-level circuit) is only switched $1/(n-1)$ times and at $1/(n-1)$ times the voltage; therefore the switching losses are lower than for a three-level converter. However, the valve stresses are also affected by inductance in the commutation loop, which increases with the level number; thus for high-level numbers it may be necessary to increase the value of the snubber capacitances, which will in turn increase the switching losses.

At the time of writing, only the three-level diode-clamped VSC topology, complemented by PWM, has been used in HVDC transmission. The main factors affecting the potential application of the multi-level configurations to HVDC transmission are as follows.

Diode clamped VSC

The main advantage of multi-level clamping is a considerable reduction in switching losses and its ability to control the harmonic content. The converter can act as a sink or generator of reactive power, but in the case of a DC link this cannot be achieved at both ends with complete independence from each other. Other negative issues are the large number of clamping diodes required (almost proportional to the square of the level number), the difficulty of maintaining the balance of DC capacitors when the power angle is not exactly equal to $\pm 90°$, the unequal power rating of the power switches and the complexity involved in providing redundant switches to increase reliability.

It is, thus, unlikely that this configuration will play an important part in future HVDC transmission.

Capacitor-clamped VSC

Capacitor clamping shares the main advantages of the diode clamping alternative. It has some flexibility in the choice of switching combinations and, therefore, more control of the clamping capacitor current, thus keeping the clamping capacitor voltage at the required level.

However, the number of clamping capacitors required is almost proportional to the square of the level number, the current rating of the clamping capacitors is high (as they have to carry the load current), and when the converter is used to control the active power, the control strategies to maintain the capacitor voltages at the required level are very complicated.

Again, this configuration, although effective in the drives industry, is unlikely to find strong support for HVDC application.

Cascaded H-bridge VSC

The number of cells required to generate the multi-level waveform is reduced by the three-level output of the individual cells. Also, the use of a modular cell simplifies maintenance and reduces manufacturing costs.

However, the number of isolated DC voltage sources (cells) needed is high and thus, in its present form, this alternative is not suitable for HVDC transmission.

6.6.1 A Cost Comparison of Alternative Configurations for Use in HVDC

This section describes a test study carried out [15] on the multi-level alternatives for possible use in HVDC transmission. The scheme ratings for the comparison were 300 MW, 150 kV, which at the time of the study represented the limit of the market for VSC power transmission technology.

Preliminary considerations, both practical and economical, led to the exclusion of multi-level topologies with more than four levels for the 300 MW power rating. The alternative configurations considered were two-level, three-level diode-clamped and four-level floating capacitor converters. The switching frequencies for each of the three converter models used in the study were selected to achieve a harmonic performance equivalent to that of a two-level converter switching at 1050 Hz. This frequency was chosen as a compromise between two limiting factors. At higher switching frequencies, the switching losses become more significant, whereas converters operating at low frequencies require the use of larger, more expensive DC capacitors and AC filters. For each topology the issues analysed were the capital costs, capitalised losses, DC capacitor volume, commutation inductances and the footprint of the converter station.

Using as a reference the costs of the two-level converter (defined as one per unit), Figure 6.21 shows the cost comparison of the three alternative schemes. The different duties of the semiconductor devices in the three topologies have been taken into account in the evaluation of the total capital cost. In the two- and three-level topologies the higher switching losses associated with the higher switching frequency for each device reduce the current capability of the semiconductor, thus requiring higher rated devices. Despite this fact, the figure shows that the two-level converter requires lower capital cost than the other two alternatives.

However, the capitalisation (at 3000 euros/kW) of the switching power losses makes the two-level scheme the most costly of the three. In fact capitalisation of losses makes the four-level floating capacitor the most economical solution. However, if the value placed on the loss capitalisation is reduced to 1000 euros/kW the three topologies yield practically the same total cost.

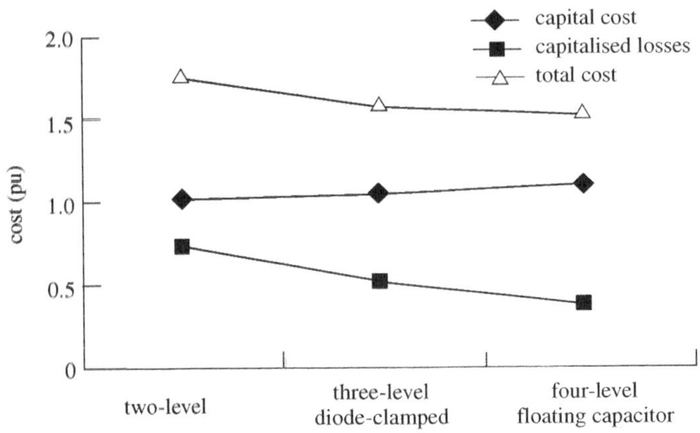

Figure 6.21 Comparison of system cost (Anderson, B.R., Xu, L., Horton, P.J. and Cartwright, P. (2002), 'Topologies for VSC Transmission', *Power Engineering Journal*, June, reproduced by permission of the IET.)

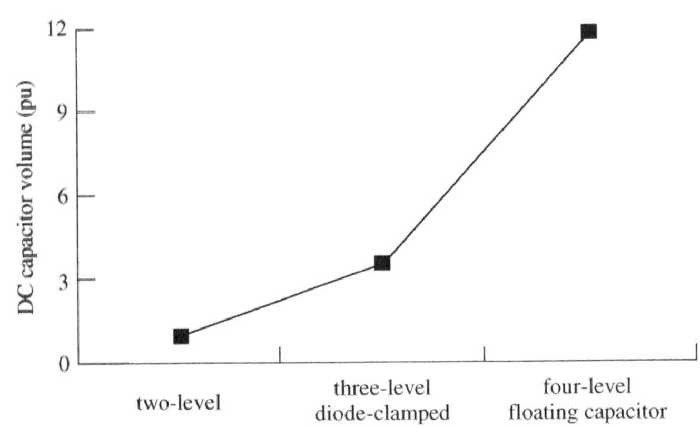

Figure 6.22 Comparison of DC capacitor volume (Anderson, B.R., Xu, L., Horton, P.J. and Cartwright, P. (2002), 'Topologies for VSC Transmission', *Power Engineering Journal*, June, reproduced by permission of the IET.)

Figure 6.22 shows the DC capacitor volume required to limit the voltage ripple to 5 %. Clearly the four-level floating capacitor converter requires considerably higher DC capacitor volume, mainly due to the additional floating capacitors, plus the fact that the lower switching frequency used in this case needs larger capacitance values.

The commutation loop stray inductance affects the snubber capacitance requirement, which in turn increases the switching loss. However, the multi-level configurations do not seem to show a significant increase in this parameter (particularly in the case of the three-level diode-clamped scheme); this is probably due to the use of many parallel-connected capacitor units.

References

1. Nabae, A. Takahashi, I. and Akagi, H. (1981), 'A new neutral point clamped PWM inverter', *IEEE Transactions on Industry Applications*, **IA-17**(5), 518–523.
2. Lai, J.H. and Peng, F.Z. (1996), 'Multi-level converters - a new breed of power converters', *IEEE Transactions on Industry Applications*, **32**(3), 509–517.
3. Yuan, X. and Barbi, I. (2000), 'Fundamentals of a new diode clamping multi-level inverter', *IEEE Transactions on Power Electronics*, **15**(4), 711–718.
4. Jung, G. *et al.* (1991), 'A general circuit topology of multilevel inverter', IEEE Power Electronics Specialist Conference, MIT, USA, Vol. PESC 91, Record: 96–103.
5. Liu, Y.H. (2003), 'Multi-level voltage and current reinjection ac-dc conversion', PhD thesis, University of Canterbury, New Zealand.
6. Scheidecker, D. and Tennakoon, S.B. (1995), 'Optimisation of the number of levels in a multi-level advanced static VAR compensator', EPE'95, Seville, 2.494–2.498.
7. Choi, N.S. Jung, Y.C. Liu, H.L. and Cho, G.H. (1992), 'A high voltage large capacity dynamic VAR compensator using multi-level voltage source inverter', PESC'92, 538–545.
8. Hochgraf, C. and Lasseter, R.H. (1997), 'A transformer-less static synchronous compensator employing a multi-level inverter', *IEEE Transactions on Power Delivery*, **12**(2), 881–887.
9. Meynard, T. and Foch, H. (1992), 'Multi-level conversion: high voltage choppers and voltage source inverters', IEEE PESC'92, 397–403.
10. Meynard, T. and Foch, H. (1993), 'Imbricated cells multi-level voltage source inverters for high voltage applications', *European Power Electronics Journal*, **3**(2), 99–106.
11. Meynard, T. and Foch, H. (1991), 'Dispositif electronique de conversion d'energie electrique', French Patent No. 09582, International Patent PCT No. 92.00652.
12. Peng, F.Z., Lai, J.S. McKeever, J.W. and Van Coevering, J. (1996), 'A multilevel voltage source inverter with separate dc sources for static VAR generation', *IEEE Transactions on Industry Applications*, **32**(5), 1130–1137.
13. Hanson, D.J., Woodhouse, M.L., Horwill, C., Monkhouse, D.R. and Osborne, M.M. (2002), 'STATCOM: a new era of reactive compensation', *IEE Power Engineering Journal*, 151–160.
14. Mohan Mathur, R. and Varma, R.K. (2002), *Thyristor-based FACTS controllers for electrical transmission systems*, IEEE Power Engineering Series, John Wiley & Sons, Inc., New York, p. 421.
15. Andersen, B.R., Xu, L., Horton, P.J. and Cartwright, P. (2002), 'Topologies for VSC Transmission', *IEE Power Engineering Journal*, 142–150.

7

Multi-Level DC Reinjection

7.1 Introduction

In Chapter 3 (Section 3.8.2) a new concept, referred to as DC reinjection, has been described to reduce the harmonic content of the AC and DC voltage waveforms of line-commutated CSCs. The basic principle is the addition of a current waveform, derived from the DC side current, to the rectangular current of the converter bridges.

This concept has been extended in Chapter 4 (Section 4.4) to self-commutating conversion (both of the current and voltage source types), where either IGCT or IGBT switching devices are used to generate the reinjection waveform without the need for line-commutated conversion. The ideal reinjection waveform has been approximated by a symmetrical linearly increasing and decreasing AC component varying at six times the fundamental frequency.

However, the ideal reinjection waveforms required for complete harmonic cancellation, or even its symmetrical triangular approximation, are very difficult to generate from the constant DC source. A more practical solution is to replace the symmetrical triangular waveform by a symmetrical multi-step approximation, which can be easily derived from the DC voltage or current source. It is shown in this chapter that the stepped reinjection waveform provides adequate harmonic cancellation and creates a soft-switching condition during the valve commutations.

The first part of the chapter describes the self-commutating MLVR (Multi-Level Voltage Reinjection) concept and three possible ways of implementing it. These are referred to as clamp-controlled, transformer-coupled and cascaded H-bridge reinjection. The second part of the chapter describes the implementation of the reinjection concept in current source self-commutating conversion. In this case the reinjection circuit is a stepped current (instead of voltage) waveform also varying at six times the fundamental frequency.

A common feature of all multi-level reinjection configurations is the ability to achieve soft switching, in the form of zero-voltage switching (ZVS) (in VSC) or zero-current switching (ZCS) (in CSC) and, therefore, this property is discussed first.

Flexible Power Transmission: The HVDC Options J. Arrillaga, Y.H. Liu and N.R. Watson

7.2 Soft Switching in Multi-Level Reinjection Converters

Soft switching has been a popular development in power electronics due to its potential to enhance the performance and reduce the switching losses. It has been successfully applied for over two decades in the power range of up to 100 kW. The implementation has mainly taken the form of resonant and quasi-resonant circuits designed to provide a ZVS condition. Such an approach reduces di/dt in the switching device that is turning on and dv/dt across the device that is turning off.

The resonant soft-switching principle has also been proposed for high-power multi-level VSC topologies [1–7], but in these cases the extra circuitry is costly and complex. In the multi-level reinjection concept, the required ZVS or ZCS condition can be achieved naturally. This condition is synchronised with the firing control of the main switches; it is established before the switching action starts and terminated after the switching dynamic process finishes.

With reference to multi-level voltage reinjection (MLVR) VSC, the reinjection components are kept completely separate from the main power transfer path and, therefore, their current ratings are low. The on-state reinjection switches are used to clamp the main power switching devices that are changing their switching state. Under this condition, dynamic voltage sharing is achieved without a problem by the series-connected power switches under synchronous control; the energy stored in the snubbers and parasitic capacitance of the power switches can be recovered without losses and thus the snubbers can be simple and inexpensive. The main bridge can be designed as a standard unit with the appropriate voltage and current ratings, without concern for dynamic voltage balancing of the series-connected switches or parasitic energy absorption. Another positive effect is the absence of risk of short-circuiting the DC power source during the commutation between the main bridge valves in a phase leg, i.e. there is no need for a dead interval between switching off one of the main valves and switching on another in the same pole or phase leg. The power converter reliability is therefore increased.

Similarly, the reinjection components involved in the ZCS condition in the case of multi-level current reinjection (MLCR) CSC are of low voltage ratings. In this case the reinjection switches provide the ZCS condition for the main converter power switches. Thus, unlike the case of a resonant switching technique, the energy stored in the inductors and stray inductance is gradually returned to the system; therefore, the interface between the converter and the AC system can be very simple. In contrast, the AC capacitors required by conventional CSC self-commutating conversion to absorb the inductive energy stored in the AC system inductance during the switching off process can be very large. Thus in the absence of ZCS, the interface between the converter and AC system would be expensive; that is why present FACTS devices have avoided the use of self-commutating CSC and present HVDC links still favour line-commutated conversion.

7.3 Clamp-Controlled MLVR [8][9]

In this configuration the two main bridges are periodically clamped by a reinjection switching circuit, common to the three phases, to the different potentials of a subdivided DC source capacitor.

It has been shown in Chapter 6 (Section 6.2.2) that the conventional multi-level diode clamping configuration requires a total number of clamping diodes (N_{CD}) equal to $3(m-1)(m-2)$, which is almost proportional to the square of the level number (m).

In the MLVR converter the clamping structure is simplified in two ways. First, by substantially reducing the number of clamping devices (although these are now controllable switches instead of diodes), and, second, by separating the clamping circuit from the main power circuit. A generalisation of the reinjection structure is shown in Figure 7.1 for an even number of levels. This circuit functions as a controllable voltage divider and provides an m-level reinjection voltage (u_j) to the main bridges via their common (neutral point). The DC capacitor is split into $(m-1)$ cascade-connected units to provide the m accessible voltage levels. The GTO–diode pair valves connected to the capacitor bank and the DC capacitors form a controllable multi-level voltage divider. By turning on the appropriate pairs, i.e. providing a unique path from the reinjection circuit output terminal u_j to the required voltage level, the reinjection circuit output voltage can be any of the possible m voltage levels.

In the general m-level configuration of Figure 7.1, the $m/2$ paths from levels $V_{(m/2)} \ldots V_{(m-1)}$ to the output terminal (denoted by symbol u_j) are formed by the combination of the horizontal clamping switches of the upper half circuit and the vertical switches (G_{u1} to $G_{u(m-1)}$). Of these vertical switches $m/2$ (i.e. G_{u1} to $G_{u(m/2)}$) are shared by all the upper half paths. Similarly the $m/2$ paths from level $V_0 \ldots V_{(m/2)-1}$ to the output terminal are formed by the horizontal clamping switches of the lower half circuit and the vertical switches (G_{d1} to $G_{d(m-1)}$). Again switches ($G_{d(m/2)}$ to $G_{d(m-1)}$) are shared by all the lower half paths. Each clamping path in the upper half circuit consists of force-commutated switches and anti-parallel free-wheeling diodes to allow the flow of current from level nodes to the output terminal, while each clamping path in the lower half circuit consists of force-commutated switches and free-wheeling diodes to allow the flow of current from the output terminal to level nodes. This structure ensures that the voltages across switch units $G_{u(m/2+1)} \ldots G_{(m-1)}$ and G_{d1} to $G_{d(m/2-1)}$ are kept within the nominal level voltage under steady-state conditions, i.e. they are clamped to stable and equal voltage levels.

The commonly shared valves $G_{d(m/2)} \ldots G_{d(m-1)}$ and $G_{u1} \ldots G_{u(m/2)}$ are not clamped to individual voltage levels. Therefore some arrangement must be made for balancing the steady-state voltages across them; however, no dynamic balancing problem occurs, because the output voltage is changed level by level.

In an m-level converter there are $(m-2)$ inner level nodes if m is even, in which case the total number of clamping switches is

$$N_{CSme} = \left(\frac{m}{2} - 1\right)\left(\frac{m}{2}\right) = \frac{m(m-2)}{4}$$

If m is odd, the total number of clamping switches is

$$N_{CSmo} = \left(\frac{m-1}{2} + 1\right)\left(\frac{m-1}{2}\right) = \frac{m^2-1}{4}$$

Further reduction in the number of switches can be achieved using a binary grouping arrangement. The m-level configuration has m paths to the output terminal, but only m switches require individual paths; the remaining switches belong to common paths and can, therefore, be shared. The binary concept is illustrated for eight levels in Figure 7.2. In stage

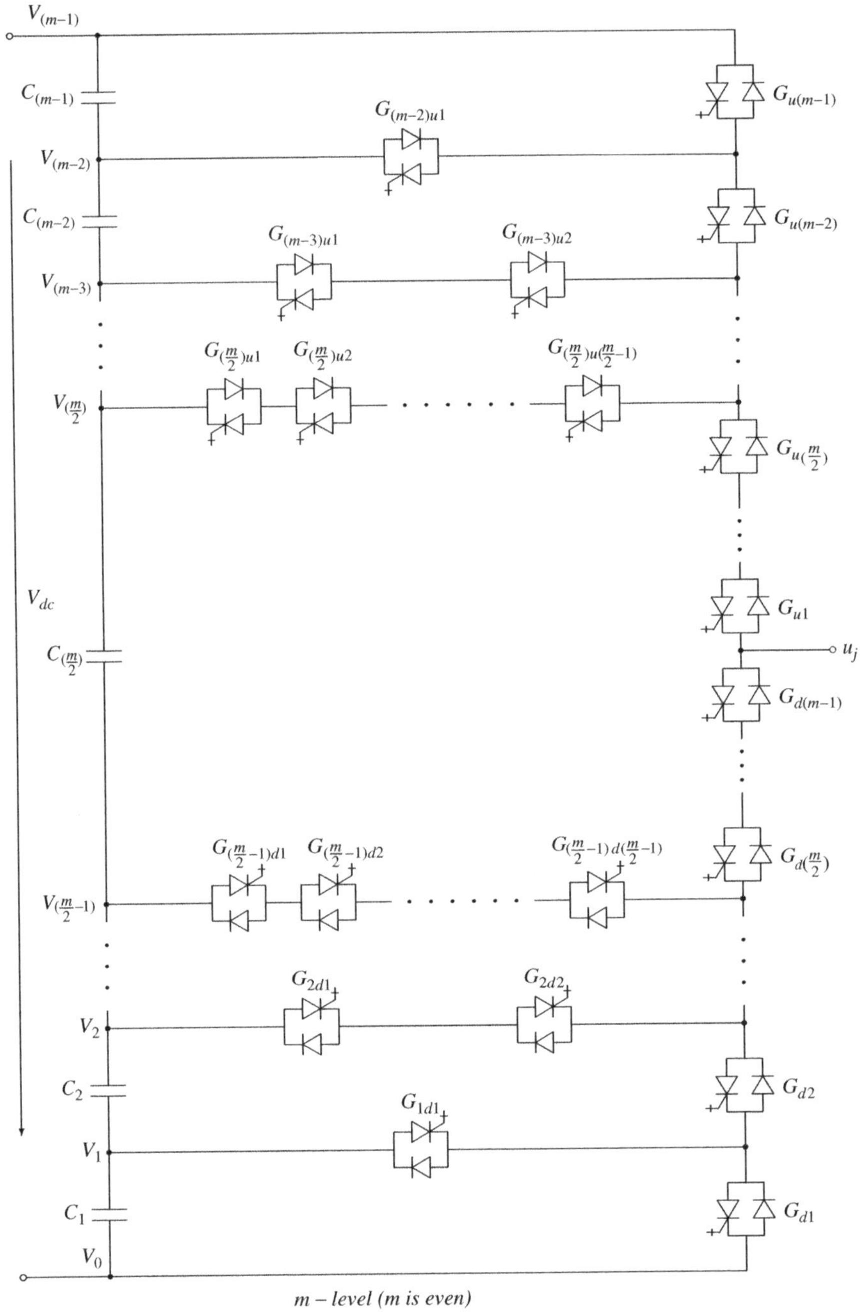

Figure 7.1 Topological structure of the clamp-controlled MLVR-VSC

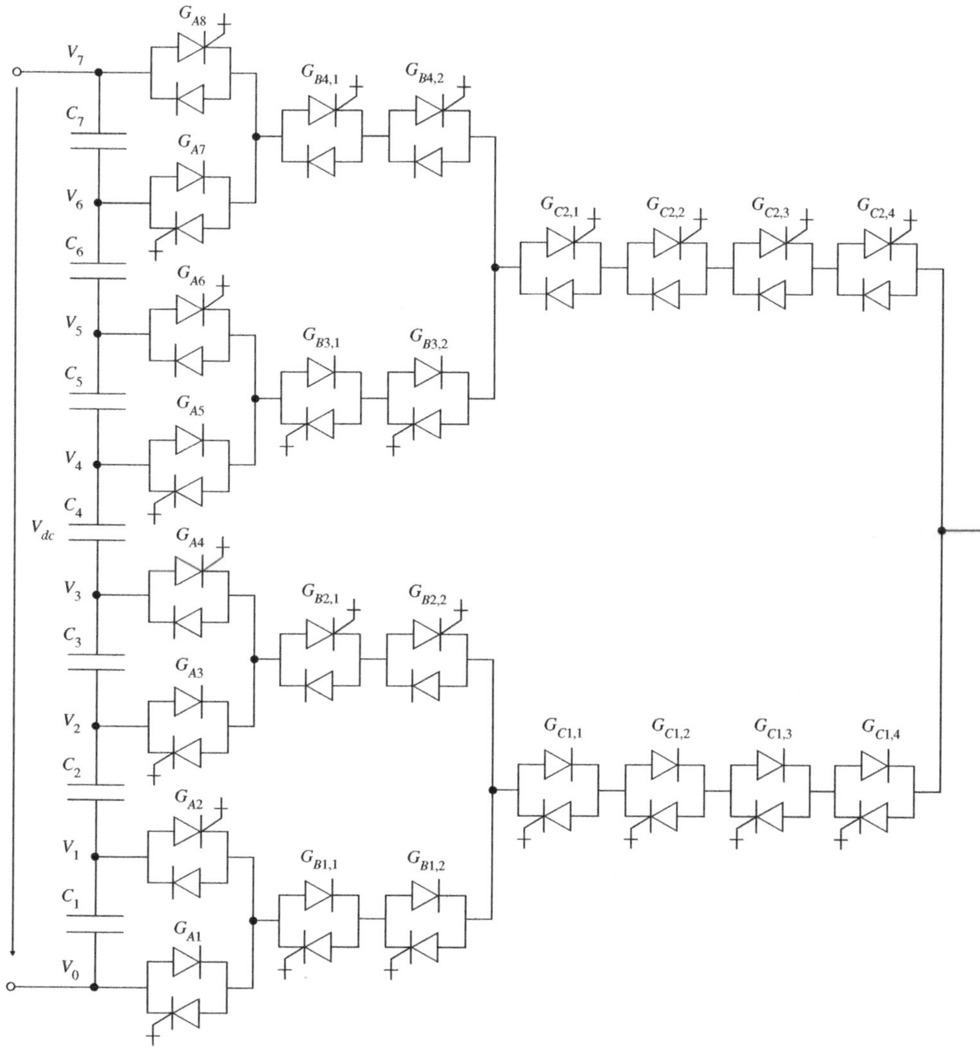

Figure 7.2 Binary-grouped, clamp-controlled multi-level configuration

A the eight paths need to be controlled individually. In the second stage the number of paths are halved; the maximum voltage difference between the two adjacent outputs in stage B is $(2 \times 2 - 1)V_L$ and therefore three switches are needed to withstand this voltage, but since one of them is already present in stage A, only two more switches are required in this stage. Similarly, stage C only needs $m/(2 \times 2)$ paths and the maximum voltage difference between adjacent groups is now $(2 \times 2 \times 2 - 1)$; therefore seven switches are needed to withstand this voltage. With the help of the switches already present in stages A and B, only four more switches are needed in stage C. The total number of switches required for the m-level configuration is thus

$$N_{Snum} = m \log_2 m$$

Obviously the switches in the stages close to the converter output terminal are used by more paths.

In Figure 7.3, which illustrates the $m = 8$ configuration, the switching symbol of every arm in the main bridges represents seven (i.e. $m - 1$) series-connected switch units. The voltages across the reinjection switch branches depend on their position in the reinjection circuit; thus each reinjection valve is represented individually in the figure. Reinjection switches S_{j4} to S_{j11} are common to the different paths, they are not directly clamped to their corresponding voltage levels and each of them needs a large resistance in parallel to balance the steady-state voltage across the units of the series chain. The dynamic voltage balance, on the other hand, can be achieved by asynchronous firing control, which ensures that the voltage increment across them is the level voltage $V_L = V_{DC}/(m - 1)$.

It must be clarified that in a high-voltage application the level voltage V_L will generally be higher than the rating of present switches and, therefore, each symbol in the diagram may itself represent a chain of series-connected switches with steady-state and dynamic voltage balancing auxiliary components.

The current ratings of the reinjection switches are smaller than those of the main bridges, because they only carry the AC components of the bridge output currents.

7.3.1 Firing Coordination

The voltages across the two main bridges in Figure 7.3 are formed by the DC voltage U_{DC} and the reinjection voltage u_j; thus to produce the required voltages the switches in the main bridges must be controlled synchronously with the reinjection switches.

With reference to the eight-level reinjection scheme, the following combinations of switches need to be turned on to provide the bidirectional current paths between the neutral point and the required levels to generate the reinjection voltage (u_j):

$v_0(S_{j1}$ to $S_{j7})$	designated as switch combination S_1
$v_1(S_{j2}$ to S_{j7} and $S_{h1})$	S_2
$v_2(S_{j3}$ to S_{j7} and S_{h21} to $S_{h22})$	S_3
$v_3(S_{j4}$ to S_{j7} and S_{h31} to $S_{h33})$	S_4
$v_4(S_{j8}$ to S_{j11} and S_{h41} to $S_{h43})$	S_5
$v_5(S_{j8}$ to S_{j12} and S_{h51} to $S_{h52})$	S_6
$v_6(S_{j8}$ to S_{j13} and $S_{h6})$	S_7
$v_7(S_{j8}$ to $S_{j14})$	S_8

In general, to generate the required reinjection waveform the m switching paths need to be controlled in the following sequence: $\ldots, S_2, S_1, S_2, \ldots, S_{m-1}, S_m, S_{m-1}, \ldots, S_2, S_1, S_2, \ldots$. The simplest firing method is achieved by using an equal interval, i.e. every reinjection switch in the sequence is ON for $30/(m - 1)°$.

7.3.2 Analysis of the Voltage Waveforms

The waveforms are described under ideal conditions, i.e. all the DC capacitances are of infinite size, the on-state switches' voltage drop is zero and their off-state impedance infinite,

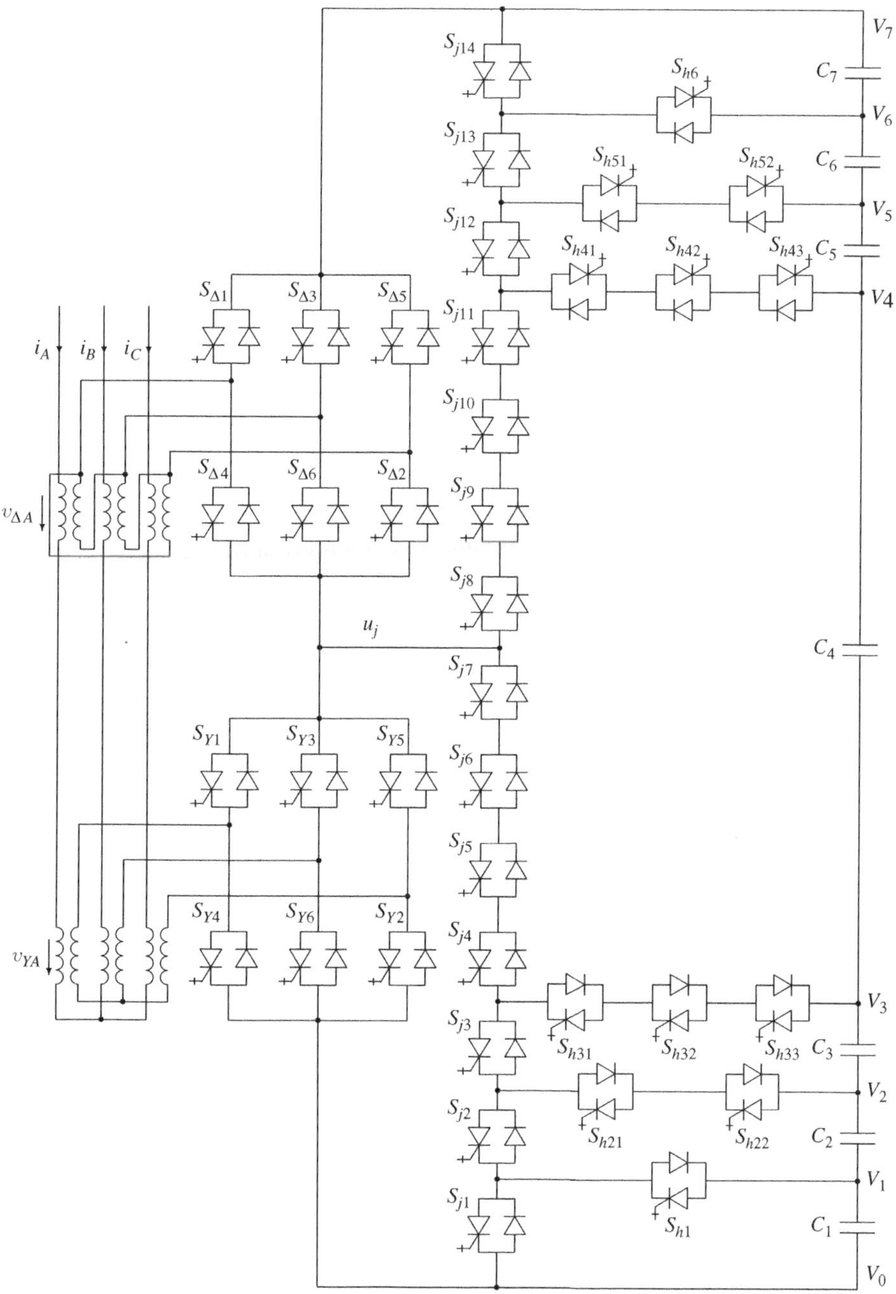

Figure 7.3 Structure of the eight-level MLVR-VSC

the free-wheeling diodes are ideal and the transformer leakage reactance is ignored. Also the voltages across the capacitors are constant and of the same value, i.e. $V_L = V_{DC}/(m-1)$.

The voltage waveforms across the main bridges, shown in Figures 7.4(a) and (b) for the eight-level configuration, illustrate that these change every $30/(m-1)^\circ$ in the sequence $0, V_L, 2V_L, \ldots, (m-1)V_L, (m-2)V_L, \ldots, V_L, 0$ and that the zero voltages (coinciding with the valve commutations) occur around the points $(0^\circ, 60^\circ, \ldots, 360^\circ)$ and $(30^\circ, 90^\circ, \ldots, 330^\circ)$ for the star – star-and star – delta-connected bridges respectively.

In Figures 7.4(a) and (b) the voltages across the bridges connected to the star–star and star–delta interface transformers vary between zero and full DC voltage every 30° in $(m-1)$ steps. The voltage across a secondary phase of the star-connected transformer (V_{Ya}), shown in Figure 7.4(c), is one-third (between 0° and 60°), two-thirds (between 60° and 120°), and again one-third (between 120° and 180°). A secondary winding of the delta-connected bridge is either directly connected to the bridge terminal through the on-state valves or short-circuited by them; thus $V_{\Delta a}$ (shown in Figure 7.4(d)) is zero (between 0° and 30°), the bridge voltage (between 30° and 150°), and zero again (between 150° and 180°). The last two voltages add together on the series-connected primary sides of the transformers to constitute the output voltage, shown in Figure 7.4(e).

The Fourier components of the output voltage $[V_A]$ are given by the expression [8]

$$V_{An} = \frac{8[1-(-1)^n]k_n V_{DC}}{3n(m-1)} \sin\left[\frac{n\pi}{12(m-1)}\right] \cos\left(\frac{n\pi}{6}\right)\left[\cos\left(\frac{n\pi}{6}\right) + \frac{\sqrt{3}}{2}\right]$$
$$\times \left\{(m-1)\sin\left(\frac{n\pi}{6}\right) + \sqrt{3}\sum_{i=1}^{m-2} i \sin\left[\frac{n\pi}{6} + \frac{in\pi}{6(m-1)}\right] - \sum_{i=1}^{m-2} i \sin\left(\frac{in\pi}{6(m-1)}\right)\right\}$$
$$\text{for} \quad m \geq 3, n = 1, 2, 3, \ldots \tag{7.1}$$

From the above expression, the fundamental peak value of the output voltage is

$$V_{A1} = \frac{16 k_n V_{DC}}{\pi(m-1)} \sin\left[\frac{\pi}{12(m-1)}\right]\left[\frac{(m-1)}{2} + \sum_{i=1}^{m-2} i \cos\left(\frac{\pi}{6} - \frac{i\pi}{6(m-1)}\right)\right] \tag{7.2}$$

and the total harmonic distortion (THD) is

$$\text{THD}v_A = \sqrt{\frac{\pi^2(4+\sqrt{3})(m-1)^2\left[1 + \frac{11-6\sqrt{3}}{13(m-1)^2}\right]}{27 \times 128 \sin^2\left[\frac{\pi}{12(m-1)}\right]\left\{\frac{(m-1)}{2} + \sum_{i=1}^{m-2} i \cos\left[\frac{\pi}{6} - \frac{i\pi}{6(m-1)}\right]\right\}^2} - 1} \tag{7.3}$$

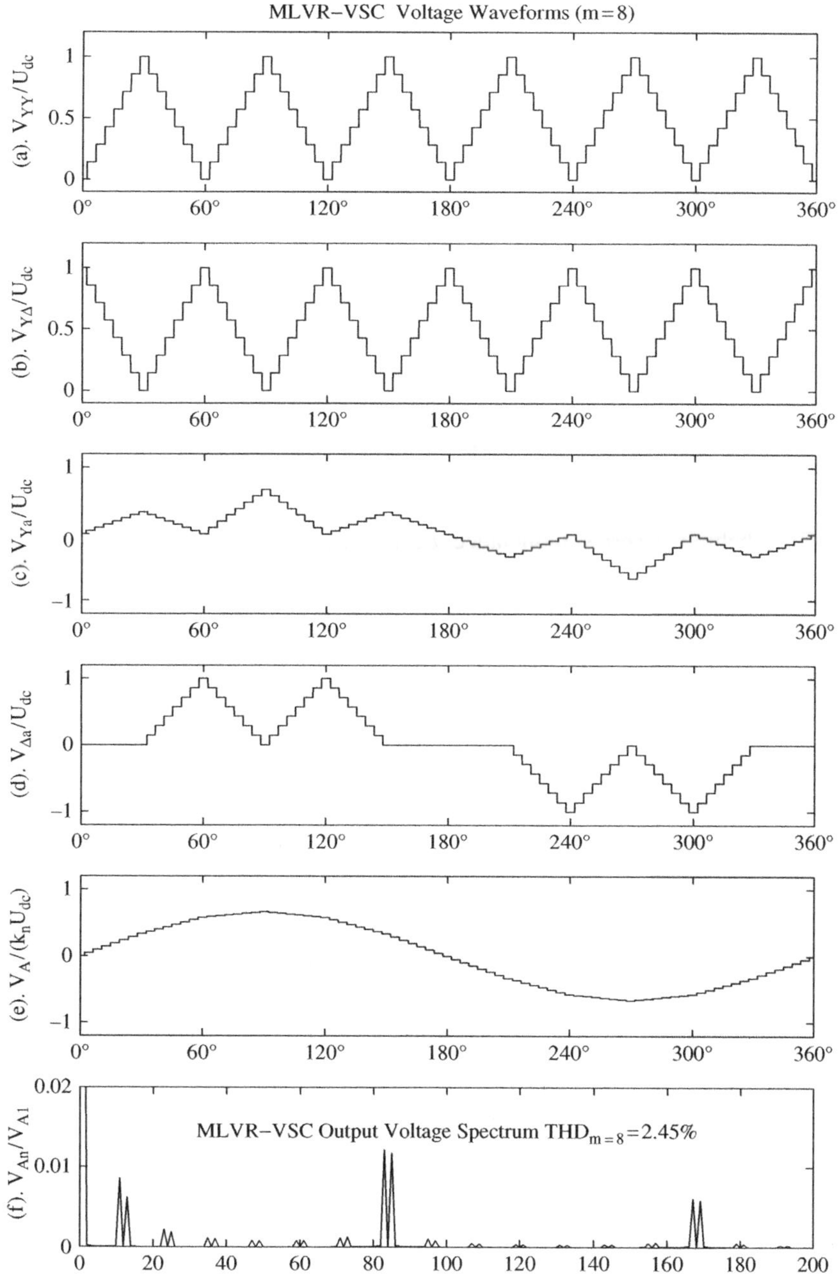

Figure 7.4 Voltage waveforms of the eight-level MLVR-VSC

Table 7.1 Output voltage THD of the MLDC and MLVR configurations

Pulse number	12	24	36	48	60	72	84
MLVR level		3	4	5	6	7	8
THD_{MLVR} (%)		7.8	5.3	4	3.3	2.77	2.5
MLDC level	3	4	5	6	7	8	9
THD_{MLDC} (%)	31.1	23.2	18.3	15.5	13.1	11.6	10.1

Table 7.1 shows the approximate output voltage THDs of the multi-level diode-clamped and multi-level reinjection alternatives for different pulse numbers.

7.3.3 Analysis of the Output Current

Figure 7.5 illustrates the VSC connected to a system represented by a three-phase voltage source behind a series inductance L_S (which includes the converter transformer leakage).

The system and converter voltage vectors are expressed as

$$V_s(\omega t) = \begin{bmatrix} v_{sA} \\ v_{sB} \\ v_{sC} \end{bmatrix} = \begin{bmatrix} V_{sm}\sin(\omega t + \delta) \\ V_{sm}\sin(\omega t - 120^\circ + \delta) \\ V_{sm}\sin(\omega t + 120^\circ + \delta) \end{bmatrix} \tag{7.4}$$

$$V(\omega t) = \begin{bmatrix} v_A(\omega t) \\ v_B(\omega t) \\ v_C(\omega t) \end{bmatrix} = \begin{bmatrix} \sum\limits_{n=1}^{\infty} V_{An}\sin(n\omega t) \\ \sum\limits_{n=1}^{\infty} V_{An}\sin(n\omega t - 120^\circ) \\ \sum\limits_{n=1}^{\infty} V_{An}\sin(n\omega t + 120^\circ) \end{bmatrix} \tag{7.5}$$

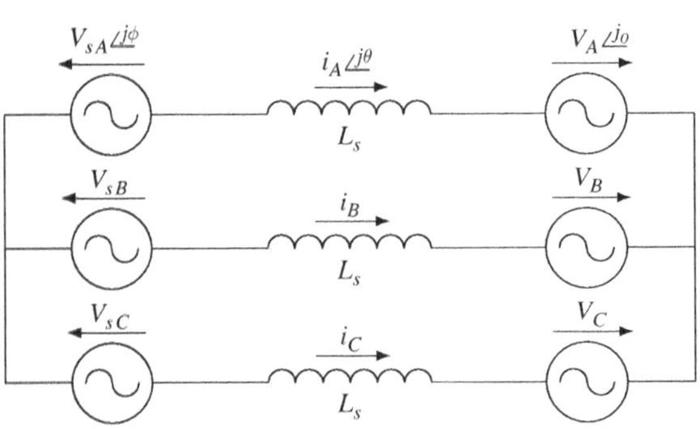

Figure 7.5 Simplified equivalent circuit of the AC system and VSC

The output current vector, derived from the circuit of Figure 7.5, has the expression

$$I(\omega t) = \begin{bmatrix} i_A(\omega t) \\ i_B(\omega t) \\ i_C(\omega t) \end{bmatrix} = \begin{bmatrix} \sum_{n=2}^{\infty} \frac{1}{nX_S} V_{An} \cos(n\omega t) \\ \sum_{n=2}^{\infty} \frac{1}{nX_S} V_{An} \cos(n\omega t - 120^\circ) \\ \sum_{n=2}^{\infty} \frac{1}{nX_S} V_{An} \cos(n\omega t + 120^\circ) \end{bmatrix}$$

$$+ \begin{bmatrix} \frac{1}{\omega L_s}[V_{A1} \cos(\omega t) - V_{sm} \cos(\omega t + \delta)] \\ \frac{1}{\omega L_s}[V_{A1} \cos(\omega t - 120^\circ) - V_{sm} \cos(\omega t - 120^\circ + \delta)] \\ \frac{1}{\omega L_s}[V_{A1} \cos(\omega t + 120^\circ) - V_{sm} \cos(\omega t + 120^\circ + \delta)] \end{bmatrix} \tag{7.6}$$

This expression shows that the harmonic content of the output current is solely caused by the harmonic components of the converter voltage.

Under perfectly balanced conditions the current waveform can be analysed using the amplitude and phase angle of the individual frequency component of a single phase. For instance, the amplitude of the fundamental component in phase A is given by

$$I_{A1} = \frac{1}{\omega L_s} \sqrt{V_{sm}^2 + V_{A1}^2 - 2V_{sm} V_{A1} \cos(\delta)} = \frac{V_{A1}}{\omega L_s} k \tag{7.7}$$

where

$$k = \sqrt{1 + \frac{V_{sm}^2}{V_{A1}^2} - 2\frac{V_{sm}}{V_{A1}} \cos(\delta)}$$

and the amplitude of the nth harmonic current

$$I_{An} = \frac{V_{An}}{n\omega L_s} \tag{7.8}$$

which is likely to be insignificant for practical levels of L_s.

7.3.4 Capacitor Voltage Balancing [10]

The DC output currents of the two main bridges are related to the converter system output by the following expressions:

$$I_{Ydc}(\omega t) = k_{Yn} \begin{cases} -i_B(\omega t) & 0 < \omega t < \pi/3 \\ i_A(\omega t) & \pi/3 < \omega t < 2\pi/3 \\ -i_C(\omega t) & 2\pi/3 < \omega t < \pi \\ i_B(\omega t) & \pi < \omega t < 4\pi/3 \\ -i_A(\omega t) & 4\pi/3 < \omega t < 5\pi/3 \\ i_C(\omega t) & 5\pi/3 < \omega t < 2\pi \end{cases} \tag{7.9}$$

$$I_{\Delta DC}(\omega t) = k_{\Delta n} \begin{cases} -i_B(\omega t + 30°) & -\pi/6 < \omega t < \pi/6 \\ i_A(\omega t + 30°) & \pi/6 < \omega t < \pi/2 \\ -i_C(\omega t + 30°) & \pi/2 < \omega t < 5\pi/6 \\ i_B(\omega t + 30°) & 5\pi/6 < \omega t < 7\pi/6 \\ -i_A(\omega t + 30°) & 7\pi/6 < \omega t < 3\pi/2 \\ i_C(\omega t + 30°) & 3\pi/2 < \omega t < 11\pi/6 \\ -i_B(\omega t + 30°) & 11\pi/6 < \omega t < 13\pi/6 \end{cases} \tag{7.10}$$

where k_{Yn} and $k_{\Delta n}$ are the turns ratios of the interface transformers.

The simplified circuit of Figure 7.6 shows the individual paths from each of the m-level nodes $(v_0, v_1, \ldots, v_{(m-2)}, v_{(m-1)})$ to the output terminal (u_j) of the reinjection converter represented by ideal switches.

The reinjection output current (i_j) is

$$i_j(\omega t) = i_{cu} - i_{cd} = i_{\Delta DC}(\omega t) - i_{Ydc}(\omega t) \tag{7.11}$$

Only one path is in the on-state at any instant and, therefore, i_j is periodically fed from each of the m paths and distributed in discrete time intervals at six times the fundamental frequency.

The currents in the $(m-1)$ capacitors are determined by the state combinations of the m ideal switch paths of Figure 7.6 and currents i_{cu} and i_{cd}, where

$$i_{cu} = i_{\Delta DC} - I_{DC} \qquad i_{cd} = i_{Ydc} - I_{DC} \tag{7.12}$$

The currents through the capacitors that are above the node connected with the on-state ideal switch path are the same (i.e. current i_{cu}), while the currents through the capacitors below that node are also the same (i.e. current i_{cd}).

Thus, using Equations (7.9), (7.10), (7.11) and (7.12), the following expression can be developed for the capacitor current i_{ci} (for $i = 1, 2, \ldots, (m-1)$):

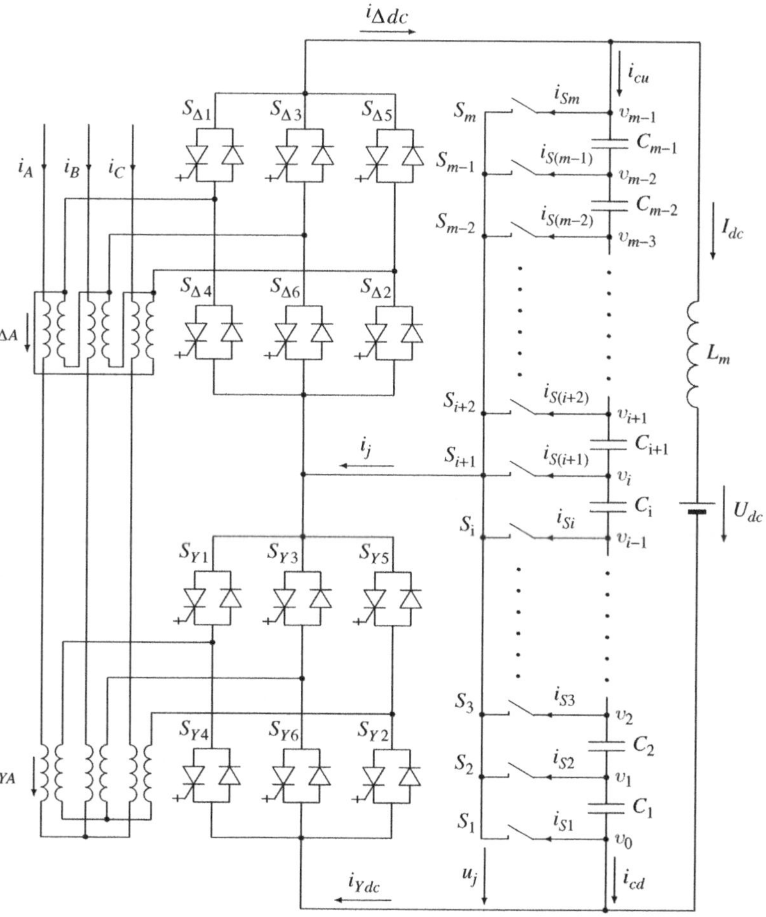

Figure 7.6 Circuit equivalent of the MLVR-VSC for system current analysis

$$i_{ci} = \begin{cases} -k_{\Delta n} i_B(\omega t + 30^\circ) - I_{DC} & -\pi/6 + \alpha < \omega t < \pi/6 - \alpha \\ -k_{Yn} i_B(\omega t) - I_{DC} & \pi/6 - \alpha < \omega t < \pi/6 + \alpha \\ k_{\Delta n} i_A(\omega t + 30^\circ) - I_{DC} & \pi/6 + \alpha < \omega t < \pi/2 - \alpha \\ k_{Yn} i_A(\omega t) - I_{DC} & \pi/2 - \alpha < \omega t < \pi/2 + \alpha \\ -k_{\Delta n} i_C(\omega t + 30^\circ) - I_{DC} & \pi/2 + \alpha < \omega t < 5\pi/6 - \alpha \\ -k_{Yn} i_C(\omega t) - I_{DC} & 5\pi/6 - \alpha < \omega t < 5\pi/6 + \alpha \\ k_{\Delta n} i_B(\omega t + 30^\circ) - I_{DC} & 5\pi/6 + \alpha < \omega t < 7\pi/6 - \alpha \\ k_{Yn} i_B(\omega t) - I_{DC} & 7\pi/6 - \alpha < \omega t < 7\pi/6 + \alpha \\ -k_{\Delta n} i_A(\omega t + 30^\circ) - I_{DC} & 7\pi/6 + \alpha < \omega t < 3\pi/2 - \alpha \\ -k_{Yn} i_A(\omega t + 30^\circ) - I_{DC} & 3\pi/2 - \alpha < \omega t < 3\pi/2 + \alpha \\ k_{\Delta n} i_C(\omega t + 30^\circ) - I_{DC} & 3\pi/2 + \alpha < \omega t < 11\pi/6 - \alpha \\ k_{Yn} i_C(\omega t) - I_{DC} & 11\pi/6 - \alpha < \omega t < 11\pi/6 + \alpha \end{cases} \tag{7.13}$$

where

$$\alpha = \frac{\pi}{12(m-1)}$$

The above equation shows that the $(m-1)$ capacitor currents are periodical time functions, such that every $\pi/3$ interval consists of two continuous current curves corresponding to the two main bridge output currents; there are six of these two curve groups in a full fundamental period for every capacitor waveform.

The voltages across the DC capacitors are determined by integrating the currents flowing through them and capacitor balancing is achieved when the integration of these currents is the same for all of them $(i = 1, 2, \ldots, (m-1))$.

The capacitors' voltage increments for every period of the fundamental frequency are given by

$$\begin{aligned}
\Delta V_{Ci} &= \frac{1}{\omega C_i}\int_{-(2i-1)\alpha}^{2\pi-(2i-1)\alpha} i_{ci} d(\omega t) \\
&= \frac{1}{\omega C_i}\left\{-2\pi I_{DC} - 2k_{\Delta n}\sum_{n=1}^{\infty}\frac{I_{Bn}[1-(-1)^n]}{n}\sin\left[\left(\frac{1}{6}\right)(n-4)\pi+\theta_{Bn}\right]\sin(n\beta_i)\right. \\
&\quad - 2k_{Yn}\sum_{n=1}^{\infty}\frac{I_{Bn}[1-(-1)^n]}{n}\sin\left[\left(\frac{1}{6}\right)(n-4)\pi+\theta_{Bn}\right]\sin\left(\frac{n\pi}{6}-n\beta_i\right) \\
&\quad + 2k_{\Delta n}\sum_{n=1}^{\infty}\frac{I_{An}[1-(-1)^n]}{n}\sin\left(\frac{n\pi}{2}+\theta_{An}\right)\sin(n\beta_i) \\
&\quad + 2k_{Yn}\sum_{n=1}^{\infty}\frac{I_{An}[1-(-1)^n]}{n}\sin\left(\frac{n\pi}{2}+\theta_{An}\right)\sin\left(\frac{n\pi}{6}-n\beta_i\right) \\
&\quad - 2k_{\Delta n}\sum_{n=1}^{\infty}\frac{I_{Cn}[1-(-1)^n]}{n}\sin\left[\frac{(5n+4)}{6}+\theta_{Cn}\right]\sin(n\beta_i) \\
&\quad \left. -2k_{Yn}\sum_{n=1}^{\infty}\frac{I_{Cn}[1-(-1)^n]}{n}\sin\left[\frac{(5n+4)\pi}{6}+\theta_{Cn}\right]\sin\left(\frac{n\pi}{6}-n\beta_i\right)\right\}
\end{aligned} \tag{7.14}$$

where $\alpha = \pi/(12(m-1))$, $\beta_i = (2i-1)\alpha$ and θ_{An} is the phase displacement between the harmonic voltage and current in phase A.

It is obvious from Equation (7.14) that under the equal on-state firing interval of the reinjection switches, the voltage increments of the $(m-1)$ capacitors depend on their position in the DC voltage divider and, therefore, capacitor balancing is not achieved.

Because the integrations of the capacitor current in each cycle are proportional to its DC component, the capacitor balance assessment is based on the DC component only. The DC components of the various currents in Figure 7.6 are given by the expressions

$$\begin{cases} I_{c(m-2)} = I_{c(m-1)} - I_{S(m-1)} \\ I_{c(m-3)} = I_{c(m-1)} - \sum_{j=m-2}^{m-1} I_{Sj} \\ \vdots \\ I_{ci} = I_{c(m-1)} - \sum_{j=i+1}^{m-1} I_{Sj} \\ \vdots \\ I_{c2} = I_{c(m-1)} - \sum_{j=3}^{m-1} I_{Sj} \\ I_{c1} = I_{c(m-1)} - \sum_{j=2}^{m-1} I_{Sj} \end{cases} \tag{7.15}$$

and

$$\begin{aligned} I_{c1} &= I_{cd} + I_{S1} \\ I_{c(m-1)} &= I_{cu} - I_{Sm} \end{aligned} \tag{7.16}$$

Equations (7.15) and (7.16) indicate that to achieve capacitor voltage balance the DC components of the $(m-2)$ inner reinjection switch currents, i.e. I_{Sj} $(j = 2, 3, \ldots, (m-1))$, are zero. Also, in an ideal 12-pulse scheme $I_{DC} = I_{cu}$ and, thus, from Equation (7.16), $I_{S1} = -I_{Sm}$, i.e. switch paths S_1 and S_m act as bypass switches for their DC components.

Capacitor voltage balance can also be assessed with reference to the charges taken away from the capacitors. Thus the DC components of the $(m-2)$ inner ideal switch currents are proportional to the charges taken away from the $(m-2)$ inner-level nodes every period. If these charges are all zero, the capacitor voltages are kept balanced.

From Equation (7.15) the DC component differences between the first capacitor (used as a reference) and other capacitors are given by

$$I_{ci} - I_{c1} = \sum_{j=2}^{i} I_{Sj} \qquad \text{for} \quad i = 2, 3, \ldots, (m-1) \tag{7.17}$$

These differences determine the capacitor voltage increment or decrement after every period; thus, when imbalance occurs, adjustments can be made to the capacitor voltages by regulating the DC components of the $(m-2)$ inner reinjection switch currents. The adjustments are implemented by slight variations of the on-state intervals [10].

The converter system power angle has a strong influence on the capacitor voltage balance. In the absence of adjustments, the capacitor current differences cannot be zero, except when $\theta = \pm 90°$. Figure 7.7 illustrates the on-state interval adjustments required by the eight-level MLVR to cancel the DC components of the reinjection switch path currents for the particular case when $\theta = 30°$.

Figure 7.8 shows the total harmonic distortion of the eight-level MLVR output voltage versus the possible range of the power factor angle θ.

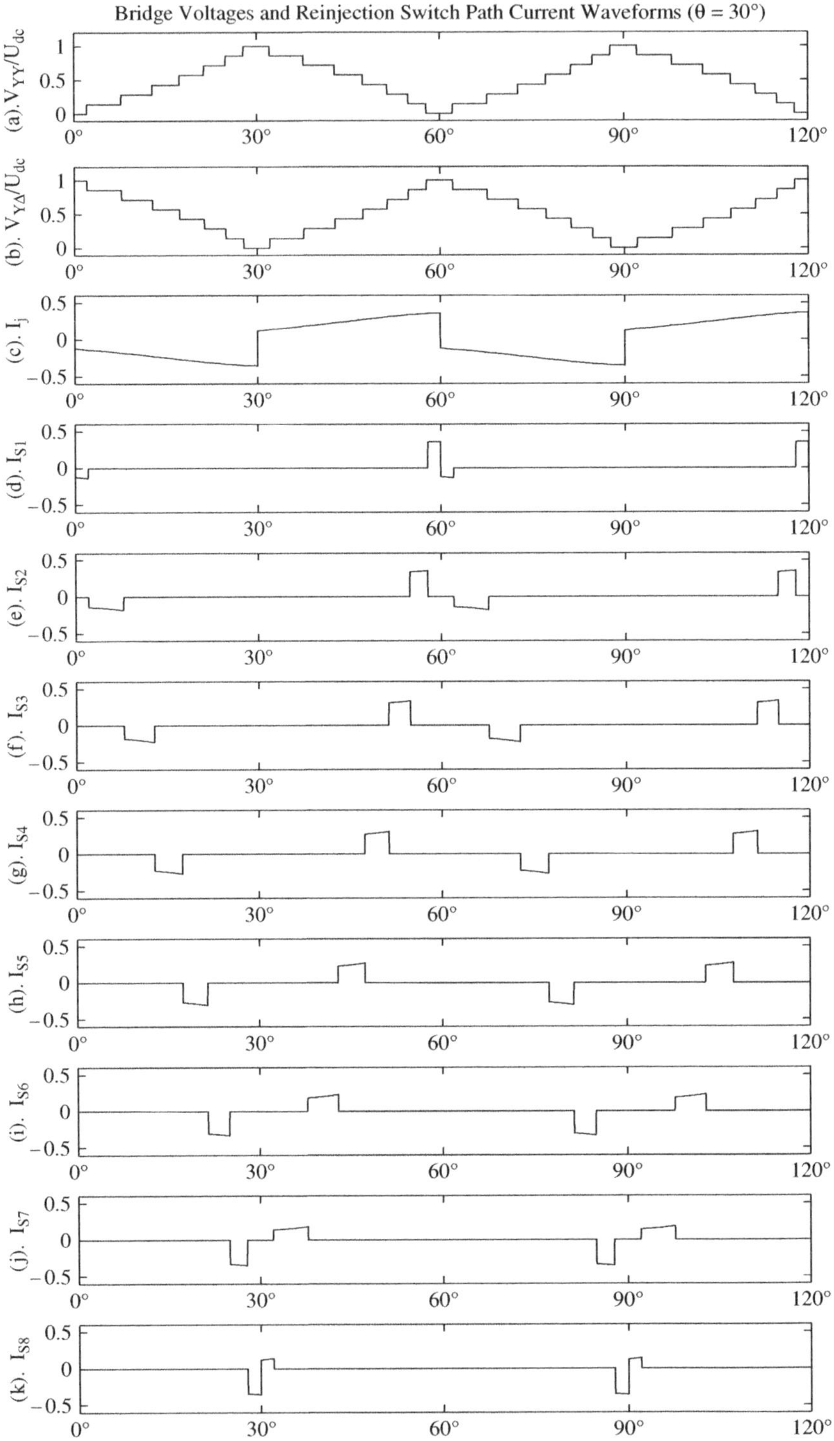

Figure 7.7 Reinjection waveforms under capacitor voltage balance control

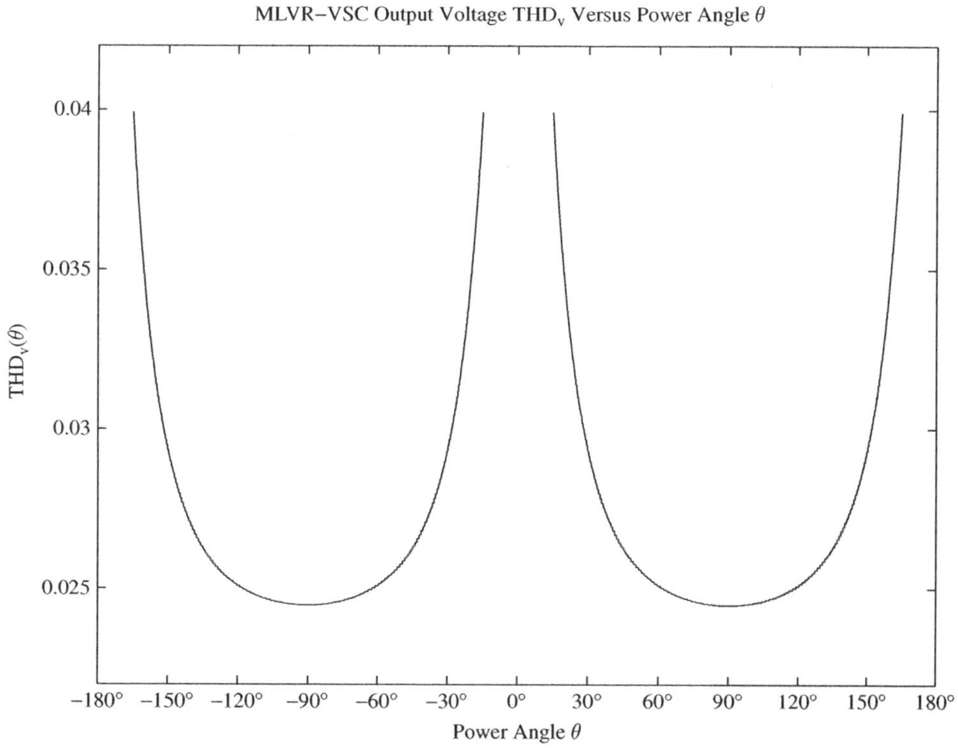

Figure 7.8 Voltage THD variation versus power angle

In the presence of asymmetry and distortion, the reinjection current may not be a periodical function of six times the fundamental frequency and, thus, the DC components have to be calculated on the full fundamental period. It is still possible to maintain capacitor voltage balancing by appropriate on-state adjustments of the reinjection switch paths, not at every one-sixth period, but for the full-cycle period of the fundamental frequency. Therefore, in this case the capacitor voltage will fluctuate more than in the case of perfectly symmetrical operation [4].

7.3.5 Dynamic Performance

The configuration of Figure 7.3, with a large resistor placed across the DC capacitor to take into account the converter losses, is used to test the operation of the MLVR scheme as a reactive power compensator. With reference to the simplified circuit of Figure 7.5, the variables to be controlled are the reactive current component and the DC side voltage (V_{DC}). The latter is controlled by changing δ, which controls the active power transfer required to charge or discharge the DC capacitors. Then the DC capacitor voltage determines the amplitude of the AC output voltage. The following equations apply to the circuit of Figure 7.5:

$$P = \frac{1}{X_S}(3V_S V_A \sin(\delta)) \qquad Q = \frac{1}{X_S}(3V_S V_A \cos(\delta)) - \frac{1}{X_S}(3V_S^2)$$
$$I_{Re} = \frac{1}{X_S}\sin(\delta) \qquad I_{Im} = \frac{1}{X_S}(V_A \cos(\delta)) - \frac{1}{X_S}V_S \tag{7.18}$$
$$I_{DC} = \frac{P - P_L}{V_{DC}} \qquad V_{DC} = \frac{1}{C}\int I_{DC}dt$$

The above equations lead to the block diagram of Figure 7.9(a), used for the control of active and reactive power, and Figure 7.9(b), used for the control of I_{Re} and I_{Im}.

The closed-loop control structure of a reactive power compensator is shown in Figure 7.10. The reactive power order is first converted to an imaginary current order I_{Im}^* while the actual I_{Im} is derived from measurements of the three-phase voltages and currents. The current error is passed to a proportional–integral (PI) controller to generate the required phase angle order for the control of the DC side voltage and output current and thus the reactive power.

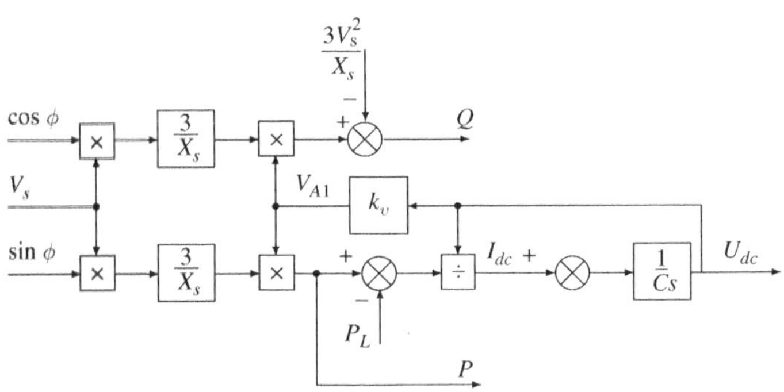

(a)

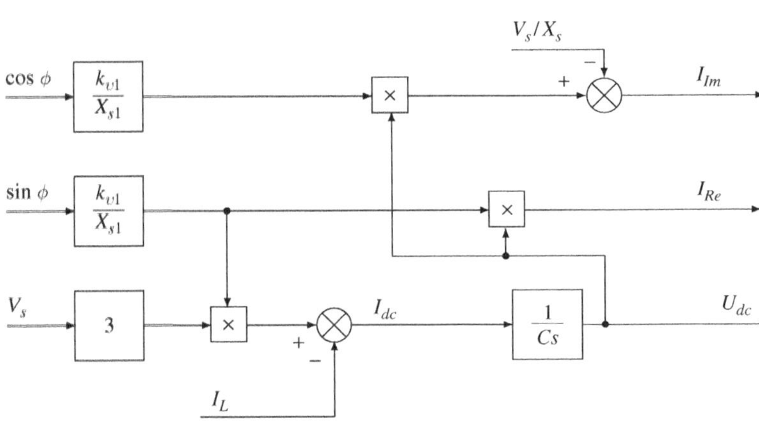

(b)

Figure 7.9 Block diagram of the MLVR-VSC control; (a) control of the real and reactive power; (b) control of the real and imaginary current components

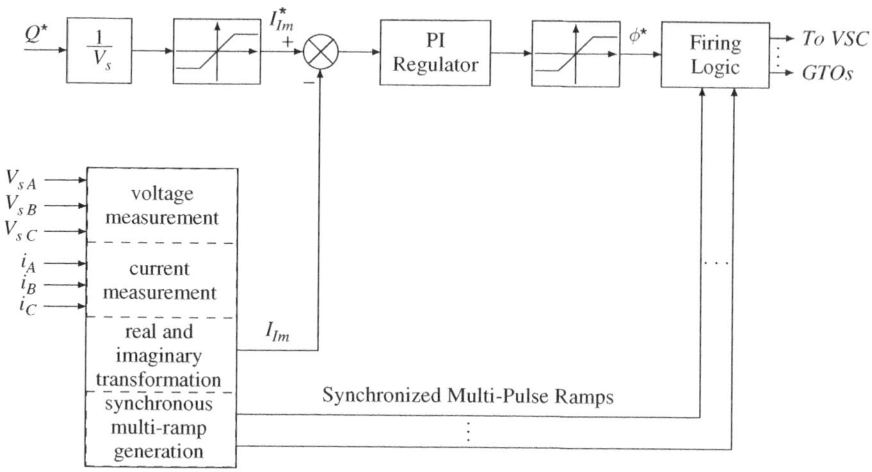

Figure 7.10 Closed-loop control structure of the MLVR-VSC

The following information is used for the test system to verify the steady-state waveforms and simulate the dynamic performance of the converter (the EMTDC/PSCAD program is used for this purpose): reinjection level number equal to 8; rated power 100 MVA; rated voltage 100 kV; AC system reactance 10 % (based on transformer rating); load resistance 10 kΩ. The steady-state waveforms, illustrated in Figure 7.11, are very much as predicted by the theory.

The dynamic response to changes in reactive power order (Q_{ref}) is illustrated in Figure 7.12(a); Q_{ref} changes from 0 to 1 per unit after 0.025 s, from 1 to −1 per unit after 0.21 s and from −1 to 1 after 0.41 s. In each case Q reaches the new state in under 0.05 s. The corresponding variations of capacitors ($V_1 \ldots V_6$) and DC voltage (V_{DC}), shown in Figures 7.12(b) and (d) for two different time scales, clearly indicate that capacitor voltage balance is maintained under steady and dynamic conditions. Finally, the variation of output voltage (V_A) and current (I_A), shown in Figures 7.12(c) and (e) in two different time scales, indicates that the operation takes place smoothly from fully inductive to full capacitive within two cycles and with very little harmonic distortion.

7.4 Transformer-Coupled MLVR [11]

The transformer-coupled MLVR, shown in Figure 7.13, uses two single phase reinjection bridges, each fed from one-half of the DC capacitor. These bridges are connected to an isolating transformer with two series-connected secondaries that produce a five-level reinjection voltage waveform at six times the fundamental frequency. Its main attraction is that it is free from capacitor balancing problems. However, the application of this technique beyond the five-level configuration is not envisaged, because the generation of an m-level reinjection waveform by this method would require $(m-1)/2$ reinjection transformers.

To explain the operating principle, the converter system is assumed to consist of ideal switches and transformers and contain unlimited capacitance on the DC side.

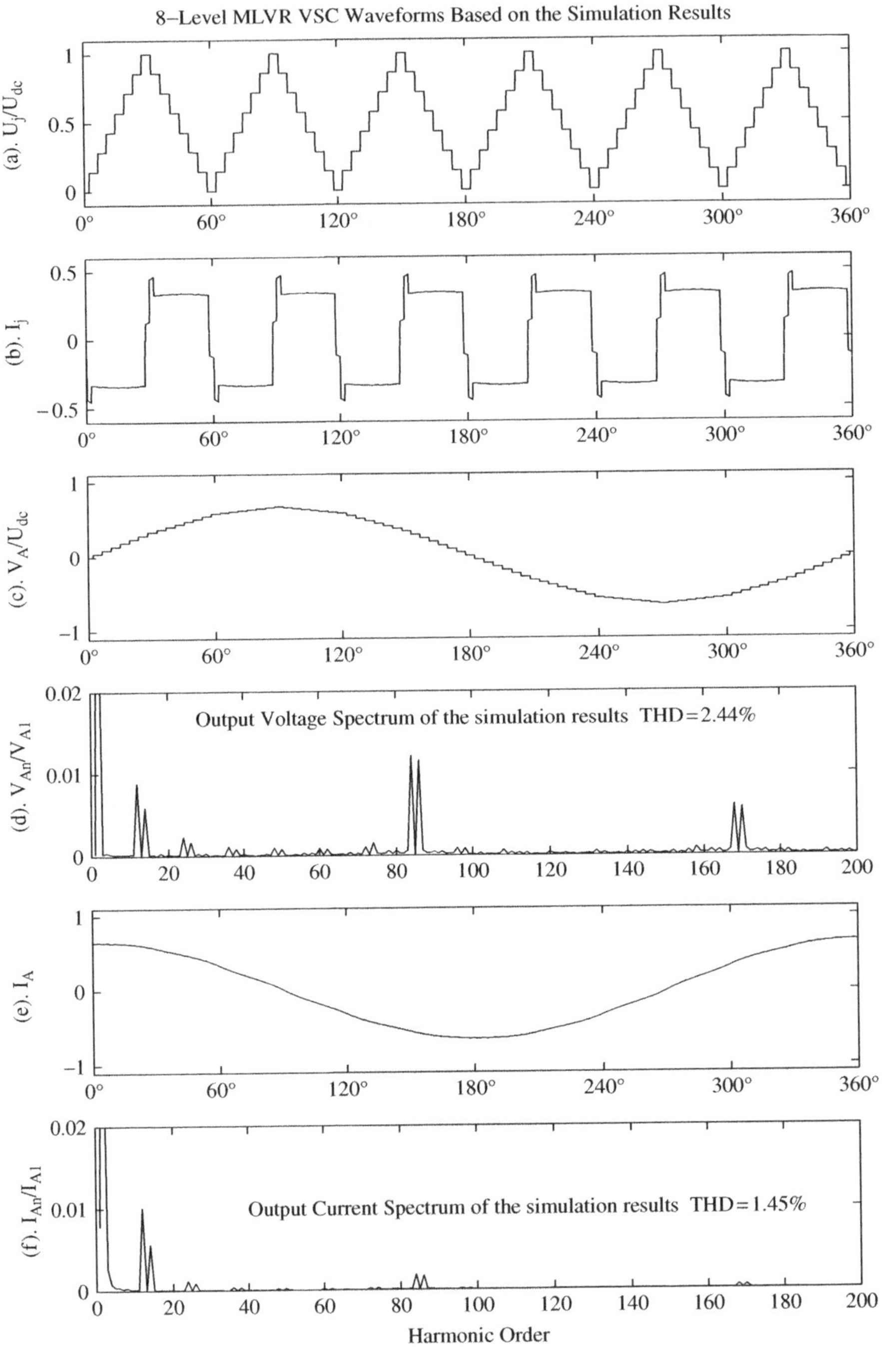

Figure 7.11 MLVR-VSC simulated waveforms for capacitive operation

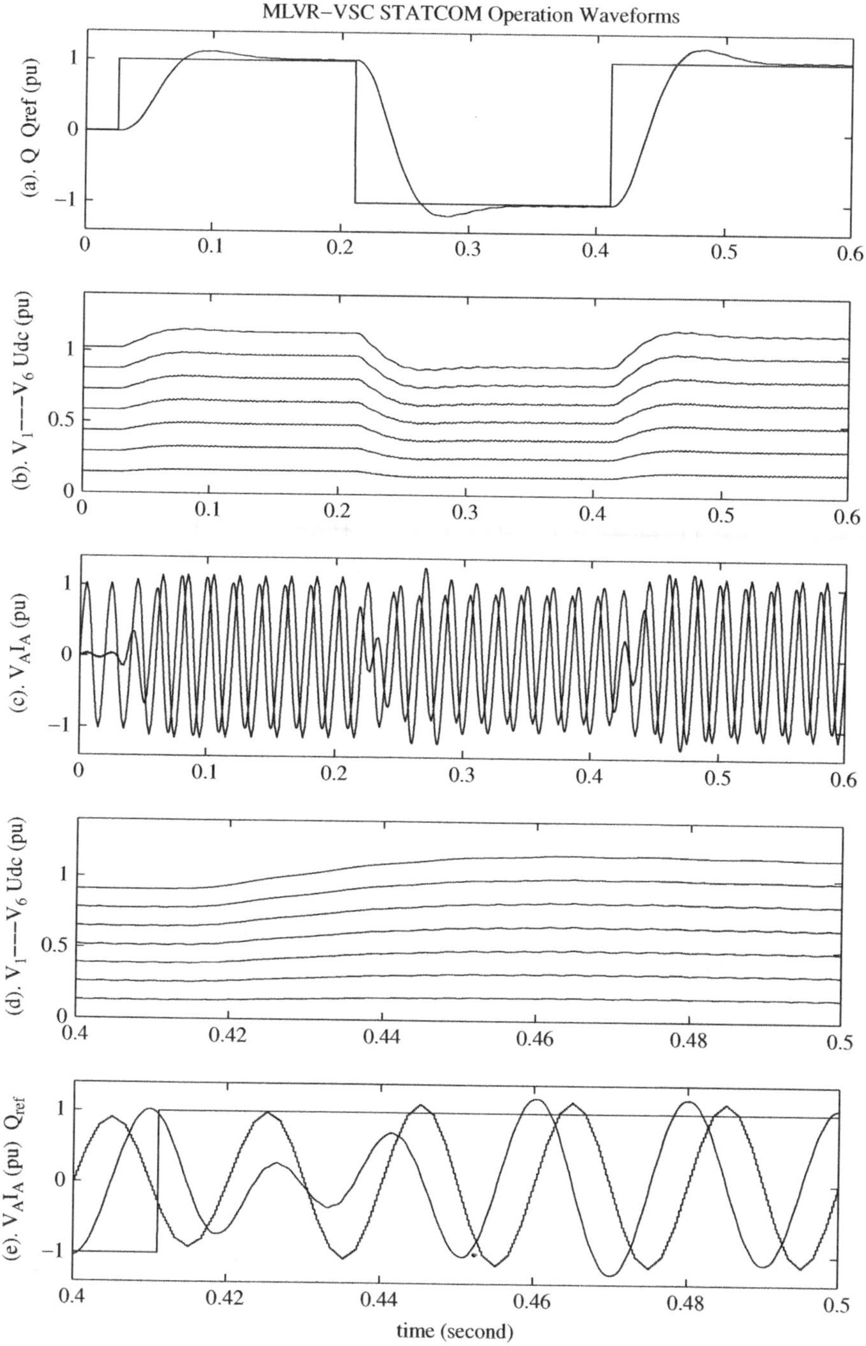

Figure 7.12 MLVR-VSC dynamic response to step reactive power orders

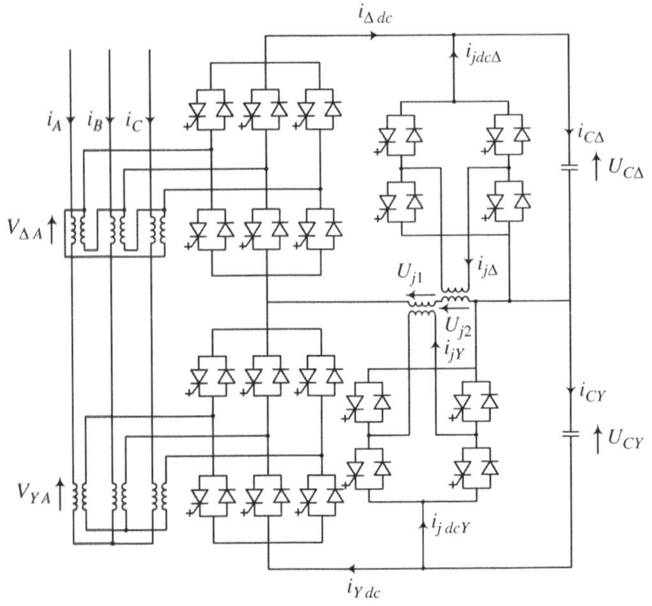

Figure 7.13 Transformer-coupled reinjection VSC

The turns ratios of the interface transformers of the 12-pulse converter ensure that the DC voltages across the upper ($U_{C\Delta}$) and the lower capacitors (U_{CY}) are the same, i.e.

$$U_{C\Delta} = U_{CY} = U_C \tag{7.19}$$

With reference to Figure 7.13, the voltages across the main star-connected and delta-connected bridges are determined by those of the reinjection bridges (U_{j1} and U_{j2}), i.e.

$$U_j = U_{j1} + U_{j2} \tag{7.20}$$

$$V_Y = U_{CY} + U_j = U_C + U_j \tag{7.21}$$

$$V_\Delta = U_{C\Delta} - U_j = U_C - U_j \tag{7.22}$$

Therefore adjustment of U_j can shape the AC output waveforms of both bridges simultaneously. The reinjection voltage U_j is determined by the turns ratios of reinjection transformers and the switching pattern of the reinjection bridges. Three possible levels (shown in Figures 7.14(a) and (b)) are achieved for the U_{j1} and U_{j2} voltages, i.e. zero voltage (short-circuiting the primary winding of its reinjection transformer), k_jU_C and $-k_jU_C$ (forward and reverse connecting the primary winding of the reinjection transformer across the DC capacitor); k_j, the turns ratio, is optimised to permit operation in all four quadrants without capacitor balancing problems. Figure 7.14(c) shows that U_j is a five-level waveform, which rises to its maximum in 30° and falls to the original state in another 30°; thus U_j is a pulse train of six times the fundamental frequency.

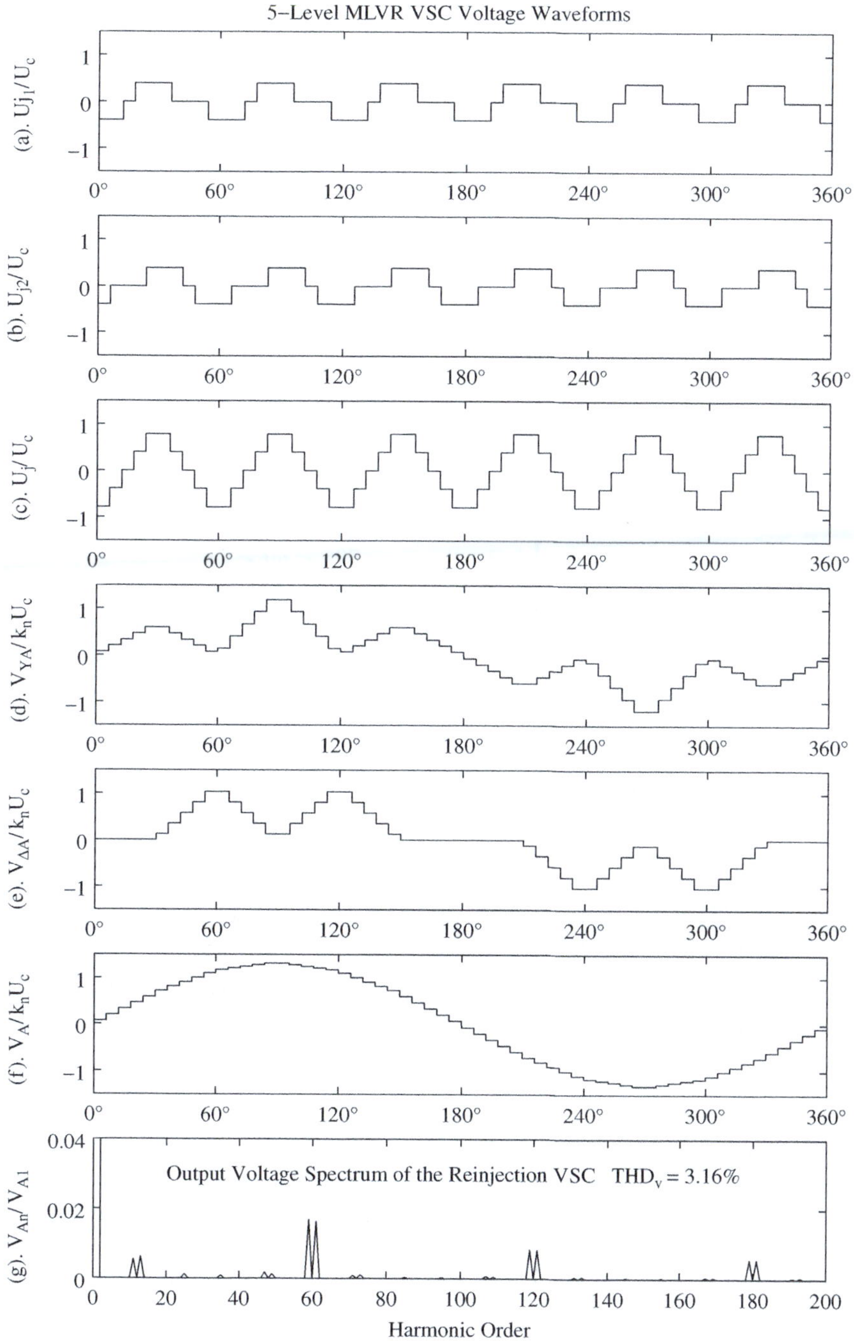

Figure 7.14 Voltage waveforms of the transformer-aided reinjection VSC

V_{YA}, the phase voltage across the primary windings of the star – star-connected interface transformer, shown in Figure 7.14(d), has the following components in its quarter cycle (k_n is the transformer turns ratio):

$$\frac{V_{YA}(\omega t)}{k_n U_C} = \begin{cases} (1-2k_j)/3 & 0^\circ < \omega t < 6^\circ \\ (1-k_j)/3 & 6^\circ < \omega t < 12^\circ \\ 1/3 & 12^\circ < \omega t < 18^\circ \\ (1+k_j)/3 & 18^\circ < \omega t < 24^\circ \\ (1+2k_j)/3 & 24^\circ < \omega t < 36^\circ \\ (1+k_j)/3 & 36^\circ < \omega t < 42^\circ \\ 1/3 & 42^\circ < \omega t < 48^\circ \\ (1-k_j)/3 & 48^\circ < \omega t < 54^\circ \\ (1-2k_j)/3 & 54^\circ < \omega t < 60^\circ \\ 2(1-2k_j)/3 & 60^\circ < \omega t < 66^\circ \\ 2(1-k_j)/3 & 66^\circ < \omega t < 72^\circ \\ 2/3 & 72^\circ < \omega t < 78^\circ \\ 2(1+k_j)/3 & 78^\circ < \omega t < 84^\circ \\ 2(1+2k_j)/3 & 84^\circ < \omega t < 90^\circ \end{cases} \tag{7.23}$$

Its Fourier content is given by

$$V_{YA}(\omega t) = \sum_{n=1}^{\infty} \frac{4}{3} \cos^2(n\pi/6) b_n k_n U_C \sin(n\omega t) \tag{7.24}$$

where

$$b_n = \frac{[1-(-1)^n]}{n\pi} \left\{1 + 2k_j \left[8 \sin\left(\frac{n\pi}{6}\right) \sin\left(\frac{n\pi}{12}\right) \cos\left(\frac{n\pi}{30}\right) \cos\left(\frac{n\pi}{60}\right) - 1\right]\right\} \tag{7.25}$$

Similarly, the following expression provides the Fourier content of $V_{\Delta A}$, the phase voltage across the primary windings of the star–delta-connected interface transformer shown in Figure 7.14(e):

$$V_{\Delta A}(\omega t) = \sum_{n=1}^{\infty} \frac{2}{\sqrt{3}} \cos\left(\frac{n\pi}{6}\right) b_n k_n U_C \sin(n\omega t) \tag{7.26}$$

The primary side output voltage $V_A(\omega t)$, shown in Figure 7.14(f), is derived from the addition of $V_{YA}(\omega t)$ and $V_{\Delta A}(\omega t)$.

The peak value of the nth harmonic and the fundamental components of the output voltage are

$$V_{An} = \frac{2k_n U_C}{\sqrt{3}} \cos\left(\frac{n\pi}{6}\right) \left[1 + \frac{2}{\sqrt{3}} \cos\left(\frac{n\pi}{6}\right)\right] b_n \tag{7.27}$$

$$V_{A1} = \frac{4k_n U_C}{\pi} \left\{1 + 2k_j \left[4 \sin\left(\frac{\pi}{12}\right) \cos\left(\frac{\pi}{30}\right) \cos\left(\frac{\pi}{60}\right) - 1\right]\right\} \tag{7.28}$$

The total harmonic distortion of the phase output voltage is

$$\mathrm{THD}_V = \sqrt{\frac{2V_{ARMS}^2}{V_{A1}^2} - 1} \tag{7.29}$$

and the reinjection transformer turns ratio required to minimise THD_V is

$$k_j = (7+4\sqrt{3})\left[4\sin\left(\frac{\pi}{12}\right)\cos\left(\frac{\pi}{30}\right)\cos\left(\frac{\pi}{60}\right)-1\right] = 0.3927$$

which results in a minimum THD_V of 3.16 %.

Finally, Figure 7.14(g) shows the spectrum of the output voltage as a percentage of the fundamental component.

7.5 Cascaded H-Bridge MLVR

The neutral point clamping concept [12] has gained general acceptance for its inherent capacitor balancing capability. However, this property is only available to the three-level configuration, which, therefore, needs to be complemented by PWM.

This section describes a cascaded H-bridge reinjection configuration that combines the benefits of three-level neutral clamping (in so far as the DC capacitor balancing is concerned) and the reduced harmonic content and lower dv/dt of the multi-level schemes. Unlike the presently used H-bridge multi-level conversion concept, where the cascaded H-bridge structure constitutes the three-phase power converter, the proposed scheme uses the H-bridges in a separate lower rated reinjection circuit.

In the diode-clamped multi-level scheme, the number of diodes increases almost proportionally to the level number squared, whereas the number of switches in the H-bridge reinjection circuit increases only in direct proportion to the magnitude of the DC voltage (i.e. does not depend on the level number).

The cascaded H-bridge MLVR achieves the ZVS condition without carrying DC and, therefore, the power rating of this circuit is a fraction of the main converter rating. Finally the facility with which this configuration can be designed for high voltages makes it perfectly suited for use in HVDC transmission.

7.5.1 Basic Structure and Waveforms

Figure 7.15 shows the proposed configuration. Except for the added H-bridge chain between the middle node of the DC capacitors and the neutral point of the two main bridges, it is the standard configuration of a 12-pulse VSC.

To produce an m-level voltage waveform across the main bridge DC terminals the reinjection circuit requires $(m-1)/2$ identical single phase H-bridges, each having a capacitor pre-charged to a voltage level equal to $V_{DC}/(m-1)$.

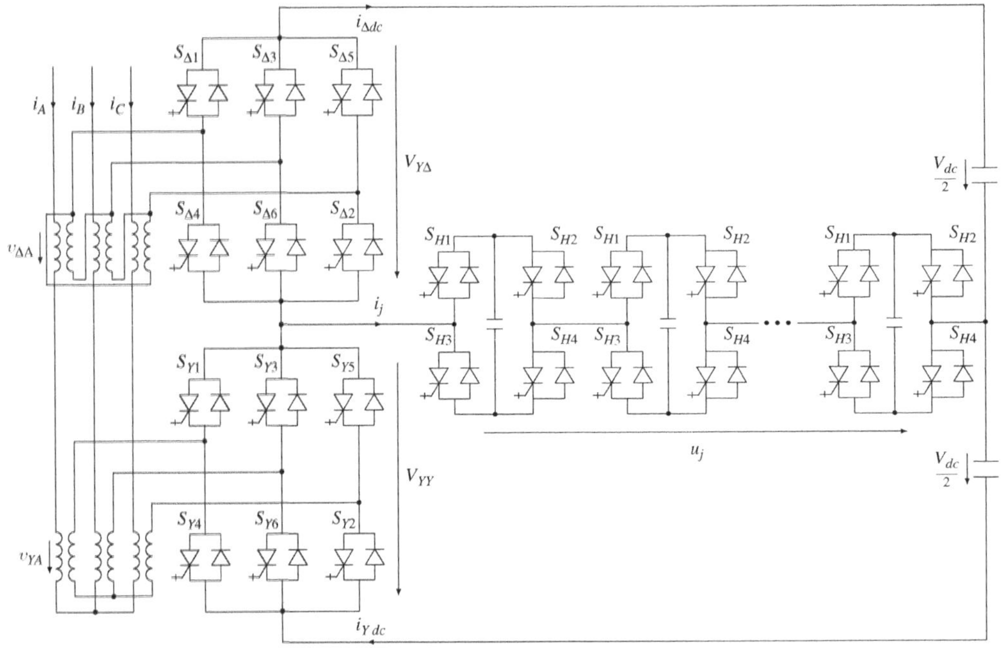

Figure 7.15 MLVR H-bridge VSC configuration

Since each H-bridge generates three different output states, i.e. $-V_{DC}/(m-1)$ (state S_{-1}), 0 (state S_0) and $+V_{DC}/(m-1)$ (state S_{+1}), an appropriate combination of the $(m-1)/2$ bridge outputs can generate any one of the possible m voltage levels, i.e.

$$-\frac{V_{DC}}{2}, -\frac{(m-3)V_{DC}}{2(m-1)}, \ldots, -\frac{V_{DC}}{m-1}, \frac{2V_{DC}}{m-1}, \ldots, \frac{V_{DC}}{2}$$

To achieve the linear reinjection waveform proposed in Section 4.5.1, the H-bridges are fired periodically at six times the fundamental frequency and the firings are synchronised with those of the two main bridges. The multi-level reinjection voltage is added to the two DC capacitor voltages to shape the voltages across the two main bridges into equal linear staircase waveforms, but with a 30° phase difference. Finally, the series combination of the two main bridge voltages produces a high-quality output voltage waveform at the converter terminals.

The theoretical reinjection waveforms produced by the cascaded H-bridge configuration are not different from those of the previous MLVR configurations, as Figure 7.16 shows for a seven-level case. Graph (a) shows that the reinjection voltage varies periodically at six times the fundamental frequency in synchronism with the main bridge valves' commutations. Graphs (d) and (c) illustrate the voltage waveforms across the interface transformer phase windings for the delta and star connections respectively. Graph (c) contains the voltage across the star-connected bridge, which is zero in the regions when the reinjection voltages equals $-V_{DC}/2$, thus providing a soft-switching condition for the commutations in this bridge. Similarly, graph (d) shows that when the reinjection voltage equals $V_{DC}/2$, the voltage across the delta-connected bridge is zero, again providing a soft-switching condition for the commutations in that bridge. Graph (e) shows the output voltage waveform and (f) its frequency spectrum.

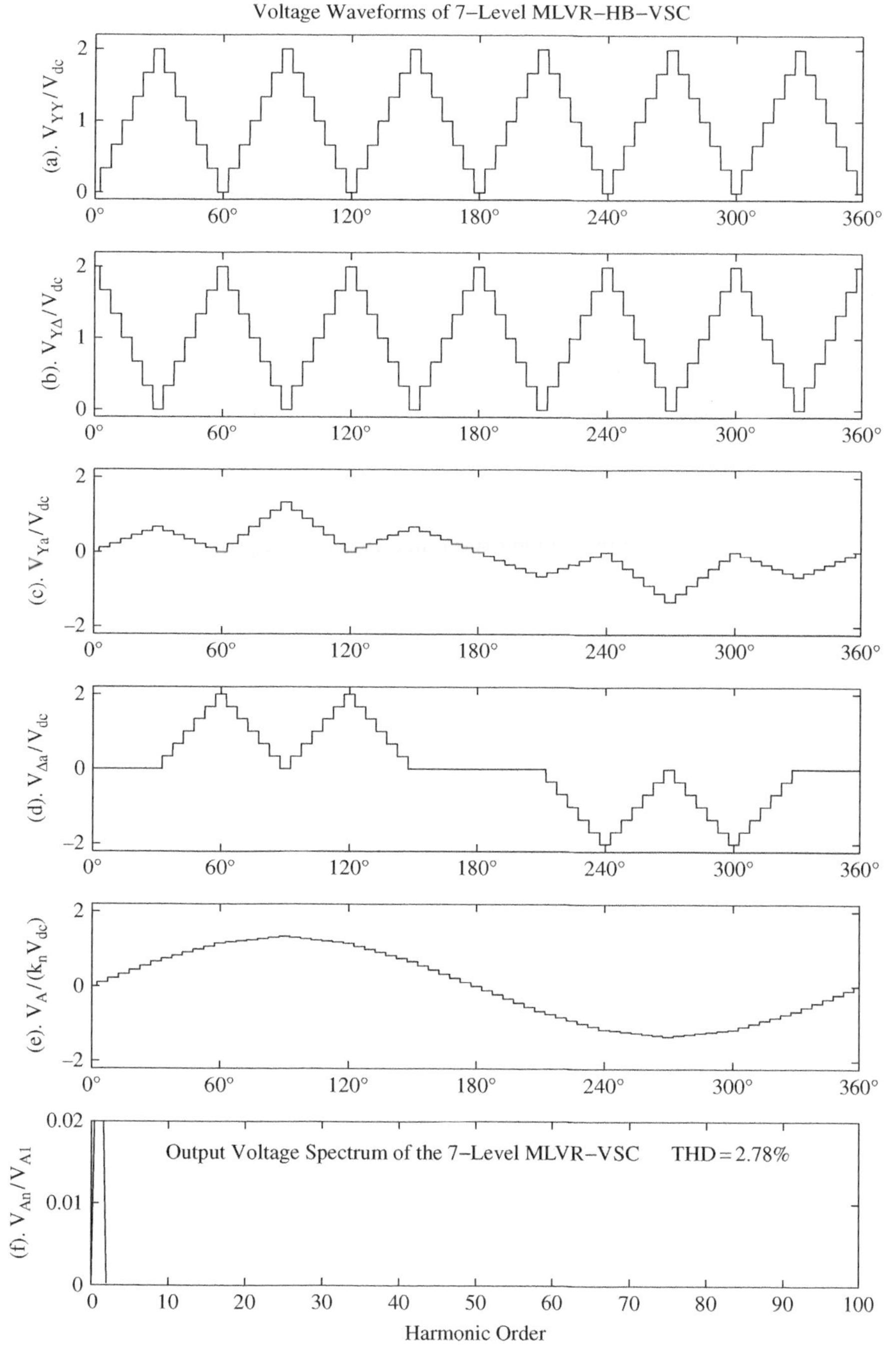

Figure 7.16 Voltage waveforms for the seven-level H-bridge MLVR configuration

It must be emphasised that the current rating of the extra H-bridge chain of the reinjection circuit is only a fraction of the main converter phase current, as the DC components of the balanced two main bridges cancel each other. The H-chain functions like a harmonic filter placed on the DC side (thus eliminating the risk of harmonic resonances); it also provides a ZVS condition, dynamic voltage balance for the main bridge series-connected switches and reduces the dv/dt stress of the power switches and components. Moreover, the scheme should permit the use of hybrid main valves consisting mostly of thyristors and a few self-commutating switches.

7.5.2 Switching Pattern of the Reinjection Bridges

As explained earlier, following a change of switching state in either of the two main bridges, the reinjection waveform is increased or decreased in m step levels of equal width with a period of $\pi/6$. In the configuration of Figure 7.15, following a switching state change in the star–star bridge, the reinjection voltage is increased step by step. Similarly, after a change of switching state in the star–delta bridge, the reinjection waveform is decreased step by step. In each case the reinjection voltage reference is the voltage from the neutral point of the two main bridges to the common node of the two DC capacitors.

To generate the required multi-level waveform, while achieving the lowest switching frequency (i.e. the reinjection voltage frequency), the $(m-1)/2$ H-bridges can be triggered in a variety of patterns. The preferred pattern for high-voltage application should use the lowest switching frequency (i.e. six times the fundamental) as well as provide self-balancing capacitor voltage capability.

Figure 7.17 illustrates a firing pattern where the output states of the H-bridges are all symmetrical about the vertical line at $\pi/6$, while the widths of the states (S_{+1}, S_0 and

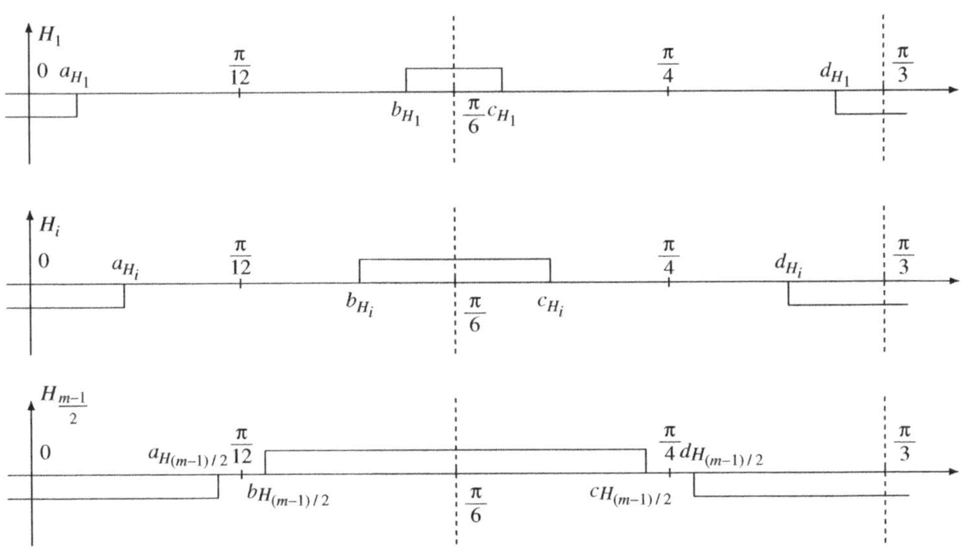

Figure 7.17 Switching pattern of the reinjection bridges

S_{-1}) of the individual H-bridges are not equal. If the $(m-1)/2$ H-bridges are numbered $H_1, H_2, \ldots, H(m-1)/2$, the four state changing points in the $\pi/3$ period of bridge H_k are

$$
\begin{aligned}
a_{H_k} &= \frac{k\pi}{6m} \\
b_{H_k} &= \frac{\pi}{6} - \frac{k\pi}{6m} \\
c_{H_k} &= \frac{\pi}{6} + \frac{k\pi}{6m} \\
d_{H_k} &= \frac{\pi}{3} - \frac{k\pi}{6m}
\end{aligned}
\tag{7.30}
$$

The firing sequence of the four switches in the kth bridge can be arranged as shown in Figure 7.18. $S_{H_{k1}}$ is changed from off-state to on-state at point b_{H_k}, and from on- to off-state at d_{H_k}; $S_{H_{k2}}$ is changed from on-state to off-state at point a_{H_k}, and from off to on-state at c_{H_k}; $S_{H_{k3}}$ is changed from on-state to off-state at point b_{H_k}, and from off to on-state at d_{Hi}; $S_{H_{k4}}$ is changed from off-state to on-state at point a_{H_k}, and from on to off-state at c_{H_k}. With this arrangement there are always two of the four switches in on-state and the kth H-bridge output states are determined by two switches in on-state, i.e.

$H_k = S_{-1}$ is determined by $S_{H_{k2}}$ and $S_{H_{k3}}$ in on-state

$H_k = S_0$ is determined by $S_{H_{k3}}$ and $S_{H_{k4}}$ in on-state

$H_k = S_{+1}$ is determined by $S_{H_{k1}}$ and $S_{H_{k4}}$ in on-state

$H_k = S_0$ is determined by $S_{H_{k1}}$ and $S_{H_{k2}}$ in on-state

$H_k = S_{-1}$ is determined by $S_{H_{k2}}$ and $S_{H_{k3}}$ in on-state

It is clear that the state $H_k = S_0$ can be generated either by $S_{H_{k1}}$ and $S_{H_{k2}}$ or $S_{H_{k3}}$ and $S_{H_{k4}}$ in on-state. With the firing sequences of Figure 7.18 for the four switches of the H-bridge, the firing frequency is six times the fundamental and the two zero-output stages are used equally. These two properties, low-frequency switching and equal rating of the switches, make this arrangement attractive for HVDC application.

7.5.3 Design of the Cascaded H-Bridge Chain

In the cascaded H-bridge configuration, the use of sufficiently large-level numbers to provide the waveform quality required for high-power and high-voltage applications is not a problem. The main considerations in the design of the reinjection chain are:

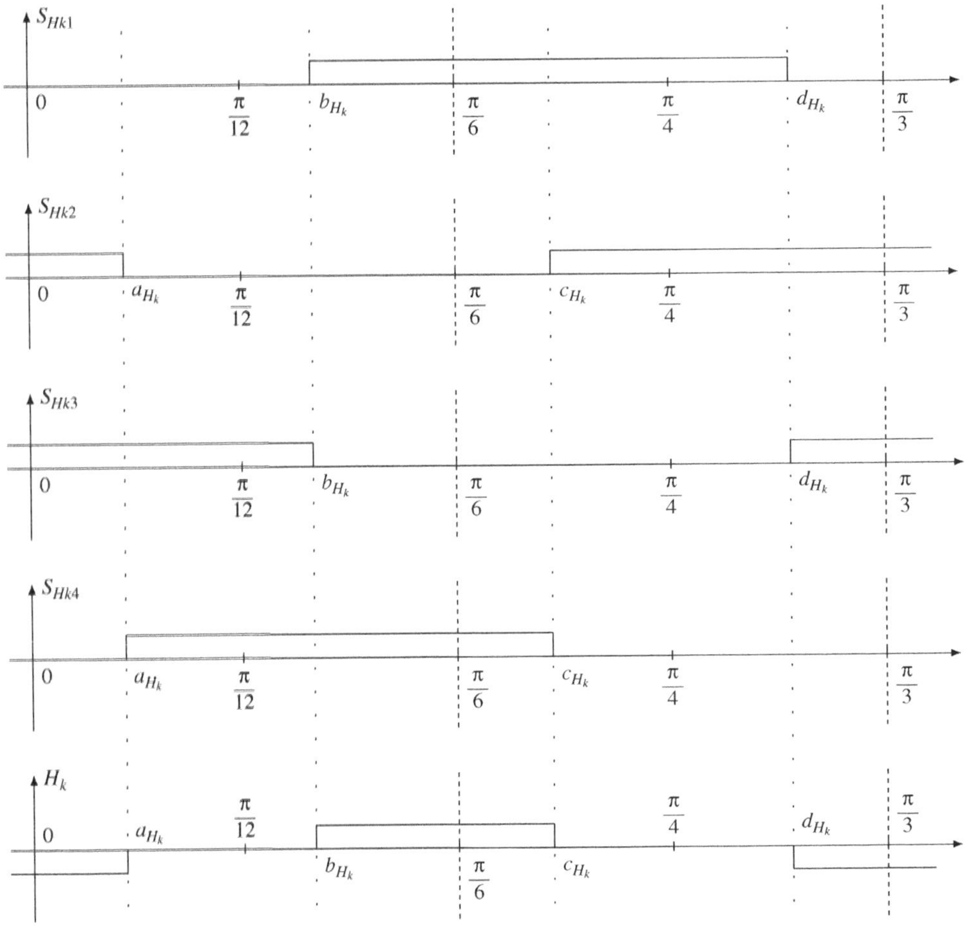

Figure 7.18 Switching sequence for the kth reinjection bridge

- achieve high voltage ratings;
- reduce dv/dt to an acceptable value;
- improve operational reliability;
- simplify the design of the main bridge and other power components.

The maximum voltage rating of an H-bridge with currently available IGCT of IGBT switching devices, without the use of series-connected chains, is under 10 kV. The ratio of DC voltage to reinjection voltage of this configuration is two. Therefore, adopting a conservative figure of 5 kV per bridge, a 50 H-bridge chain will provide 250 kV from a DC voltage of 500 kV.

The level-by-level increase or decrease of the voltage across the main converter valves with a period of repetition equal to one-twelfth of the fundamental frequency, plus the presence of zero-voltage regions during the commutations, should make it possible to construct the bridge arm using only a few forced commutated switches, the rest being thyristors. The forced commutated switches would be used to switch off the arm branch current and sustain the first level of voltage, while the by then recovered thyristors would sustain the subsequent level-by-level voltage increases. So the higher the level number (and therefore the lower the voltage increments), the fewer the number of forced commutated switches required.

Instead of single device H-bridges, the reinjection chain could be constructed using a few H-bridges of higher voltage rating valves consisting of series-connected switches and capacitors. This solution, however, would impose high dv/dt stresses across the main bridge switches and other converter components (such as cable and transformers) and would create voltage balancing problems for the series-connected switches of the high-voltage-rated H-bridges, because each H-bridge is a two-level VSC. The total number of switches and capacitors required for chains consisting of high- or low-voltage H-bridges are the same, but the capacitor size of each low-rated H-bridge is smaller than the size of each capacitor in the high-voltage H-bridge alternative; the reinjection current is distributed either to a large number of individual capacitors or to a number of capacitors grouped in series, and the former requires a lower rms current rating than the latter.

Also, the addition of redundant switches, as required for operational reliability, would result in higher costs if fewer higher voltage H-bridges were used.

In the multi-level clamped configurations discussed in Section 7.3, when the converter voltage rating is much higher than that of the individual switches, a high-level number would be needed to reduce the voltage balancing problems in the series-connected switching device. However, the number of auxiliary clamping switches increases quickly with the level number and such a solution is less cost effective. Moreover, the benefit of a high-level number is not significant for very high-level numbers. For instance, between 11 and 101 levels of the reinjection configurations, THD only reduces from 2 % to 1.1 %.

As indicated earlier in the section, the number of switches needed for the H-bridge converter configuration is independent of the level number. For instance, for the 500 kV DC voltage conversion under discussion, the number of main bridge valves has the same structure (each formed by 100 switches in series) whether the selected level number is 11 or 101. The reinjection branch of the 11-level configuration is formed by five high-voltage-rated H-bridges in series, each bridge arm consisting of 10 switches in series and 10 capacitors in series on its DC side; the reinjection branch of the 101-level configuration is formed by 50 H-bridges in series, each connected to a single capacitor. Therefore, under suitable control conditions, high-level numbers and waveform quality are not a problem in the H-bridge configuration.

7.5.4 Capacitors' Balancing

As shown in Figure 7.15, the H-bridge chain is connected between the middle node of the DC capacitors and the neutral point of the two main bridges. No independent sources are used to charge and discharge the H-bridge capacitors, although they can be pre-charged to an appropriate level before the operation starts. Therefore, special measures are needed to ensure that the H-bridge capacitors follow the voltages across the main converters as well as keep them balanced.

Equalisation of the H-bridge chain voltage and the DC capacitor voltage

To simplify the description let us assume that the level number is infinite; then the reinjection voltage becomes a triangular waveform expressed by

$$u_j(\omega t) = \begin{cases} \dfrac{U_{DC}/2}{\pi/12}(\omega t - \pi/12) & 0 < \omega t < \pi/6 \\ -\dfrac{U_{DC}/2}{\pi/12}(\omega t - \pi/4) & \pi/6 < \omega t < \pi/3 \end{cases} \tag{7.31}$$

For a high-level number the AC output voltage of the reinjection converter system is very close to a sinusoidal wave and with sufficient inductance in series with the ideal AC source the converter AC output current will be practically a sine wave.

Under balanced conditions the converter output fundamental voltages are

$$\begin{bmatrix} v_A(\omega t) \\ v_B(\omega t) \\ v_C(\omega t) \end{bmatrix} = \begin{bmatrix} V_{ph}\sin(\omega t) \\ V_{ph}\sin(\omega t - 2\pi/3) \\ V_{ph}\sin(\omega t + 2\pi/3) \end{bmatrix} \tag{7.32}$$

where V_{ph} is the peak magnitude of the phase voltage.

The AC output phase currents of the converter are

$$\begin{bmatrix} i_A(\omega t) \\ i_B(\omega t) \\ i_C(\omega t) \end{bmatrix} = \begin{bmatrix} \sqrt{2}I_{SR}\sin(\omega t + \theta) \\ \sqrt{2}I_{SR}\sin(\omega t - 2\pi/3 + \theta) \\ \sqrt{2}I_{SR}\sin(\omega t + 2\pi/3 + \theta) \end{bmatrix} \tag{7.33}$$

where θ is the power factor angle (the phase angle of the AC with respect to the converter output voltage), which is considered positive when the current leads the voltage, and I_{SR} is the fundamental (rms) component of the converter phase current output.

These currents are coupled to the two main converters by the interface transformers and then converted to the main bridge DC sides, i_{Ydc} and $i_{\Delta DC}$. The latter vary at six times the fundamental frequency and are expressed as

$$i_{Ydc}(\omega t) = \sqrt{2}k_n I_{SR} \begin{cases} -\sin(\omega t + \theta - 2\pi/3) & 0 < \omega t < \pi/6 \\ -\sin(\omega t + \theta + 2\pi/3 & \pi/6 < \omega t < \pi/3 \end{cases} \tag{7.34}$$

$$i_{\Delta DC}(\omega t) = \sqrt{2}k_n I_{SR} \begin{cases} -\sin(\omega t + \theta - \pi/2) & 0 < \omega t < \pi/6 \\ \sin(\omega t + \theta + \pi/6) & \pi/6 < \omega t < \pi/3 \end{cases} \tag{7.35}$$

where k_n is the star–star interface transformer turns ratio.

The reinjection H-bridge current is

$$\begin{aligned} i_j(\omega t) &= i_{YDC}(\omega t) - i_{\Delta DC}(\omega t) \\ &= 2\sqrt{2}k_n I_{SR} \begin{cases} \sin(\pi/12)\cos(\omega t + \theta + 5\pi/12) & 0 < \omega t < \pi/6 \\ -\sin(\pi/12)\cos(\omega t + \theta + \pi/4) & \pi/6 < \omega t < \pi/3 \end{cases} \end{aligned} \tag{7.36}$$

Using the expressions for the reinjection voltage and current, the power transferred to the reinjection branch is

$$P_{jHBs} = \frac{3}{\pi}\int_0^{\pi/3} i_j(\omega t)u_j(\omega t)d(\omega t) = \frac{36\sqrt{2}}{\pi^2}k_n I_{SR} V_{DC}\left[\tan\frac{\pi}{12} - \frac{\pi}{12}\right]\cos(\theta) \tag{7.37}$$

Equation (7.37) indicates that the reinjection branch absorbs energy during the rectification and delivers energy during the inversion processes respectively. Thus the H-bridge capacitors would be overcharged during rectifier operation and discharged during inverter operation, unless an appropriate charge is transferred from or sent to the main converter capacitors. For this configuration to operate properly, the sum of all the DC capacitor voltages across the H-bridge chain must equal the top and bottom DC voltages across the DC capacitors of the two main converters.

Equation (7.37) describes the reinjection average power in terms of the AC and DC voltage, but it is more convenient to express the power in terms of either DC voltage and current or AC voltage and current.

If the level number is infinite and the DC voltage V_{DC}, the fundamental rms output voltage of the converter phases is

$$V_{Arms1} = \frac{12\sqrt{2}}{\pi^2}\tan(\pi/12)k_n V_{DC} \tag{7.38}$$

Thus the DC current of the two main bridges is

$$I_{DC} = \frac{3V_{Arms1}I_{SR}\cos(\theta) - P_{jHBs}}{V_{DC}} = \frac{3\sqrt{2}}{\pi}k_n I_{SR}\cos(\theta) \tag{7.39}$$

Similarly, if the added voltage of all the capacitors across the H-bridge chain is $V_{DC}/2$, the equivalent DC in the H-bridge chain is

$$I_{jdc} = \frac{P_{jHBs}}{V_{DC}/2} = \frac{24}{\pi}\left[\tan\frac{\pi}{12} - \frac{\pi}{12}\right]I_{DC} = 2k_{jp}I_{DC} \tag{7.40}$$

In the absence of DC side load current, the DC Current (I_{DC}) will only be used to deliver or take charges from the two main capacitors; similarly, the equivalent DC current I_{jdc} will deliver to or take charges from all the capacitors in the H-bridge chain. To ensure equality between the voltages across the two main capacitors and all the series capacitors of the H-bridge chain, the following condition must be satisfied:

$$\frac{1}{C_m}\int_{-\infty}^{t} I_{DC}dt = \frac{1}{C_{jBequ}}\int_{-\infty}^{t} I_{jdc}dt = \frac{2k_{jp}}{C_{jBequ}}\int_{-\infty}^{t} I_{DC}dt \tag{7.41}$$

or

$$C_{jBequ} = 2k_{jp}C_m$$

where

C_m represents the capacitance of each of the two main capacitors

C_{jBequ} is the total equivalent capacitance of the H-bridge chain
$k_{jp} = (24/\pi)(\tan \pi/12 - \pi/12)$ is the ratio of the powers transferred by the main converter and the reinjection circuit.

Equation (7.41) indicates that if C_{jBequ} is equal to $2k_{jp}C_m$ and the converter DC current only flows through the two main capacitors, the voltage equality is assured; this condition is met by the STATCOM application.

However, for other applications the DC output current of the main converter bridges may not exclusively go into the two main capacitors. The part of the current going into other parts of the circuit depends on the converter operation, and particularly the load demand. As the latter is unpredictable, voltage equality is not guaranteed purely by an appropriate design of the capacitances of the two groups of capacitors. Instead, to achieve equality between the added H-bridge DC capacitor voltage and the DC voltage across the two main capacitors, two charge-passing paths have to be opened and closed periodically and they must be synchronised with the H-bridge firing sequence.

When the H-bridge chain outputs the positive maximum voltage level ($V_{DC}/2$) the bypassing state of the star–delta bridge (i.e. the two arms of a pole phase being in the on-state) is used as the charge-passing path. This forces the summed voltages of the H-bridge capacitors to be equal to the voltage across the top main capacitor connected to the star–delta bridge. Similarly, when the H-bridge chain outputs the negative maximum voltage level ($-V_{DC}/2$) the bypassing state of the star–star bridge is used as the charge-passing path. This forces the summed voltages of the H-bridge capacitors to be equal to the voltage across the bottom main capacitor connected to the star–star bridge

However, the bypassing states can only be used for this purpose when the H-bridge outputs are at their maximum positive and negative levels, otherwise this action would create a short-circuit condition in the converter. Moreover, the periodical use of the bypassing states also achieves voltage balancing of the two main capacitors, a necessary condition for the reinjection configuration to operate properly.

Balancing of the H-bridge capacitors

To generate the required reinjection voltage waveform, the differences between the capacitor voltages of the individual H-bridges must be kept within a specified narrow margin. The voltage across an H-bridge capacitor is determined by the current passing through it and the switching pattern selected for the bridge.

The fundamental voltage component generated by the converter (i.e. the added emf*s* generated by the fluxes of the two interface transformers of the double bridge configuration) is given by Equation (7.32) and the fundamental phase currents by Equation (7.33). These currents are transferred by the interface transformers to the converter sides and then, through the converter switches, to the DC side. If the currents are perfectly balanced and their harmonic content negligible, a full cycle of the reinjection current is represented by Equation (7.36).

Once the triggered pattern has been chosen, the switches in the H-bridge chain turn on and off to generate the required reinjection voltage waveform. The reinjection current, also controlled by the switching action, charges or discharges the individual H-bridge capacitor voltages depending on whether the H-bridge output is positive or negative; when the H-bridge is bypassed the current through the capacitor is zero.

The specified firing pattern can cause the capacitor voltages to drift. This condition is next analysed for the firing pattern shown in Figure 7.17.

The ith H-bridge is switched to state S_{-1} before the origin, to S_0 at a_{Hi}, to S_{+1} at b_{Hi}, to S_0 at c_{Hi} and to S_{-1} again at d_{Hi}. The switching instants occur at

$$\begin{aligned} a_{Hi} &= i\pi/6m \\ b_{Hi} &= [(m-i)\pi]/6m \\ c_{Hi} &= [(m+i)\pi]/6m \\ d_{Hi} &= [(2m-i)\pi]/6m \end{aligned} \tag{7.42}$$

The voltage increment of the capacitor in the ith H-bridge is calculated as follows:

$$\begin{aligned} \Delta U_{Ci} &= \frac{1}{\omega C_i}\left[-\int_0^{i\pi/6m} i_j(\omega t)d(\omega t) + \int_{(m-i)\pi/6m}^{(m+i)\pi/6m} i_j(\omega t)d(\omega t) - \int_{(2m-i)\pi/6m}^{\pi/3} i_j(\omega t)d(\omega t)\right] \\ &= \frac{-8\sqrt{2}k_n I_{SR}\sin(\pi/12)\cos(\theta)}{\omega C_i}\sin\left(\frac{\pi}{12}-\frac{i\pi}{12m}\right)\sin\left(\frac{i\pi}{12m}\right) \end{aligned} \tag{7.43}$$

For the selected firing pattern, Equation (7.43) indicates that the $(m-1)/2$ H-bridge capacitor voltages are uniformly increased or decreased by an amount determined by the power factor angle θ and that the voltage increments of the first and last H-bridge capacitors are equal.

If the capacitances of all the H-bridges are equal, the differences between the voltage increments of the H-bridge capacitors are given by the expression

$$\Delta U_{CpA} = \Delta U_{Ck} - \Delta U_{Cl} = \frac{-8\sqrt{2}k_n I_{SR}\sin(\pi/12)\cos(\theta)}{\omega C_i}\sin\left[\frac{\pi}{12}-\frac{(k+l)\pi}{12m}\right]\sin\left[\frac{(k-l)\pi}{12m}\right] \tag{7.44}$$

This expression shows that the voltage increment differences between any two capacitors are only equal when $\cos(\theta) = 0$, a condition that is, again, met by the ideal STATCOM operation (i.e. ignoring internal converter losses).

For general four-quadrant operation, however, the individual capacitor voltages would not be balanced. Thus a self-balancing mechanism needs to be developed for use in HVDC transmission.

The identical H-bridges and their series connection provide the necessary conditions to achieve voltage balancing. Once the firing pattern has been chosen, such as that of Figure 7.17, at every $\pi/3$ period, all the H-bridges in the chain must generate specific output voltages to produce the required reinjection voltage waveform. However, there is no restriction on the selection of the particular units used to generate the required states. This flexibility provides the opportunity to rearrange the $(m-1)/2$ H-bridges for the next coming periods. That is, an H-bridge that was undercharged in the previous period can be forced to gain more charge in the following one and a bridge that was overcharged to receive less charge in the next period. Thus an appropriate switching arrangement must be made in the sequence to ensure that all H-bridges are equally used.

7.5.5 STATCOM Application

Compared with the static VAR compensator (SVC) the VSC-STATCOM offers superior voltage control due to its capability to act as a reactive power generator. However, the present STATCOM, based on PWM technology, has not yet displaced the conventional SVC due to higher cost and there is, therefore, room for new concepts to try and improve the STATCOM competitiveness. As explained in previous sections, the cascaded H-bridge VSC multi-level configuration is well suited for STATCOM operation, due to the lack of capacitor balancing problems, as well as providing a practically undistorted waveform, low power losses and high operating reliability.

The simple test model of Figure 7.19 is used to verify the characteristics of the H-bridge STATCOM by means of EMTDC simulation. In the test system V_{dc} (the DC side capacitor bank voltage) and δ (the phase angle difference between the converter output and the source voltages) are the control parameters. For STATCOM application, as well as the DC side voltage, either the reactive power or the reactive current can be used as the variables to be controlled; the latter, however, is the more relevant selection.

The DC side voltage is changed by δ, which controls the small active power required to charge or discharge the DC capacitors. The DC capacitor voltage V_{dc} then determines the amplitude of the MLVR-VSC AC output voltage. The STATCOM operation is determined by Equations (7.18) and the control system described in Section 7.3.5.

The test system conditions are as follows:

- reactive power rating 100 MVAR (with two 50 MVAR converter bridges);
- rated voltage 100 kV;
- nominal leakage reactance of each of the interface transformers 10 %, their turns ratios (k_n) being 1 and $\sqrt{3}$ for the star- and delta-connected transformers respectively.

To simplify the display of results the waveforms are normalised as follows:

$Q_{Base} = 100$ MVA for reactive power Q and its order Q_{ref}

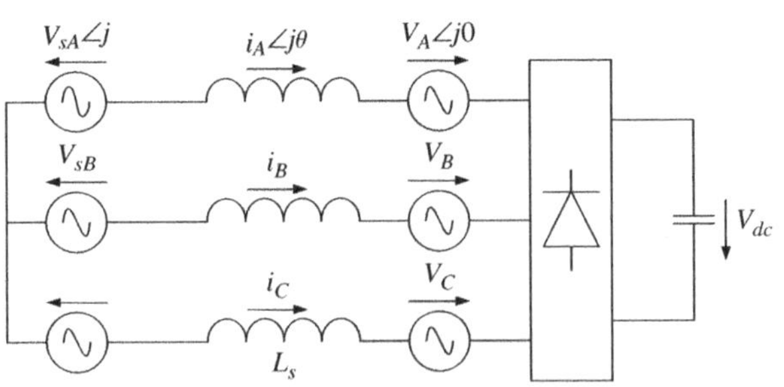

Figure 7.19 Simplified VSC test system

$V_{Base} = \sqrt{3}V_{rated}/\sqrt{2}$ for voltages on the primary side of the interface transformers (e.g. V_A)

$I_{Base} = Q_{Base}/3V_{Base}$ for currents on the primary side of the interface transformers (e.g. I_A)

$V_{dcBase} = V_{Base}/k_n$ for voltages on the converter side of the interface transformers (e.g. V_{jC1}, V_{dc}, etc.)

$I_{dcBase} = k_n I_{Base}$ for currents on the converter side of the interface transformer (e.g. I_j).

Waveform verification

Figures 7.20 and 7.21 show the test system voltage and current waveforms for a fundamental cycle in steady state at rated reactive power conditions for capacitive operation.

The variables plotted in Figure 7.20 are:

(a) U_j, the reinjection voltage

(b) V_{YY}, the normalised phase voltage of the star–star-connected bridge

(c) $V_{\Delta Y}$, the normalised phase voltage of the star–delta-connected bridge

(d) V_A, the normalised AC output voltage

(e) V_{An}/V_{A1}, the output voltage spectrum.

The variables shown in Figure 7.21 are:

(a) I_j, the reinjection current.

(b) $V_{C1} - V_{jC\Sigma}$, the normalised voltage difference between the top main capacitor and the reinjection capacitors. These waveforms are shown to illustrate that the added voltages of the reinjection capacitors follow closely (with just a small ripple) the voltages of the two main capacitors.

(c) $V_{C2} - V_{jC\Sigma}$, the normalised voltage difference between the bottom main capacitor and the reinjection capacitors. Their purpose is as explained in (b).

(d) $V_A I_A$, the normalised AC output voltage and current of the seven-level MLVR-VSC.

(e) I_{An}/I_{A1}, the output current spectrum of the seven-level MLVR-VSC.

All these waveforms are practically the same as the theoretical ones of the multi-level reinjection schemes. The simulated output voltage THD is 2.71 as compared with a theoretical value of 2.77 %.

If the converter transformers' leakage reactance is 10 %, the current harmonic content injected into the AC system is very small, i.e. 1.03 and 1.14 % for the capacitive and inductive cases respectively.

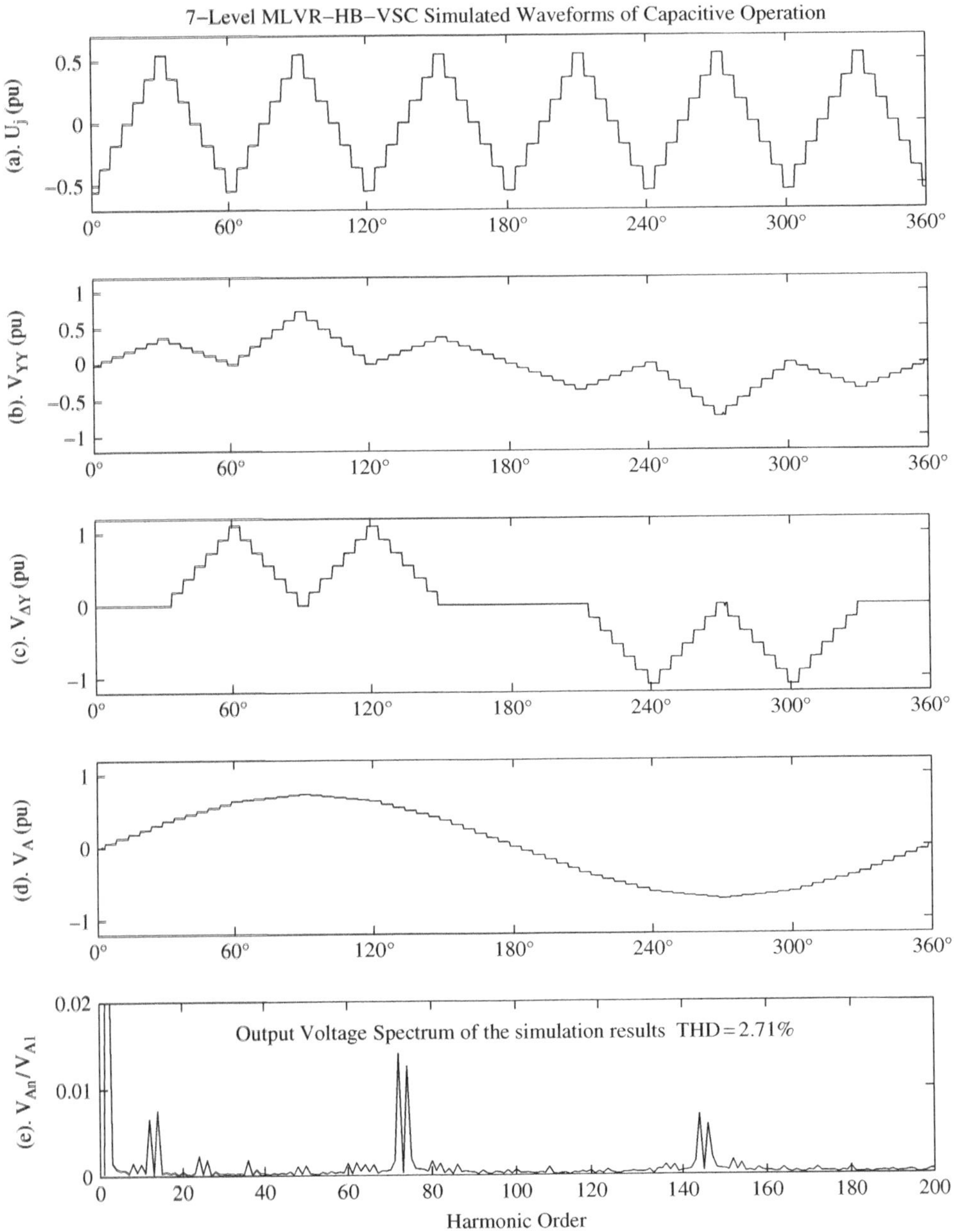

Figure 7.20 MLVR-VSC voltage waveforms for capacitive operation

Dynamic performance

Figure 7.22 illustrates the simulated dynamic performance of the seven-level MLVR-VSC operating as a STATCOM and shows the following normalised waveforms:

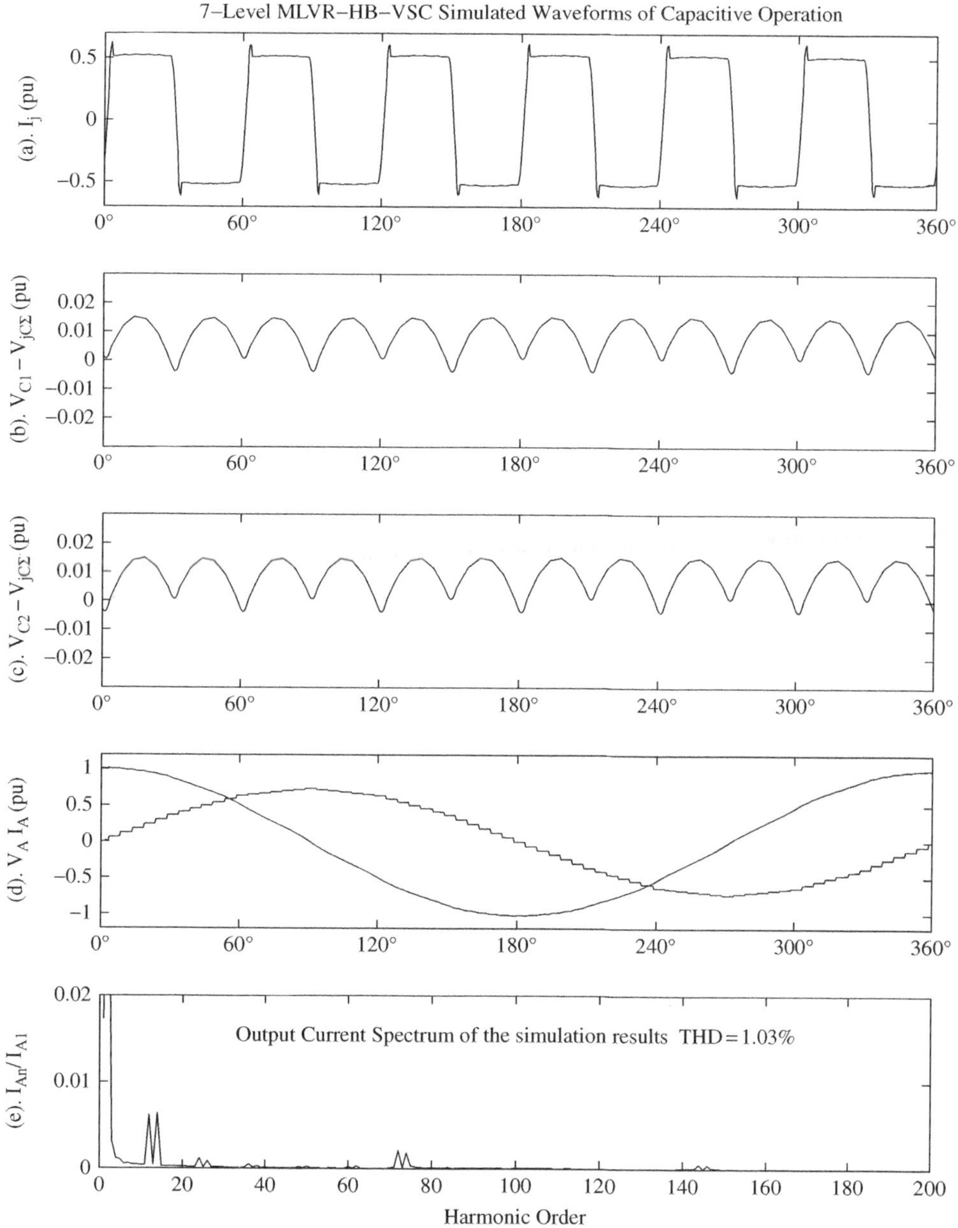

Figure 7.21 MLVR-VSC waveforms for capacitive operation

(a) Q, Q_{ref}, P, the generated (Q) and ordered (Q_{ref}) reactive powers (the capacitive being set to positive and the inductive to negative) as well as the active power (P) of the converter (generation being set to positive and absorption to negative).

(b) $V_{jC\Sigma}$, V_{C1}, V_{C2}, the total reinjection capacitor voltage and main capacitor voltages.

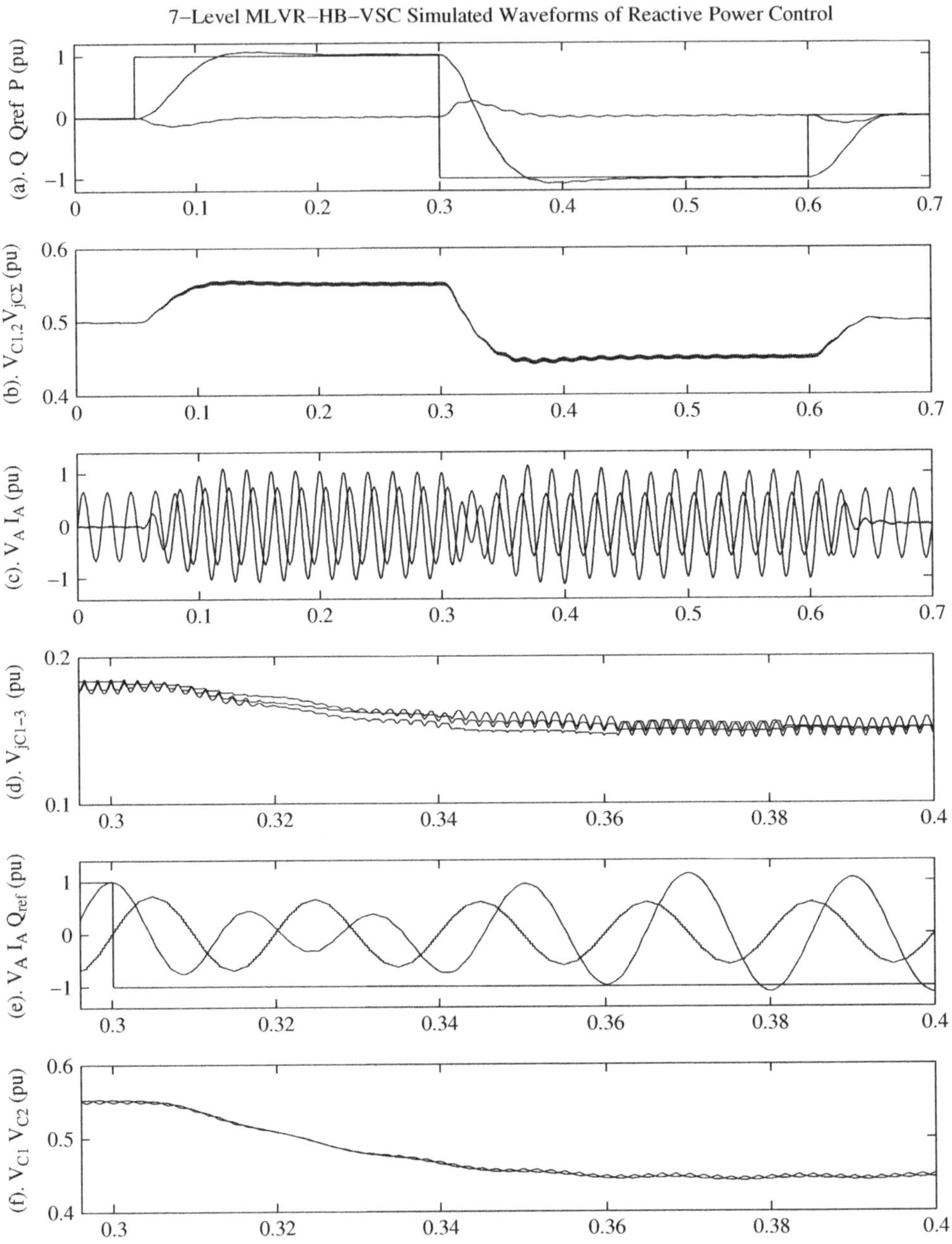

Figure 7.22 MLVR-VSC dynamic responses under step reactive orders

(c) $V_A I_A$, the AC output voltage and current of the seven-level MLVR-VSC.

(d) V_{jC1}, V_{jC2}, V_{jC3}, voltages of the reinjection capacitors in a reduced time scale (from 0.3 to 0.4 seconds) to show in greater detail the dynamic response.

(e) V_A, I_A, Q_{ref}, the AC output voltage and current in the reduced time scale (Q_{ref} is added to indicate the instant of the ordered reactive power change).

(f) $V_{C1} V_{C2}$, the main capacitor voltages in the reduced time scale.

Referring to Figure 7.22(a), the following dynamic performance is observed:

(i) The reactive power order changes from 0 to 1 pu after 0.05 seconds and the generated reactive power reaches the required value at 0.11 seconds, the maximum overshoot being about 5 %.

(ii) Q_{ref} is changed from 1 to -1 pu at 0.3 seconds; the generated reactive power reaches the new value at 0.36 seconds and the maximum overshoot is under 8 %.

(iii) Q_{ref} is changed from -1 pu to zero at 0.6 seconds; the reactive power returns to zero at 0.66 seconds and no overshoot occurs.

The waveforms in Figure 7.22(b) show that:

(iv) The voltages of the two main capacitors are kept balanced in the steady state and dynamic regions.

(v) The added voltages of the reinjection capacitors on the DC sides of the H-bridge chain follow closely the voltages of the two main capacitors both in the steady-state and dynamic regions.

The waveforms in Figure 7.22(c) show that the output voltage and current response to the reactive power changes (from full-scale inductive to full-scale capacitive in approximately three cycles) is smooth and without significant overcurrent or overvoltage. The waveforms in Figure 7.22(d) demonstrate that the reinjection capacitor voltages experience a slight drift in the dynamic region, but after a few cycles they become balanced again. The reduced scale waveforms of Figure 7.22(e) show that the harmonic contents of the output voltage and current waveforms are very low and change smoothly during the dynamic periods. Similarly those of Figure 7.22(f) clearly indicate that the capacitor voltages are kept balanced in the dynamic periods.

7.6 Summary of Main Characteristics of MLVR Alternatives

In the multi-level voltage source configurations described in Chapter 6, the auxiliary switches and/or capacitors required to increase the number of steps of the output voltage waveform are an integral part of each of the main power circuit phases and have to be rated accordingly. The switching components of the multi-level reinjection schemes are kept separate from the power circuit and are shared by the three phases; this reduces considerably the voltage and/or current ratings, as well as the number of switches and high-voltage connections.

Multi-level reinjection combines the benefits of multi-level conversion and soft switching, and requires fewer extra switching components.

The three- and five-level transformer-coupled configurations have no capacitor balancing problems, because the DC capacitor is not subdivided into clamping sections. However, each arm of the reinjection bridges is connected across the full DC voltage, which makes it unattractive for high-voltage application.

The active clamping concept permits the use of any level number without the need for reinjection transformers and the number of extra switches is substantially reduced with respect to the conventional multi-level diode clamping. Moreover, the controllability of the clamping circuit can be used to eliminate the capacitor voltage balancing problem. However, the use of high-level numbers is still impractical, as the number of clamping switches increases more than linearly with the level number.

Cascaded H-bridge voltage reinjection is the most likely multi-level contender for use in high-voltage applications. This alternative achieves the multi-level output and ZVS condition without carrying DC. Thus in the absence of DC power, the power rating of the reinjection circuit is a fraction of the main converter rating. Moreover, the number of switches required by the low-rated H-bridge structure is directly proportional to the voltage rating (as compared with the conventional diode-clamped configuration, where the number of extra switches increases with the square of the pulse number).

Another feature of H-bridge reinjection is its dynamic voltage balancing capability. In this respect, the free adjustable width of the ZVS period enables the H-bridge chain to exchange charge with the main bridge capacitors; the charge exchanges between the main bridge and H-bridge chain force their capacitors to be equal periodically; this effect is particularly important under asymmetrical source conditions. The identity of the H-bridges, their series connection and the possibility of redundant state combinations provide the self-balancing mechanism for the H-bridge capacitor chain voltages. This self-balancing mechanism is less effective when the number of H-bridges is high.

7.7 Multi-Level Current Reinjection (MLCR) [13, 14]

7.7.1 Structure and Operating Principles

MLCR can be considered as the dual of MLVR-VSC. Thus the waveforms developed for the MLVR configuration, and illustrated in Figure 7.4, also apply to the MLCR configuration. In this case current, instead of voltage, constitutes the DC source.

Unlike MLVR that requires asymmetrical switches (with unidirectional voltage blocking and bidirectional current capability), MLCR requires symmetrical switches (with bidirectional voltage blocking and unidirectional current capability). Thus if IGBT switches were to be used in MLCR conversion, a diode would have to be connected in series with the IGBT, and the extra power loss may not be acceptable in high-voltage applications. Therefore, symmetrical switches, such as the GTO and IGCT, are more appropriate devices for MLCR conversion, particularly given the lower switching frequencies involved (as compared with those of PWM conversion).

The use of switches with turn-off capability provides the reinjection circuit with the opportunity to self-commutate, instead of using the ripple voltage as the commutating voltage (the solution described in Section 3.8.2), and thus to place at will the position of the reinjection current pulses.

The same topological structure and basic control strategy of the DC ripple reinjection concept described in Section 3.8.2 for line-commutated conversion can be used in self-commuting CSC (this has been explained in Section 4.5).

When the AC system voltage is perfectly symmetrical, the 12-pulse voltage ripple is of low amplitude and therefore the inductance (L_m) required to suppress the DC current fluctuation is very low. In practice, however, larger values of L_m are needed to cope with the presence of some system asymmetry.

The required levels of current reinjection are produced by tapping the reinjection transformer. Compared with VSC, where equal size capacitors are required to share the DC voltage, the reinjection transformer taps in the CSC case can be arranged more flexibly to derive the AC output current waveforms.

Figure 7.23 shows the $(m+1)$-level MLCR configuration based on the 12-pulse series-connected converter; in this figure the reinjection switches are shown as GTOs. The output currents of the bridges I_{BY} and $I_{B\Delta}$ are shaped by the reinjection currents I_{jY} and $I_{j\Delta}$ into $(m+1)$-level waveforms and produce a $12m$-pulse equivalent output current waveform on the primary of the interface transformer.

There are two important differences regarding the double bridge MLCR system with respect to that of the single bridge. One is the operating frequency, which is now six

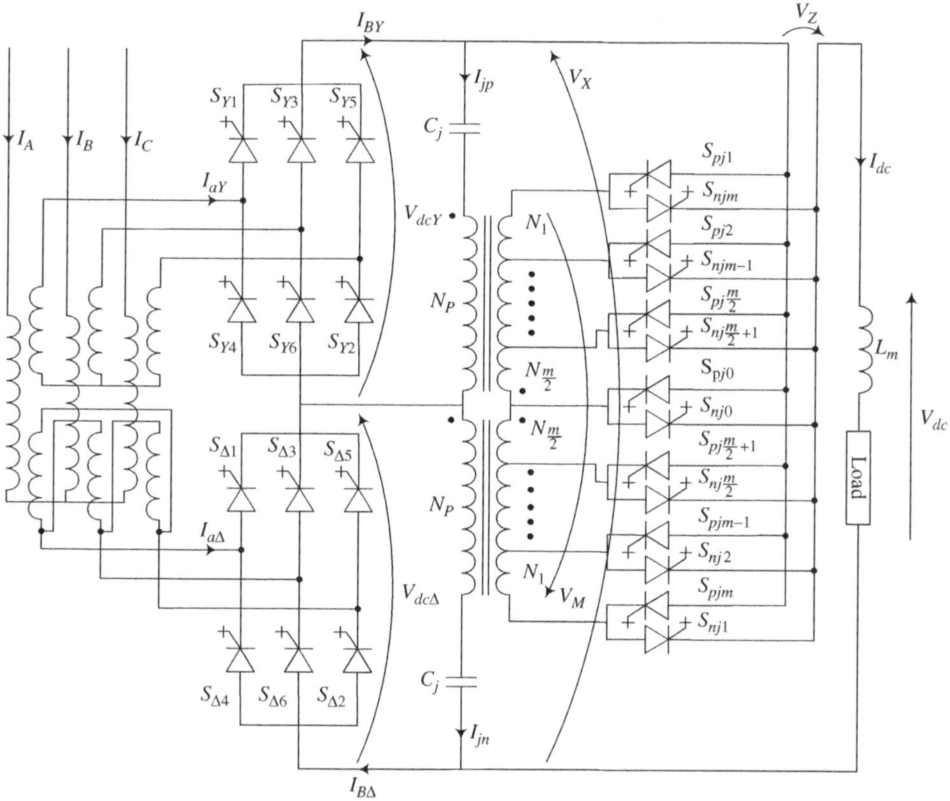

Figure 7.23 Structure of the series MLCR-CSC

times (instead of three) the fundamental; the other is the location of the reinjection point, which is now the midpoint between the series-connected bridges, instead of the transformer neutral. The primaries (N_p) of two single phase transformers are connected across the bridges' terminals and each of their multi-tapped secondaries (N_k) is periodically connected in series with the DC line, the repetition period being six times that of the fundamental frequency; this is achieved by firing simultaneously two opposite-conducting GTOs of the symmetrically placed taps on both sides of the reinjection transformer secondaries (e.g. S_{pj1} and S_{nj1}).

Similarly, Figure 7.24 shows the $(m+1)$-level MLCR configuration based on the 12-pulse parallel-connected converter. The multi-tapped reactor assisted by the switching action of the reinjection switches distributes I_{DC} to the two bridges in $(m+1)$-level waveforms.

The main property of the MLCR scheme is its capability to control the position, magnitude and duration of the reinjection steps. If these parameters are optimised to achieve maximum harmonic cancellation, for every pair of taps symmetrically placed with respect to the two reinjection transformer secondaries, the pulse number is doubled, with the midpoint tap and short-circuiting switch pair (S_{pj0}–S_{nj0}) adding an extra multiplication factor. Thus the five-level configuration shown in Figure 7.25 can achieve 60-pulse conversion (i.e. 5(reinjection level number) × 6 (reinjection frequency ratio) × 2 (number of bridges). An added bonus of the reinjection current is that the converter valve currents are forced to a very low value (i.e. to an almost ZCS condition) during the commutations, which simplifies the design of the snubber circuits. The ZSC condition does not apply to the reinjection switches, but due to the unidirectional nature of the current, the snubbers required can be of the simple resistor–capacitor–diode (RCD) type.

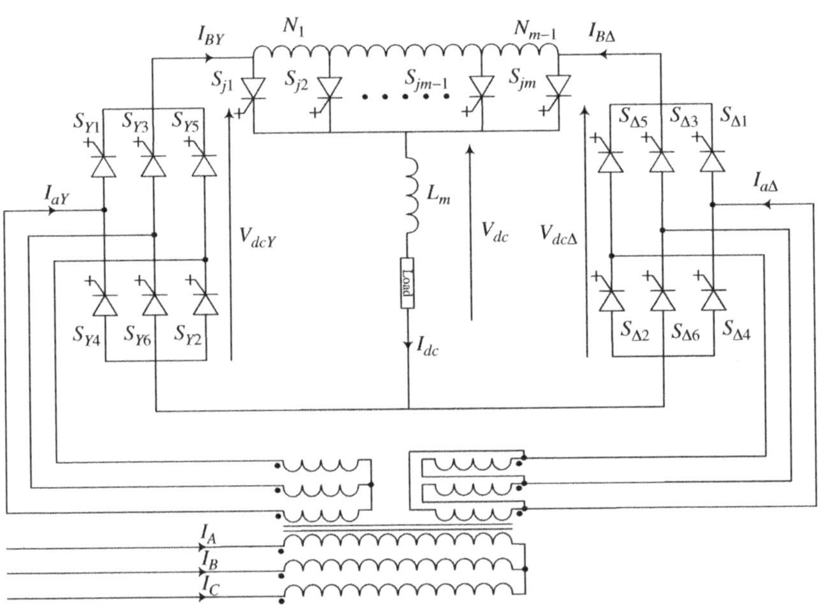

Figure 7.24 Structure of the parallel MLCR-CSC

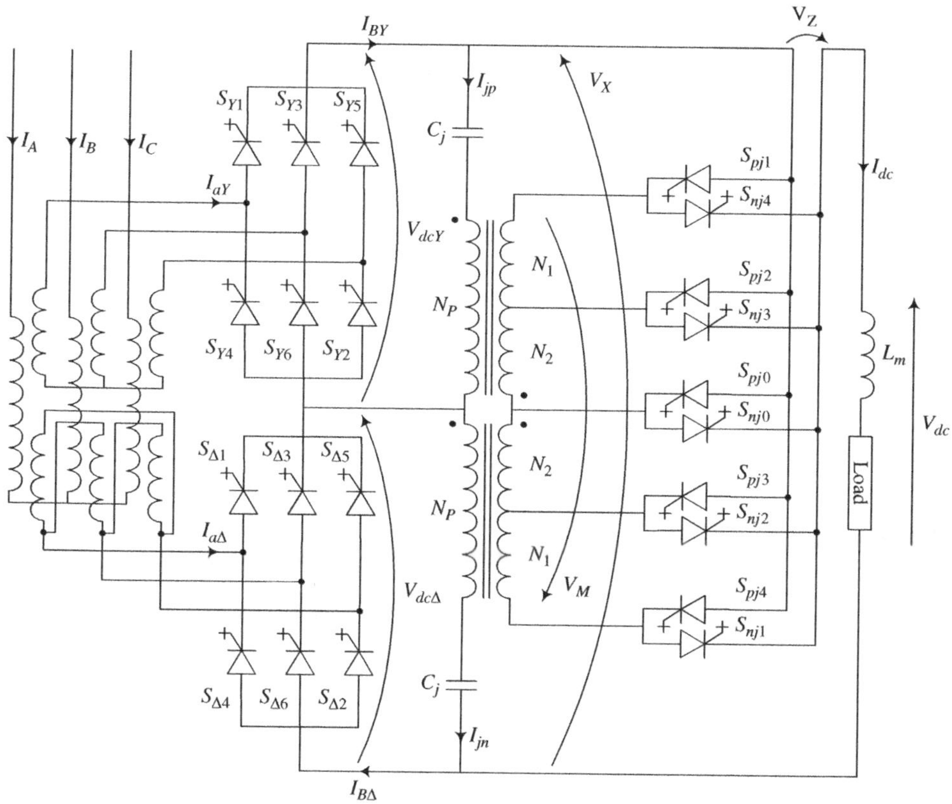

Figure 7.25 Five-level reinjection series-connected CSC

7.7.2 Self-Commutating Thyristor Conversion [15]

In the pulse multiplication scheme described above, the magnitude and duration of the reinjection steps were optimised to achieve maximum harmonic cancellation, which, as already explained, for the five-level reinjection configuration, provides 60-pulse conversion. However, this reinjection waveform does not completely cancel the converter valves' currents during the commutation and, therefore, turn-off devices are still needed to provide self-commutation.

It is shown next that the use of a non-optimal reinjection waveform (from the harmonics viewpoint) can force a ZCS condition for an interval of 6° (or 333 μs at 50 Hz) during the commutation. This should permit the outgoing thyristor to recover its blocking capability and thus make the conventional thyristor converter self-commutating. To achieve the ZCS condition the quality of the AC and DC voltage waveforms is somewhat reduced (from the optimum 60 to 48 pulse for the five-level reinjection scheme). If necessary larger zero-current regions can be achieved at the expense of further reduction in the harmonic reduction capability.

Let us consider the circuit of Figure 7.25 operating in the steady state, with valve S_{Y1} conducting. When the reinjection current forces the current of valve S_{Y1} to zero, none of the valves connected to the common cathode (CC) conduct and DC will continue to flow via the reinjection path. However, the next level of the multi-step reinjection current should force a change in DC current. This is prevented by the large DC reactor that develops the necessary transient emf (with negative polarity on the bridge CC bus) to ensure that the anode of S_{Y3} becomes positive with respect to its cathode irrespective of the potential of the AC voltage. Therefore, provided that valve S_{Y1} has by then recovered its blocking capability, it is possible to turn on valve S_{Y3} to provide a new path for the converter current. Thus the converter can commutate without the assistance of a turn-off pulse or a line commutating voltage, i.e. it can be of the conventional thyristor type.

It is therefore possible to achieve self-commutation, as well as pulse multiplication, using conventional thyristors for the converters and GTOs (or IGCTs) for the reinjection switches. This alternative gives the thyristor converter similar flexibility as a forced commutated VSC, i.e. the ability to control both the DC voltage and current, with leading or lagging power factor, as well as reducing the harmonic content. In other words, it combines the benefits of the robust and efficient conventional converter and the controllability of the advanced self-commutated technology. This is an important breakthrough that should give greater flexibility to thyristor-based HVDC transmission.

Theoretical waveforms of the self-commutating thyristor converter

The circuit diagram of the self-commutating thyristor converter is the same as that of the GTO (or IGCT) configuration shown in Figure 7.25, the only difference being the replacement of the symbol of the self-commutating switches of the main converter by that of the thyristor.

The following theoretical waveforms are shown in Figure 7.26 for the five-level reinjection scheme of the modified Figure 7.25, for the case when the converter bridges operate with a firing angle of −45° (i.e. supplying reactive power, even though the switches are thyristors):

(a) and (b) $I_{B\Delta}$ and I_{BY} are the DC currents of the bridge converters modified by their respective reinjection currents I_{jn} and I_{jp}.

(c) and (d) $I_{ca\Delta}$ and I_{aY} are the phase 'a' currents in the secondary windings of the delta- and star-connected windings respectively.

(e) I_A is the phase 'a' current on the primary side of the converter transformer, i.e. the output current of the converter.

(f) I_{An} is the harmonic spectrum of the output current. It shows that the 47th and 49th harmonic orders are prevalent, which indicates that the converter is on 48-pulse operation. The THD of the output current waveform is 4 %.

On the DC side, the reinjection circuit increases the pulse number of the voltage waveform by a factor of four, i.e. to 48 pulses per cycle (for the five-level reinjection configuration).

It follows that the combination of a conventional thyristor converter and a self-commutated multi-level reinjection circuit can provide reactive power, as well as active

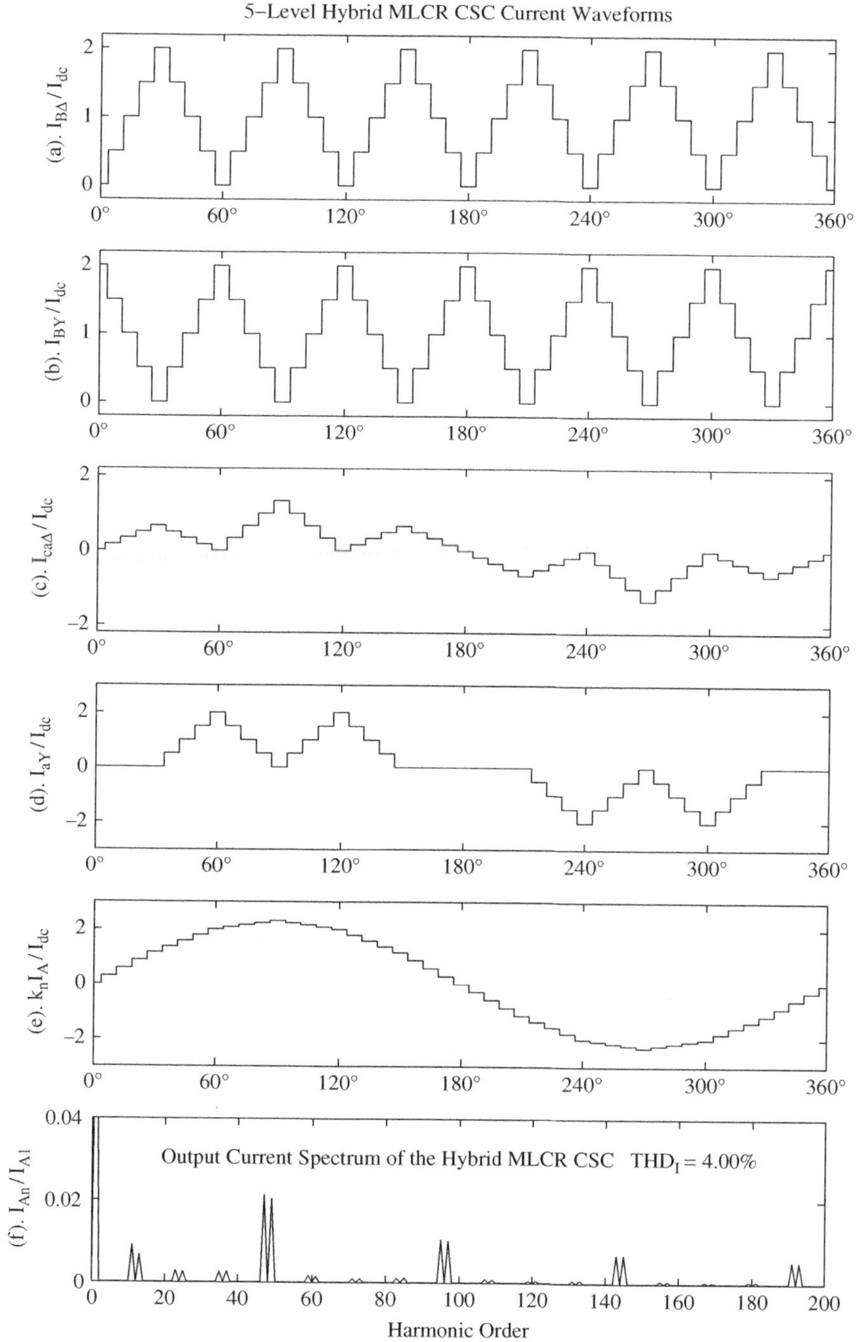

Figure 7.26 Theoretical current waveforms of the five-level reinjection CSC

power controllability. It also transforms the conventional CSC waveforms into multi-step AC and multi-pulse DC voltage waveforms. This applies equally to rectifier and inverter operation and to variable frequency supplies. The power conversion efficiency is high, because the rectified harmonic power is reinjected into the DC system. Moreover, the larger number of taps of the multi-level arrangement does not greatly increase the total current rating of the reinjection switches, because the average currents of the individual reinjection switches reduce in inverse proportion to their number; of course the rms currents of the individual switches (and therefore the total rms current) will be higher than their average values.

7.7.3 EMTDC Verification

The converter system shown in Figure 7.25 (with the main converter switches replaced by thyristors) has been modelled in the EMTDC/PSCAD package. As in the theoretical model, the converter transformers are connected to an ideal three-phase voltage source and their leakage reactances are 5 % (based on 100 MW/100 kV). A series combination of resistance and smoothing inductance is connected on the DC side to provide a nominal 100 MW of DC power, while injecting 100 MVAR (which corresponds to a firing angle of −45°) to the AC system.

Figures 7.27 and 7.28 show the steady-state waveforms on the AC and DC sides respectively. As in the theoretical case, the simulated output current (graph 7.27(e)) and its spectrum (graph 7.27(f)) show predominantly 48-pulse operation, with 4.65 % of THD, as compared with the 4 % predicted by the theoretical analysis.

The observed small content of 12-pulse related harmonics (i.e. 11 and 13 orders on the AC side and 12 on the DC side) was also present in the theoretical results. Its levels, however, are small (in the region of 1 %) and thus fall within prescribed harmonic standards.

The main differences between the theoretical and simulated waveforms relate to the presence of snubber components in the PSCAD converter model, which introduce some waveform distortion in the current waveform. Also, though not shown in the figures, di/dt of the reinjection current steps introduces high-frequency oscillations in the converter terminal voltage and a small shunt capacitor is needed to eliminate them.

Clearly the PSCAD simulation confirms that the series-connected double bridge thyristor converter commutates naturally even at negative firing angles, that it is capable of generating reactive power and that the AC and DC voltage waveforms are perfectly acceptable without the need for harmonic filters on either side of the converter.

Dynamic performance

EMTDC simulation is also used to illustrate the dynamic behaviour of the five-level MLCR-CSC shown in Figure 7.25, under the following conditions: power rating 100 MVA (provided by two 50 MVA bridges); voltage rating 100 kV; interface transformer leakage reactance 10 %; inductance (L_m) of the DC side tapped reactor 2 H.

The reactor, used as a reinjection autotransformer, has five symmetrically arranged taps, and produces linear symmetrical current reinjection waveforms.

The basic control structure of the MLCR-CSC for the control of the active and reactive power is shown in Figure 7.29, and that used for the real and imaginary current components

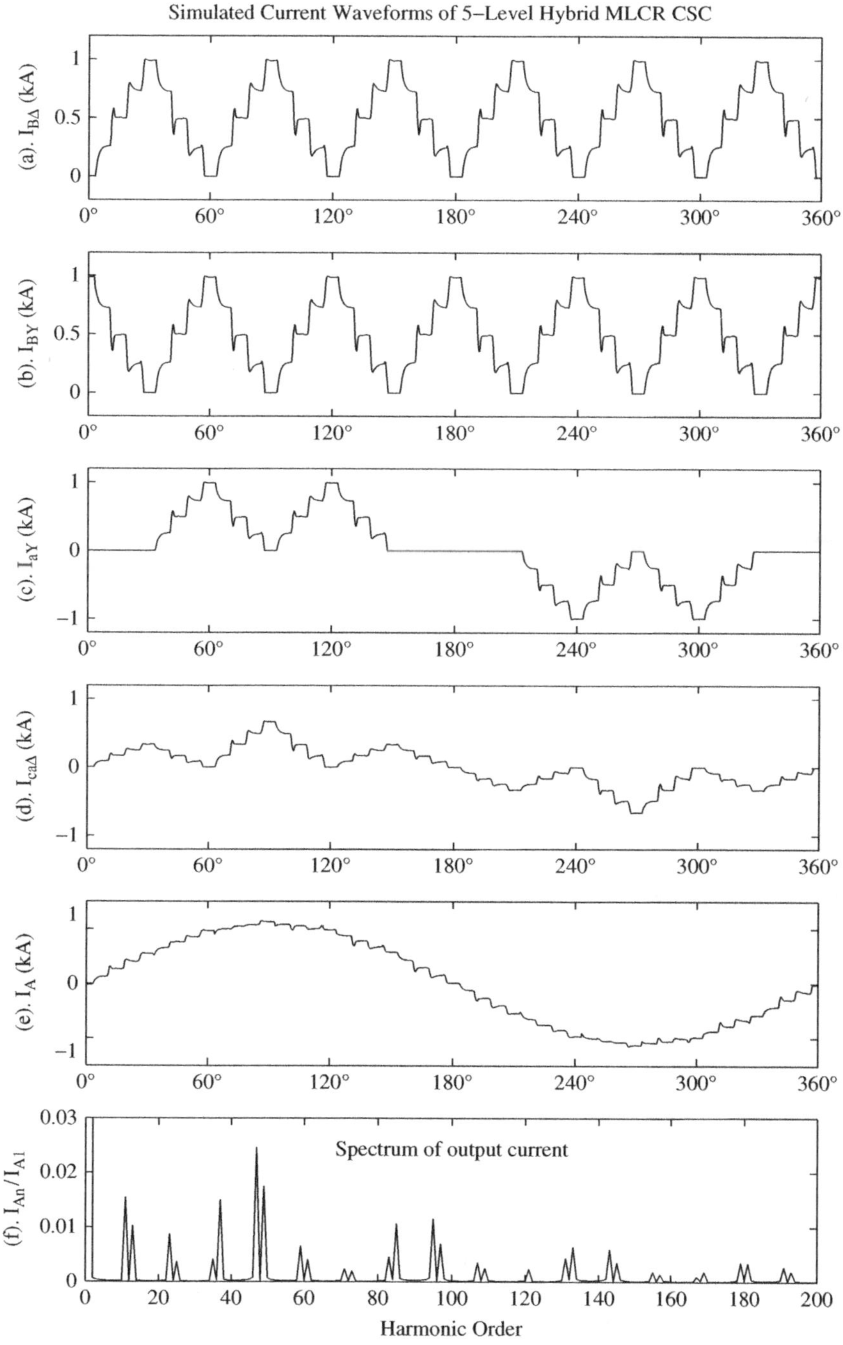

Figure 7.27 EMTDC simulation of the current waveforms of the CSC

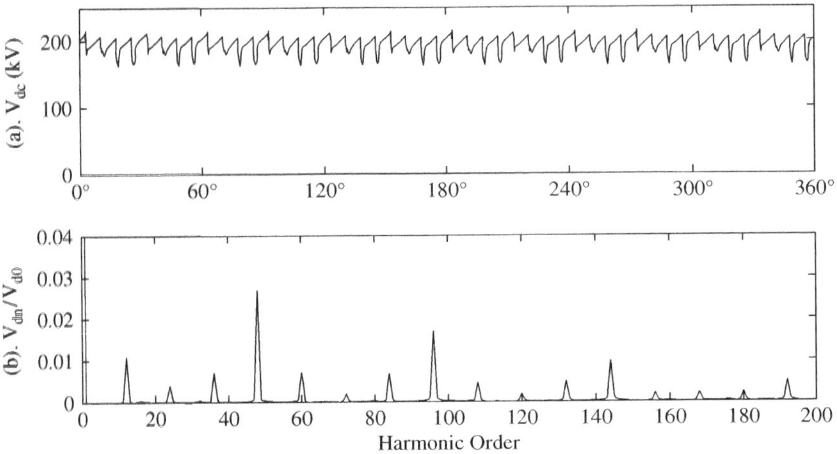

Figure 7.28 EMTDC simulation of the DC voltage waveform of the CSC

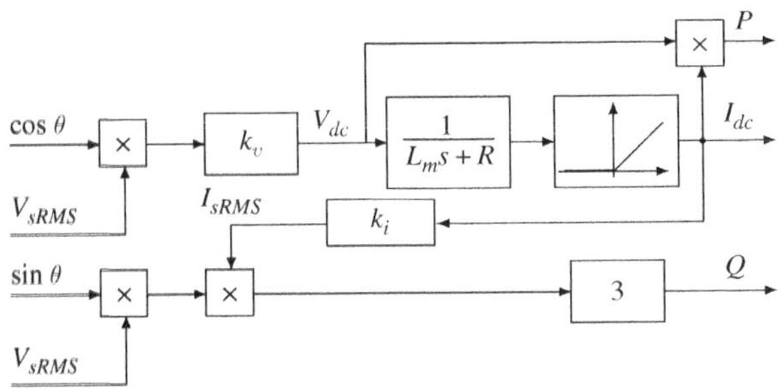

Figure 7.29 MLCR-CSC block diagram for the control of the real and reactive powers

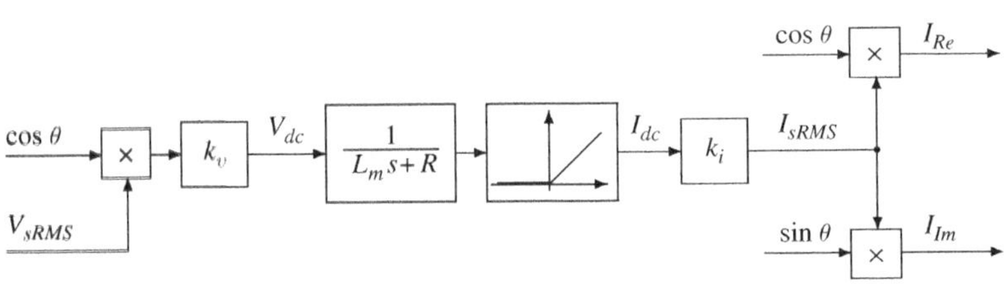

Figure 7.30 MLCR-CSC block diagram for the control of the real and imaginary currents

is shown in Figure 7.30. The test carried out assumes that the converter is acting exclusively as a reactive power controller (i.e. in the STATCOM mode).

Unlike the case of PWM-VSC, due to the low fundamental-frequency-related switching nature of MLCR, it is not possible in this case to exercise independent control of the amplitude and power angle. The only variable that controls the MLCR operation is θ (the power factor angle). A similar situation occurred in the case of the MLVR-VSC configuration described earlier, where δ, the phase angle difference between the converter output voltage and the AC source voltage, was the unique control variable. However, in the latter case only a relatively small variation of δ was needed to achieve four-quadrant operation, whereas in MLCR this condition requires a variation of θ in the range of $\pm180°$. Thus the relationships between θ and $[P, Q]$ are highly non-linear. This is shown by the following expressions derived from the block diagrams of Figures 7.29 and 7.30, where V_t is the rms value of the converter terminal (phase-to-phase) voltage:

$$P = V_{DC}I_{DC} = \frac{k_v^2 V_t^2 \cos^2(\theta)}{L_m s + R}$$

$$Q = -3V_t \sin(\theta).k_i I_{DC} = -\frac{3k_v k_i V_t^2 \sin^2(\theta)}{L_m s + R}$$

$$I_{Re} = \frac{k_v k_i V_t \cos^2(\theta)}{L_m s + R}$$

$$I_{Im} = -\frac{k_v k_i V_t \sin(\theta)\cos(\theta)}{L_m s + R}$$

For the test case (STATCOM operation), only a small variation of θ around $\pm90°$ is needed to control the DC current.

The amplitude increment or decrement of the DC load branch current I_{DC} depends on the polarity of the DC voltage across the load branch, which is proportional to the cosine function of θ ($V_{DC} = k_v V_t \cos(\theta)$), if the converter voltage V_{tS} is kept constant and the losses in the smoothing reactor are ignored.

However, the amplitude increment or decrement of the unidirectional DC cannot be solely determined by the polarity of the power angle increment around the $\pm90°$ mark, because for $+90°$

$$dV_{DC} = d[k_v V_t \cos(90° + \Delta\theta)] \approx -k_v V_t d(\Delta\theta)$$

whereas for $-90°$

$$dV_{DC} = d[k_v V_t \cos(-90° + \Delta\theta)] \approx k_v V_t d(\Delta\theta)$$

Thus the power angle increment $\Delta\theta$ around $\pm90°$ has to be coordinated with the converter operating condition to generate the appropriate polarity of $\Delta\theta$.

The simulation results shown in Figure 7.31 include:

(a) the generated Q and Q_{ref};

(b) the average DC current and voltage;

(c), (d), (e) the three-phase voltages and currents;

(f) the DC voltage and three-phase currents (with the time scale reduced).

The normalised bases for the waveforms in Figure 7.31 are

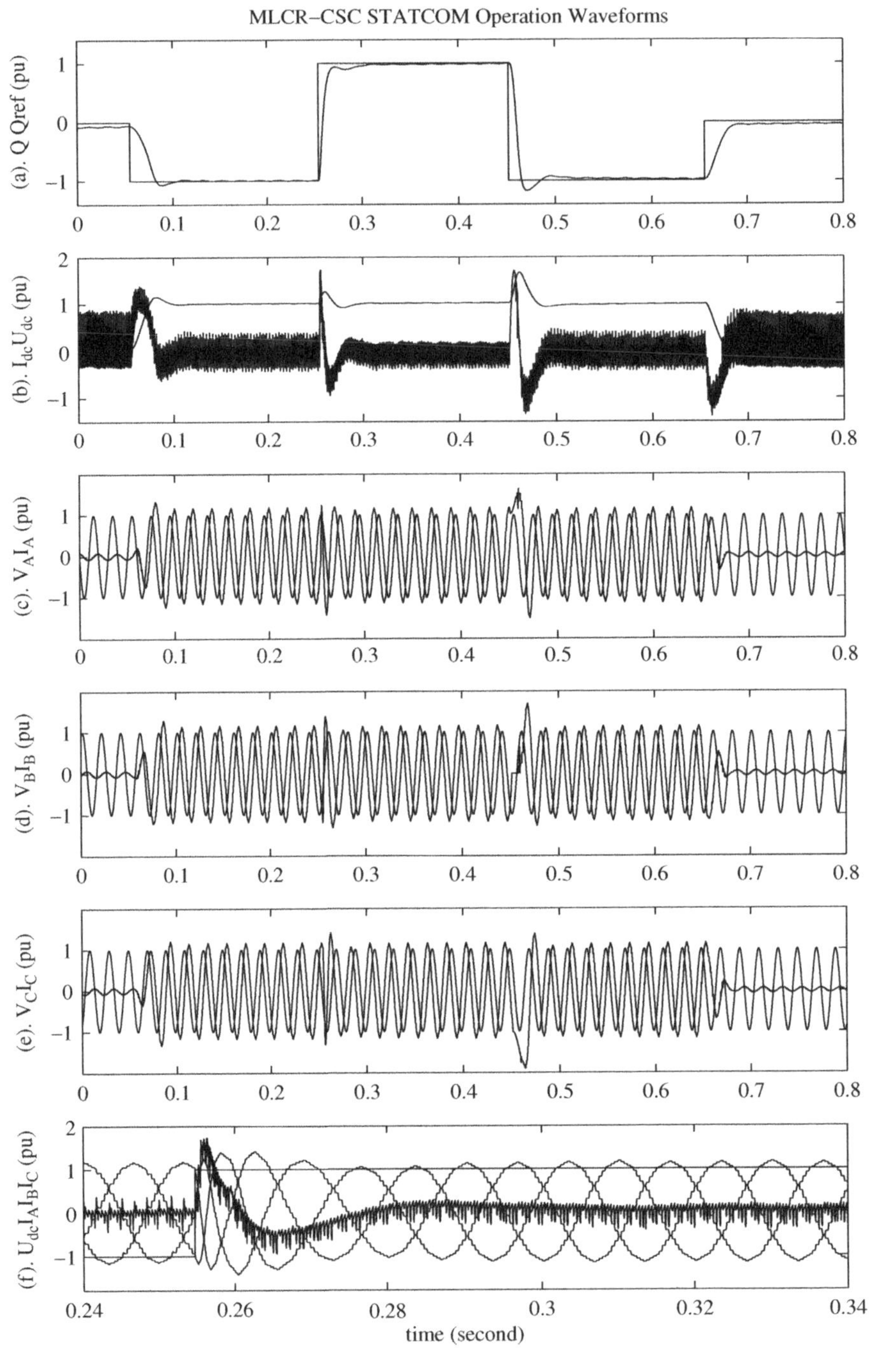

Figure 7.31 Dynamic responses of the MLCR-CSC STATCOM

100 MVA for Q and Q_{ref}

$$\frac{100\sqrt{2}}{\sqrt{3}} \text{ kV for } V_{DC} \text{ and } V_A, V_B, V_C$$

$$\frac{\sqrt{3}}{\sqrt{2}} I_{SR} = \frac{1}{\sqrt{2}} \text{ kA for } I_{DC} \text{ and } I_A, I_B, I_C$$

The reactive power order is changed four times (in full step) in a total time interval of 0.8 s. By an appropriate choice of parameters for the PI controller, MLCR responds to these orders very quickly and without steady error.

7.8 MLCR-CSC Versus MLVR-VSC

In common with other multi-level schemes, both MLCR and MLVR avoid high dv/dt rates.

The MLCR alternative is better suited to large HVDC transmission schemes for the following reasons:

- The provision of a close to ZCS condition of the valves in the main bridges eliminates the need for the interfacing capacitor, normally required to absorb the inductive energy stored in the AC system during the commutation period.
- Moreover, MLCR offers the interesting possibility of using conventional thyristor bridges, because the reinjection current waveform can be controlled to force the bridge currents completely to zero during the commutation process.
- The absence of shunt capacitance at the converter terminals eliminates the risk of harmonic parallel resonances, a common occurrence in line-commutated conversion.
- The problem of capacitor balancing of MLVR configurations does not apply to the MLCR alternatives.
- MLCR is capable of supplying balanced current under symmetrical and asymmetrical AC source conditions, even during faults.
- The MLCR circuit is placed in the main power flow path and, therefore the reinjection switches are rated for low voltage and high current.
- The interface transformer of a six-pulse converter MLVR requires a path for the triple fluxes generated by triple harmonic voltage components. Such is not the case with MLCR, which generates triplen harmonic current but no triplen harmonic voltage components; therefore, there is no special requirement for the interface transformer in this case. This, however, is not an important issue for high power applications because the 12-pulse configuration is always the preferred option. For the 12-pulse MLVR the interface transformers have to be separated and connected in series, whereas in the 12-pulse MLCR the two bridges can be powered by a common transformer with a set of primary windings and two sets of isolated secondary windings.

- The power losses in a DC reactor are much higher than those in a DC capacitor. However, the size of the DC reactor can be reduced due to the high frequency voltage across the reactor (for a level number m, the DC voltage ripple frequency of the MLCR is $12(m-1)$ for the 12-pulse system).

- MLCR does not suffer from high currents during DC faults, because the rate of rise of current during disturbances is limited by the DC smoothing reactor and the fault current can be suppressed by control action. Instead, the MLCR experiences high overvoltages across the DC reactor induced by the fault current. In MLVR, on the other hand, the capacitor *di/dt* is high during DC faults and is more likely to damage the switching devices.

References

1. Yuang, X., Stemmler, H. and Barbi, I. (1999), 'Evaluation of soft switching techniques for neutral clamped (NPC) inverters', Power Electronics Specialists Conference (PESC'99), 2: 659–664.
2. Yamamoto, M., Sato, S., Hiraki, E. and Nakaoka, M. (2002), 'Auxiliary active resonant commutated snubber-assisted 3-level 3-phase voltage source soft-switching inverter', Proceedings of the Power Conversion Conference (PCC'02), 3: 1245–1250.
3. Turpin, C., Deprez, I., Forest, F., Ricardeau, F. and Meynard, T.A. (2002), 'A ZVS imbricated cell multilevel inverter with auxiliary resonant commutated poles', *IEEE Transactions on Power Electronics*, **17**(6), 874–882.
4. Teichmann, R. and Bernet, S. (1999), 'A multilevel ARCP voltage source converter topology', The 25th Annual Conference of the IEEE Industrial Electronics Society, IECON'99, 2: 602–607.
5. Cho, J.G., Back, J.W., Yoo, D.W., Won, C.Y. and Rim, G.H. (1998), 'Zero voltage switching three-level auxiliary resonant commutated pole inverter for high power applications', *IEE Proceedings, Electric Power Applications*, **145**(1), 25–32.
6. Dijkhuizen, F.R., Duarte, J.L. and Van Groningen, W.D.H. (1998), 'Multilevel converter with auxiliary resonant commutated pole', Conference Record of the IEEE Industry Applications Society Annual Meeting, 2: 1440–1446.
7. Yuan, X., Orglmeister, G. and Barbi, I. (2000), 'ARCPI resonant snubber for the neutral point clamped inverter', *IEEE Transactions on Industry Applications*, **36**(2), 586–595.
8. Liu, Y.H., Arrillaga, J. and Watson, N.R. (2004), 'Multi-level voltage reinjection - a new concept in high voltage source conversion', *IEE Proceedings, Generation, Transmission and Distribution*, **151**(3), 290–298.
9. Liu, Y.H., Arrillaga, J. and Watson, N.R. (2004), 'A new STATCOM configuration using multi-level dc voltage reinjection for high power applications', *IEEE Transactions on Power Delivery*, **19**(4), 1828–1834.
10. Liu, Y.H., Arrillaga, J. and Watson, N.R. (2005), 'Capacitor voltage balancing in multi-level voltage reinjection (MLVR) converters', *IEEE Transactions on Power Delivery*, **20**(2), 1728–1737.
11. Perera, B.L., Liu, Y.H., Arrillaga, J. and Watson, N.R. (2005), 'A five-level reinjection scheme for high pulse voltage source conversion', *IEE Proceedings, Electric Power Applications*, **152**(2), 209–216.
12. Nabae, A., Takahashi, I. and Akagi, H. (1981), 'A new neutral point clamped PWM inverter', *IEEE Transactions on Industry Applications*, **IA-17**(5), 518–523.
13. Perera, L.B., Liu, Y.H., Watson, N.R. and Arrillaga, J. (2005), 'Multi-level current reinjection in double-bridge self-commutated current source conversion', *IEEE Transactions on Power Delivery*, **20**(2), 984–991.

14. Perera, L.B., Liu, Y.H., Watson, N.R. and Arrillaga, J. (2005), 'Multi-level current reinjection self-commutated HVDC converter', *IEE Proceedings, Generation Transmission and Distribution*, **152**(5), 607–615.
15. Arrillaga, J., Liu, Y.H., Perera, L.B. and Watson, N.R. (2006), 'A current reinjection scheme that adds self-commutation and pulse multiplication to the thyristor converter', *IEEE Transactions on Power Delivery*, 21(3), 1593–1599.

8

Line-Commutated CSC Transmission

8.1 Introduction

As explained in Chapter 3, CSC is the more practical solution for use with line-commutated conversion (LCC) and has, therefore, been the configuration adopted in conventional (thyristor-based) HVDC transmission. The capacity installed worldwide has already reached 75 GW in 92 projects, at voltage levels of up to ±600 kV, and new schemes continue to be added.

DC transmission results in lower losses than equivalent AC lines, but because of the converter plant required at the terminals, it is only viable when the distance involved is long and the amount of power to be transferred large. However, substantial progress made in the ratings and characteristics of the thyristor valve has reduced the so-called break-even distance and increased the competitiveness of DC schemes. Moreover, HVDC links, with or without transmission distance, can be justified for the interconnection of AC systems of different frequencies or different frequency control philosophies. A detailed discussion of the place of HVDC in modern power systems is given in [1] and [2].

With the development of reliable and more flexible self-commutating conversion, CSC-LCC technology is encountering increasing competition from PWM-VSC transmission (discussed in Chapter 10) for low and medium power ratings. Despite its relatively low flexibility in respect to the control of reactive power, the thyristor-based line-commutated CSC transmission is still the only practical option for very large-power transmission schemes.

The main recent developments in LCC transmission are taking place in India and China, which are becoming the world's larger users of HVDC. In China the first stage of the Three Gorges scheme, a 1060 km, 3000 MW link completed in 2004, includes the largest HVDC converter (1500 MW, 500 kV). The final HVDC capacity of this project, expected to be achieved by 2009, is 18 200 MW.

Flexible Power Transmission: The HVDC Options J. Arrillaga, Y.H. Liu and N.R. Watson

8.2 The Line-Commutated HVDC Converter

The main electrical components of each pole of the HVDC converter station are shown in the circuit diagram of Figure 8.1 [3]. All the components enclosed within the thick rectangle are located inside the valve building.

The switching components on the AC side of the converters, i.e. isolators and circuit breakers, are of conventional design. The switches on the DC side consist of minimum oil circuit breakers, capable of interrupting small currents for the switching of the neutral bus load and for the changeover from single pole metallic return to bipolar operation. Also, to limit inrush currents and overvoltages during transformer energisation, the converter breakers are provided with pre-insertion resistors.

The measuring equipment, including a voltage divider, current measuring transducers and current transformers, provides the necessary input signals for the control and protection circuits.

A typical layout for a 1000 MW bipolar HVDC station [4], shown in Figure 8.2, illustrates that the major proportion of the space is taken up by the external plant components and particularly the capacitors used as part of the harmonic filters and for voltage support.

Figure 8.3 shows a three-dimensional view of the valve hall. The transformers (on the right of the picture) and the smoothing reactor (on the left) are placed close to the hall walls with their bushings passing through the wall. The floor area of the valves, services and control rooms is only a small fraction of the total station area. The auxiliary power equipment used for cooling and air-conditioning is placed immediately under the valve hall. The building normally contains a steel structure designed to act as a Faraday cage to reduce electromagnetic radiation from the valve hall which might cause radio interference.

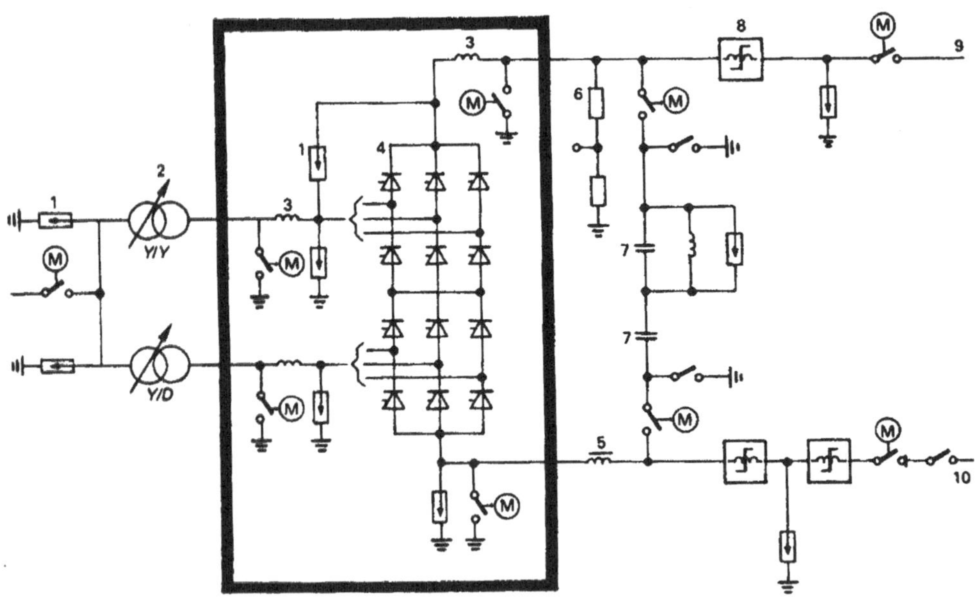

Figure 8.1 Typical circuit diagram for one pole of a converter station (Arrillaga, J. (1983), *HVDC Transmission*, IEE Power Engineering series 6, reproduced by permission of the IET.)

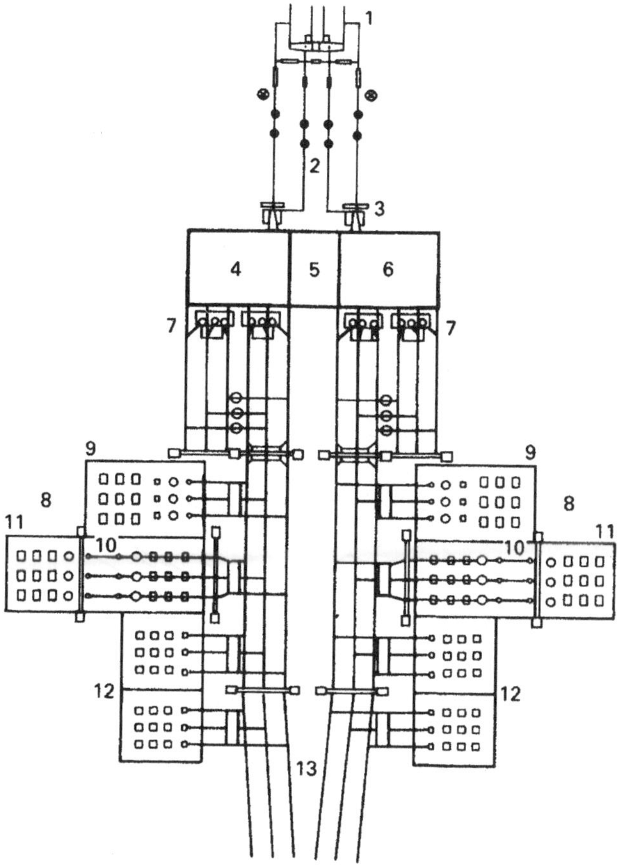

Figure 8.2 Station layout for a bipolar HVDC station: 1, DC and electrode lines; 2, DC switchyard; 3, DC smoothing reactors; 4, valve hall, pole 1; 5, service building with control room; 6, valve hall, pole 2; 7, converter transformers; 8, AC harmonic filters; 9, high-pass filter; 10, 11th harmonic filter; 11, 13th harmonic filter; 12, shunt capacitors; 13, AC switchyard (Arrillaga, J. (1983), *HVDC Transmission*, IEE Power Engineering series 6, reproduced by permission of the IET.)

The high-voltage thyristor valves

The application of semiconductor technology to HVDC transmission has been made possible by the development of reliable series-connected structures of individual devices.

A large number of thyristors in series are required to provide the complete valve with the necessary voltage rating. Series connection of thyristors requires components to be added to the valve to distribute the off-state voltage uniformly between the individual units. Thus each thyristor is served by several passive components, not only to ensure that this voltage sharing is achieved, but also to protect individual thyristors from overvoltage, excessive rate or rise of voltage (*dv/dt*) and rate of rise of inrush current (*di/dt*). The thyristor switch, together with its local voltage grading and triggering circuit, is known as a thyristor level.

The electrical circuit of a thyristor level is shown in Figure 8.4 and forms the basic 'building block' of a valve.

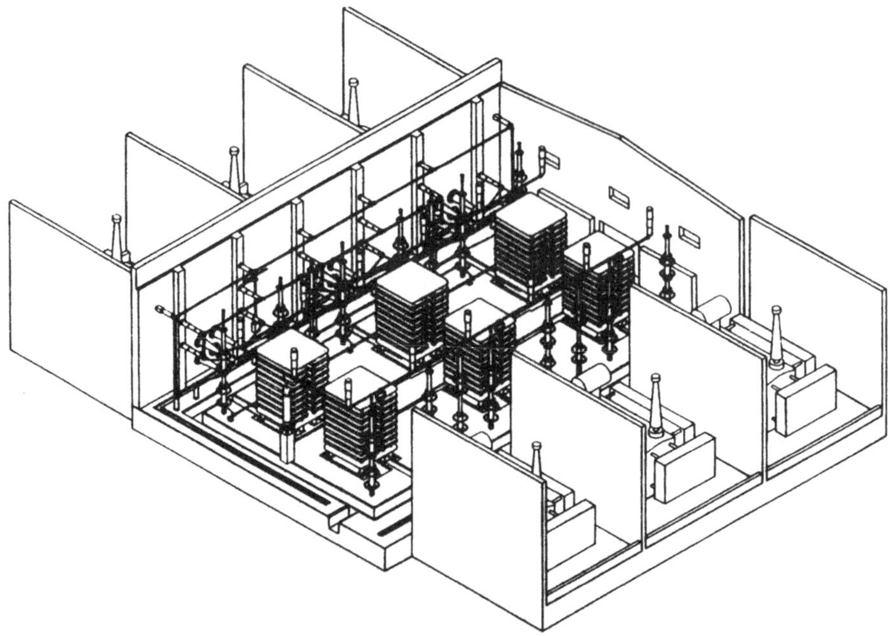

Figure 8.3 Cutaway view of the valve hall showing the locations of the major components (Arrillaga, J. (1998), *HVDC Transmission*, second edition, IEE Power and Energy series 29, reproduced by permission of the IET.)

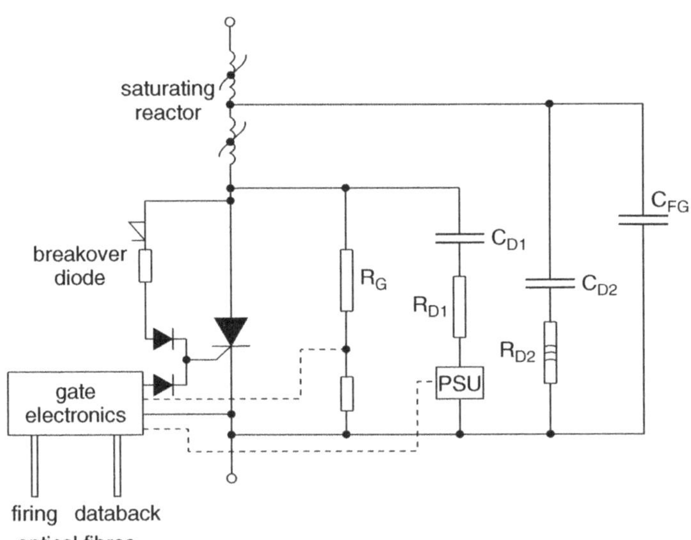

Figure 8.4 Components of a thyristor level (Arrillaga, J. (1998), *HVDC Transmission*, second edition, IEE Power and Energy series 29, reproduced by permission of the IET.)

The function of the saturating reactor is to present a large inductance in series with the stray capacitances of the external circuit. This is necessary to protect the thyristors from damage immediately after firing. However, too much inductance is itself undesirable, because it increases the reactive power absorbed by the converter. The saturating reactor avoids this problem by exhibiting high inductance only at low current. At full load current the effect of the reactor becomes negligible.

Voltage distribution is achieved by several components, acting over different frequency ranges. Direct voltage is distributed by a 'DC grading resistor' (R_G). Voltage distribution in the range from power frequency up to a few kilohertz is controlled by a complementary pair of *RC* grading circuits (R_D and C_D). This frequency range includes the natural frequency which characterises the voltage oscillation encountered when valves turn off at the end of a conduction interval. The component values are chosen to minimise the magnitude of this voltage overshoot.

Insulation failures within the converter can subject the valve to voltage oscillations of much higher frequency, at which these *RC* circuits cease to be effective. To ensure that no thyristor level experiences an unduly severe voltage during such events, a capacitive grading circuit is also included. This 'fast-grading capacitor' (C_{FG}) is arranged to discharge via part of the saturating reactor in order to limit its contribution to thyristor inrush current.

Normally the thyristor is triggered into conduction at a particular 'point on wave' determined by the control system. The command to fire the valve is sent as an optical signal from a 'valve base electronics' cubicle at earth potential to every thyristor in the valve, via individual optical fibres. The optical signals are decoded by a 'gate electronics' unit located adjacent to each thyristor, which then generates a pulse of current to trigger the thyristor. The gate electronics derives the (small) power necessary for its operation from the displacement current in the *RC* grading circuit during the off-state interval.

Thyristors could be damaged by excessive forward voltage or forward *dv/dt*. However, they can be protected by arranging for the gate electronics to trigger the thyristor into conduction independently of the central control system.

In marginal cases, some thyristors may block forward voltage while others do not. In the limit, this could result in the whole of the valve-winding voltage being applied to a single thyristor, so that the thyristor would be destroyed if it were not protected. However, this is prevented by a back-up triggering system, which consists of a series string of small overvoltage-triggered thyristors, connected from the anode to the gate of the main thyristor via a current limiting resistor. When the forward voltage across the main thyristor threatens to exceed the maximum safe value, the semiconductor elements of the break-over diode conduct, and pass a heavy pulse of current to the gate of the main thyristor, triggering it rapidly into conduction. If the thyristor-level components are suitably rated, the break-over diode can operate repetitively in the event of a failure of the gate electronics, thereby preventing consequential failure of the thyristor.

By connecting a suitable number of thyristor levels in series, a valve of the necessary voltage rating can be constructed. The number of series-connected thyristors required to meet the operating conditions for a particular scheme is determined by the protective level of the arrester and a test withstand margin (typically 15 %). Extra redundant levels are included to allow the equipment to remain in service after a small number of thyristor failures.

The power thyristors used in the valves are constructed from monocrystalline silicon wafers over 100 mm in diameter. Thyristors of this type require a very high clamping pressure to maintain adequate thermal and electrical contact between the silicon and the electrical connections and heat sinks. Removal of a thyristor is accomplished by using a hydraulic tool to separate the two heat sinks which are in contact with the particular thyristor while retaining sufficient clamping force on the other thyristor to ensure that its pressure contacts are not compromised. A particular advantage of this system is that it is not necessary to open either the power circuit or the cooling circuit in order to replace a thyristor.

The thyristors produce considerable heat loss, typically 30 to 40 W/cm^2 (or over 1 MW for a typical quadruple valve), and an efficient cooling system is thus essential. Each thyristor unit is normally provided with a double heat sink and the heat is taken away from the sinks by circulating water. High-purity water combines superb cooling with high dielectric strength. The purity of the water must be very high to remove all ionic components, which would otherwise cause the coolant to bypass the electrical insulation of the valve. The liquid coolant is distributed to every thyristor level in the valve through electrically insulating polyethylene hoses. Water cooling systems, however, require careful design to prevent leakage (which would have disastrous consequences) and corrosion.

Feedback signals make it possible to monitor the state of each individual thyristor. Microcomputers are used in the control room to process the information from the valve. A faulty thyristor is immediately detected and the exact position of the defective thyristor is reported. Since each valve contains a somewhat larger number of thyristors than are actually needed, the converter can continue to operate, even if some thyristors are defective; these are only replaced during the planned regular maintenance.

The auxiliary power needed for the thyristor firing is obtained from the voltage across the thyristor.

Self-protected light-triggered thyristors are now available and their use reduces the number of valve components, the valve-tier volume and the valve cost. It also improves the thyristor-level reliability. The electrically triggered thyristors require a local electronics unit for the generation of trigger pulses as well as protection and monitoring. All signal communication within and to the valves across potential differences is performed using light pulses transmitted by light guides (fibre optics). This applies to both the firing signals for the individual thyristors in the valve and the feedback signals from each thyristor level to the valve control equipment.

Converter transformers

The converter transformers are mostly of conventional design. The standard 12-pulse converter configuration can be obtained with any of the following arrangements:

- six single phase two-winding;
- three single phase three-winding;
- two three-phase two-winding.

Star or delta connections are exclusively used for the above configurations. In conventional transformers the insulation gap between winding and yoke is relatively small, as the winding part with potentials close to ground are situated very close to the yoke. In the case of a converter transformer this is not possible, since the potentials of its connections are determined by the combination of conducting valves at any particular instant, and the entire winding must be fully insulated. As a result of the insulation the radial leakage flux at the ends of the windings increases.

As the leakage flux of a converter transformer contains very large harmonic content, it produces greater eddy-current loss and hot spots in the transformer tank.

Apart from the normal AC insulation requirement, the converter transformer is subjected to a direct voltage depending on its position with respect to the ground.

When a short-circuit fault occurs across a valve arm, the fault current is predominantly limited by the converter transformer impedance, whose minimum value is determined by the allowable surge on-state current of the thyristor units. Thus the converter transformer leakage impedances are normally larger than those of conventional transformers. On-load tap-changing (OLTC) is normally used to reduce the steady-state demand of reactive power. The tap-changer is the most critical mechanical device of the HVDC station, the number of operations being much greater than those of OLTC used in conventional AC systems.

Other important considerations are the shipping weight and dimensional limitations, particularly in modern converter stations where the valve size is no longer a limitation.

Harmonic filters

The justification for the use of harmonic filters has already been made in Chapter 3. In the converter layout of Figure 8.2 there are two sets of harmonic filters on the AC side; they consist of tuned branches for the 11th and 13th harmonics and a high-pass branch, typically tuned to the 24th harmonic. Extra shunt capacitors are often needed, when the AC system cannot provide the necessary additional reactive power more economically. Generally, a high-pass DC filter tuned to the 12th harmonic is also placed on the DC side of the converter.

The design of the AC filters is a complex exercise that must take into consideration the harmonic impedances of the AC system under all possible operating conditions as well as the detuning effects of frequency excursions and filter component variations. The subject is discussed in some detail in [1] and a more comprehensive study of the subject can be found in [5].

A new concept referred to as ConTune AC filter [6] has been developed to improve the quality factor for best filtering performance and reduce the filter losses. In this case the tuning frequency is automatically adjusted to provide perfect tuning irrespective of the varying system and filter component conditions. This development is described in more detail in Chapter 9.

Smoothing reactors

While the current waveform on the DC side improves with increasing inductance, the control response slows down and the resonance frequency reduces, making the stabilisation of current control more difficult.

The main purpose of the smoothing reactor is to reduce the rate of rise of DC following disturbances on either side of the converter. This in turn reduces the number of commutation failures following AC voltage reductions and limits the current peak seen by the rectifying station during DC line short circuits. The second task of the reactor is to reduce the levels of voltage and current harmonics on the DC line and the transfer of non-harmonic frequencies between the two interconnected AC systems.

In back-to-back interconnections the absence of DC short circuits and the lack of interference from DC side harmonics permit a reduction in the reactor's size and even its elimination, but the other effects indicated above still need to be considered when deciding the reactor's size. A judicious choice of reactor inductance will shift the DC side series resonant frequency away from the low-order harmonic. This is particularly important to avoid a fundamental frequency resonant condition, which has been one of the causes of converter transformer saturation, as explained in Section 3.11.

To be effective, the reactor inductance must maintain its value under short-circuit currents and therefore a partial or total air-core construction is used to achieve a linear magnetising characteristic.

Although the reactor cost reduces if located on the ground side of the converter, there is a need to protect the converter from lightning surges; this requires some inductance on the high-voltage side. Thus in general the location of a single reactor on the line side proves to be a more economical solution.

8.3 HVDC Converter Disturbances

According to the origin of the malfunction, converter disturbances can be divided into three broad groups:

1. Malfunction of the valves or their associated equipment. The main types are: misfire, firethrough and backfire (only in mercury arc valves).

2. Commutation failure, the most common disturbance during inverter operation. This fault often follows other internal or external disturbances.

3. Short circuits within the converter station. Although these faults are rare, they must be taken into consideration in converter design.

Misfire and firethrough

Misfire is the failure to fire a valve during a scheduled conducting period and firethrough is the failure to block a valve during a scheduled non-conducting period. These faults are caused by various malfunctions in the control and firing equipment and their effect is more critical when they occur at the inverter end. With rectifier operation they do not constitute a serious disturbance unless they are sustained, in which case they can introduce voltage and current oscillations on the DC side.

Commutation failure

This fault is the result of a failure of the incoming valve to take over the DC before the commutating voltage reverses its polarity, taking into account the need for sufficient extinction time. A true commutation failure is due to varying conditions in the external

AC or DC circuits combined with inadequate predictive control of the inverter extinction angle. Either a low alternating voltage or high DC, or both, can prevent completion of the commutation process in sufficient time for safe commutation; in such cases the DC current is shifted back from the incoming valve to the previously conducting valve.

Moreover, due to the DC rise, caused by the temporary DC voltage collapse at the inverter end, the subsequent commutation may also be unsuccessful, thus causing a double successive commutation failure. In this case the inverter output voltage V_d reverses for nearly half a cycle.

As a result of the DC short circuit at the inverter end, the transformer is either partially or totally bypassed and the DC line current exceeds the current in the AC lines. This effect has been used to detect the occurrence of commutation failures.

In practice, following the detection of a commutation failure, the next firing instant is advanced by the constant extinction angle control. If the failure is caused by low alternating voltage following an AC disturbance, upon clearance of the disturbance the normal voltage will return and prevent further commutation failures. However, in the event of recurring commutation failures the valve group should be blocked. This action is often combined with bridge or valve group bypass in the case of a multi-group converter station.

The probability of commutation failure can be reduced by increasing the minimum extinction angle allowed in normal operation. This, however, increases the VAR compensation required and a compromise is reached where a reasonably low probability of commutation failure is acceptable. As a guideline, the minimum margin angle is fixed to avoid commutation failures during voltage reductions of up to 15 %.

Internal short circuit

Although rare, short circuits can occur at various locations of the converter station. They can be caused by maloperation of earthing switches, deteriorating insulators or surge arrester failures, particularly during transient overvoltages.

A flashover across a non-conducting valve produces a phase-to-phase short circuit with a very large overcurrent on the conducting valve.

The largest stress is produced during rectification with a small firing delay, and the worst instant is immediately after a commutation, in which case the valve current is limited only by the transformer leakage reactance and the system source impedance.

Bypass action

Many of the valve faults are of a temporary nature and can be eliminated by a temporary absence of conduction. This is achieved by a bypass scheme, such that one of the main bridge arms provides the necessary bypass for the DC. Blocking of a converter through bypass pairs involves blocking of the main firing pulses and the simultaneous injection of continuous firing pulses to a bypass pair. Resumption of normal operation simply demands the restoration of firing pulses with suppression of the blocking pulses.

8.4 Structure of the HVDC Link

Modern DC schemes are designed exclusively for 12-pulse operation. A single line diagram of a typical HVDC link [3] is illustrated in Figure 8.5. The scheme includes

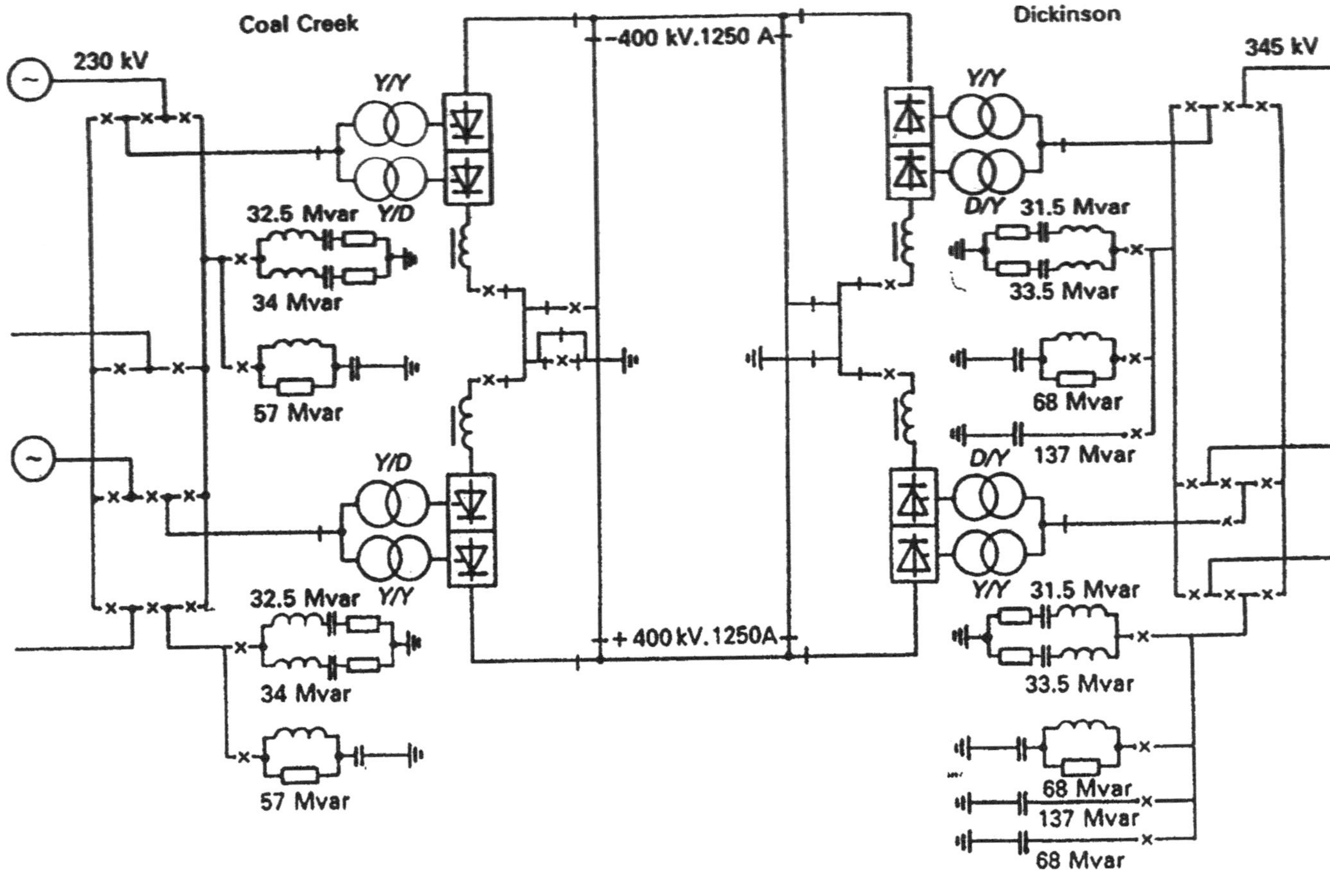

Figure 8.5 Single line diagram of a typical HVDC scheme [3] (Arrillaga, J.(1983), *HVDC Transmission*, IEE Power Engineering series 6, reproduced by permission of the IET.)

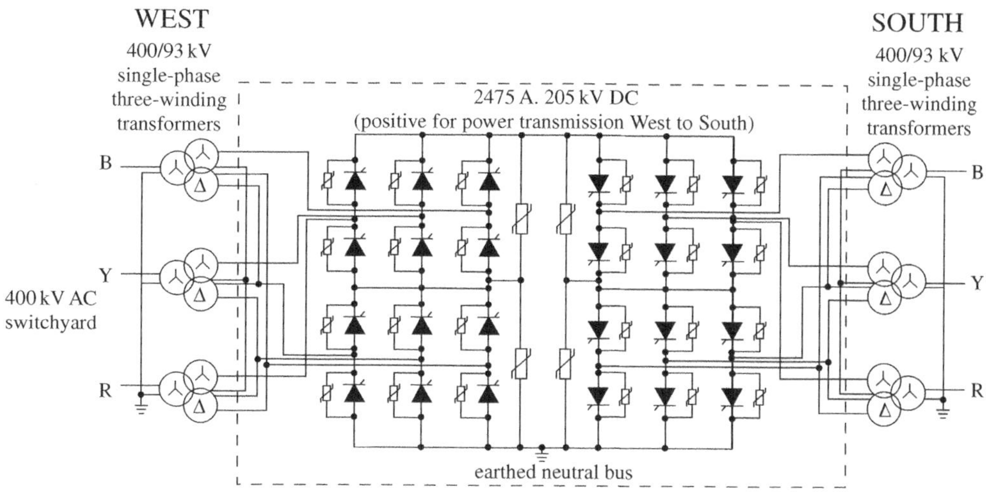

Figure 8.6 Schematic arrangement of converter equipment for one pole (Arrillaga, J. (1998), 'HVDC Transmission', second edition, IEE Power and Energy series 29, reproduced by permission of the IET.)

two valve groups at each end, each valve group consisting of two series-connected six-pulse bridges supplied from two converter transformers. The transformers are connected in star–star and star–delta, respectively, to provide the necessary 30° phase shift for 12-pulse operation.

A three-phase diagram of one pole of a DC link is shown in Figure 8.6, giving details of the converter and valve overvoltage protection. Besides the extensive protective measures incorporated at the individual thyristor level, it is necessary to provide overall valve protection against reverse overvoltage. Figure 8.6 shows that this is accomplished by connecting a gapless metal–oxide surge arrester across the valve. This surge arrester constitutes the primary protection of the valve against overvoltages of external origin. The arrester has to withstand continuous operation while subjected to the valve off-state voltage, including the periodic switching transients occurring every cycle. The arrester provides a protective level typically around 70 % higher than the peak of the normal operating voltage of the valve.

Overhead lines

The first consideration in DC transmission design is the possibility of using the ground as a conductor. Whenever this is possible, a monopolar transmission scheme will provide a very economical solution. In general, however, the use of ground as a permanent conductor is rarely permitted and a bipolar arrangement is used with equal currents in the two conductors. Normally, if one of the conductors fails, the temporary use of ground return is permitted.

The basic principles determining the dimensioning of overhead lines and hence the towers are very much the same for AC and DC transmission. Therefore the tower designs for DC transmission are very similar to those of AC lines.

However, the power carrying capability of DC lines is substantially increased with respect to that of AC and this reduces the land use and visual effect considerably. This effect is illustrated in Figure 8.7 for the 3000 MW transmission corridor of the Three Gorge transmission scheme in China.

Figure 8.7 Artist's view of the Three Gorges transmission corridor using either DC or AC lines (Reproduced by permission of ABB.)

For DC lines the number of insulators required is usually determined by the DC voltage withstand level under contaminated conditions rather than on switching surge or impulse flashover performance unless the pollution levels are very low.

The need for ground wires is less justified in DC lines, because (with bipolar lines) only half of the power is affected by a line fault, and the fault current can be kept under control without the need for mechanical switching. However, ground wires are often used because they relieve some of the harmonic currents flowing through the ground and thus reduce telephone interference.

The choice of conductors depends largely on corona and field-effect considerations. Those effects, however, are not expected to be a special problem for DC voltages up to ±1200 kV. The critical design parameters with respect to corona are radio interference and audible noise, although the corona losses will obviously play a part in the economic choice of conductor. The existence of ion currents under the lines is a characteristic of HVDC transmission, but the magnitude of touch potentials (from an insulated person touching a ground object) appears to be considerably lower under DC lines than those which occur under corresponding AC voltage lines.

Cable transmission

In contrast to overhead line insulators, in which the breakdown can occur through flashover on the outside, cable breakdown occurs due to puncture through the insulation and this is where DC has a major advantage over AC. Mainly due to the absence of ionic motion in the DC cable insulation, the working stress of oil-impregnated paper-insulated cables can be as high as 30 to 40 kV/mm under DC compared with 10 kV/mm under AC, but the great majority of those in operation are in the 25 to 30 kV range, which still provides a substantial advantage over AC cable insulation. Due to thermal limitations, the power rating of AC cables only increases approximately in proportion to the voltage, while the charging current increases with the distance and with the square of the voltage. Consequently, unlike overhead transmissions, AC transmission by cable uses relatively low voltages, except for very short distances (mainly on submarine crossings). Also, considerable currents circulate in the sheaths and reinforcing materials, which increase the thermal losses; these losses are often reduced by cross-bonding of cable sheaths, an expensive solution. Forced cooling of the conductors is thus necessary, even at moderate power levels, and cooling stations are also needed along the route.

Moreover, solid insulated DC cables can be manufactured for much higher voltages, with considerable reduction in the number of cables needed for a given power rating; this results in less repair problems and way-leave requirements.

Apart from the reduction in losses, resulting from the absence of any appreciable current in the sheaths and reinforcing materials, DC cables are subjected to less overcurrent stresses.

Limitations in DC cable ratings have often introduced restrictions on converter station configurations. For instance, the 2000 MW Cross-Channel HVDC link between England and France employs two bipolar links rather than a multi-bridge configuration. The scheme uses eight DC cables rated at 270 kV and 2000 A, buried in four trenches dug into the sea bed to avoid damage by anchors.

As the number of HVDC projects using submarine cables increases, more economical designs (requiring increased operating voltages) are being developed, because the cost of DC cables as a proportion of the total cost of the schemes is high. The transient overvoltage stresses, although of major concern, can be controlled with overvoltage limiting devices.

There are three types of paper-insulated DC cables, oil-filled, gas-pressurised and mass-impregnated, often referred to as solid cables. XLPE is a fourth type, at the time of writing only used in one installation, at 150 kV DC.

The great bulk of cables is of the mass-impregnated type, in operation at voltages up to 450 kV DC. This design of cable is usually cheaper to manufacture than either oil-filled/pressurised or gas-pressurised cable and is not restricted in length by the need for intermediate pressure/feeding stations. A cross-section of the cable used in the New Zealand DC link upgrade (rated 350 kV and 500 MW) is shown in Figure 8.8.

Oil-pressurised and gas-pressurised cables can withstand higher insulation working stresses than mass-impregnated cables, of the order of up to 50 % higher. Whereas oil-filled cables have generally been limited in length to a few tens of kilometres, a flat-sided, oval-shaped, oil-filled cable 120 km long has been developed for 400 kV DC, across land (without intermediate pressure/feeding stations) and installed in a commercial project (Kontek). In this design of cable, two insulated cores (operated in parallel) are arranged within a common lead sheath and outer reinforcing tapes and insulation, which forms the flat-sided oval construction. This

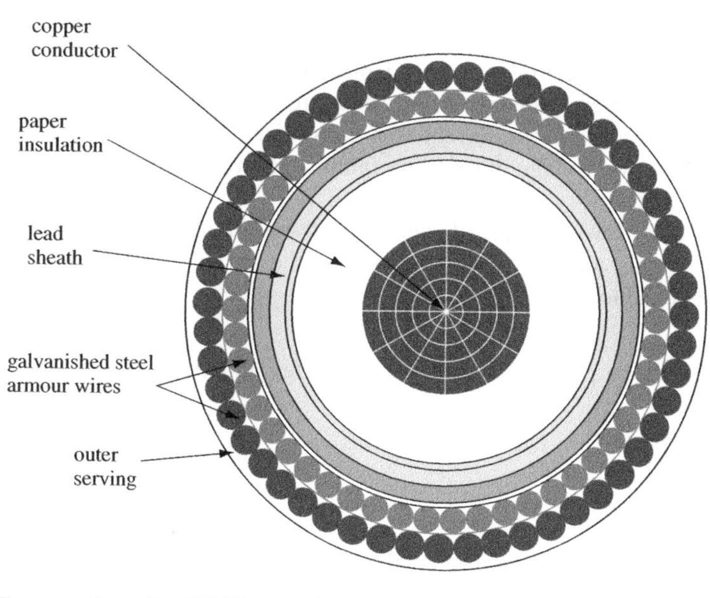

Figure 8.8 Cross-section of a 350 kV mass-impregnated cable (Arrillaga, J. (1998), *HVDC Transmission*, second edition, IEE Power and Energy series 29, reproduced by permission of the IET.)

allows volumetric expansion of the cable and limits the internal pressure rise to a fraction of that of single core round-cable design.

While the overall length of present DC cable connections spans from around 30 km to 250 km, there is no known limitation for lengths above 250 km. A contract has been recently placed with ABB for a 580 km submarine interconnection between Norway and The Netherlands (NorNed) with a rating of 700 MW at ±450 kV. The maximum sea depth is currently 500 m, but tests have been made on cables for 2000 m depth; apart from the heavy armour required for deep cables, no other limitations are presently known in this respect. On the other hand, in shallow waters, cables must be protected against mechanical damage. The cost advantage of mass-impregnated, paper-insulated cables and their applicability to submarine installations at great depth makes them very suitable for these applications.

Earth electrodes

The resistivity of the upper earth layer is typically of the order of 4000 Ω m and, therefore, the electrodes cannot be placed directly in contact with earth of such high resistance. They are not placed at the converter station to prevent some current entering the converter transformers via the primary neutral causing DC saturation. Other technical reasons influencing the siting of the electrodes are:

- the possibility of DC ripple interfering with power systems, telephone systems, railways, etc.;
- metallic corrosion caused by DC in equipment in contact with earth, such as cable sheaths and pipes.

Consequently the earth electrodes must be placed in an area of sufficient thickness and conductivity and sufficiently distant from urban areas, pipelines, etc. An electrode line is normally laid to carry any earth current to a convenient distant point with low resistivity. To reduce transient overvoltage during line faults a separate right of way is normally used for the electrode line.

With electrodes of low resistance located in low-resistivity earth layers resulting in low DC electric potentials and potential gradients at the surface of the earth, the following advantages can be achieved:

- the electrode can be placed closer to the converter station;
- shorter electrode line;
- reduced power losses;
- reduced interference in the vicinity of the electrode;
- further opportunities to use monopolar HVDC transmission.

Kimbark [7] and Uhlmann [8] have given considerable coverage to the experience gathered on the subject of earth resistivity and current distribution in the ground. Such information has been used in the design of electrodes.

8.5 DC System Configurations

The simplest HVDC scheme, shown in Figure 8.9(a), is a monopolar configuration with ground return. It consists of a single conductor connecting one or more 12-pulse converter units in series or parallel at each end and uses either sea or earth return. An electrode is required at each end of the line. Because of magnetic interference and corrosion problems, ground return is rarely permitted and metallic return is used instead. Both configurations

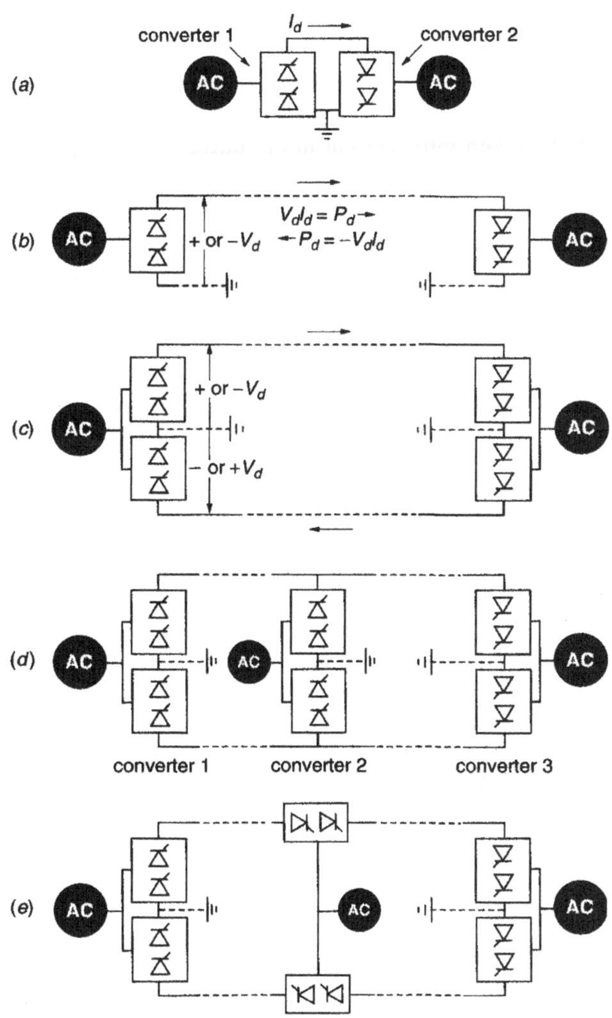

Figure 8.9 Basic HVDC configurations (Arrillaga, J. (1998), *HVDC Transmission*, second edition, IEE Power and Energy series 29, reproduced by permission of the IET.)

require a DC smoothing reactor at each end of the HVDC line, usually located on the high-voltage side and, if the line is overhead, DC filters are normally required.

A bipolar HVDC system (shown in Figure 8.9(b)) consists of two 12-pulse converter units in series with electrode lines and ground electrodes at each end and two conductors, one with positive and the other with negative polarity to ground for power flow in one direction. For bidirectional flow, the two conductors reverse their polarities. With both poles in operation, the imbalance current flow in the ground path can be held to a very low value.

The extension of HVDC technology to multi-terminal configurations has been discussed for over four decades, the interest peaking in the 1980s. However, only one fully multi-terminal scheme was constructed for commercial operation. Its object was to convert the Hydro-Quebec–New England link (commissioned in 1986) into a five-terminal scheme with the addition of three further terminals. However, the original two-terminal link (between Des Cantons and Comerford) was never integrated into the multi-terminal DC network because of anticipated performance problems. No other (higher than three) multi-terminal scheme has been considered since then using CSC. Accepting that three terminals is the only practical multi-terminal option, when at the planning stage a third terminal extension is to be made in the future, this must be taken into account in the initial cost comparison between the AC and DC alternatives, because the tapping is equivalent to reducing the transmission distance, which makes the DC solution less competitive.

Normally the power rating of a proposed third-terminal tap is relatively small compared with the main transmission link and, therefore, the high transmission voltages of the bipolar interconnection will require expensive converter equipment for the parallel connected third terminal (shown in Figure 8.9(c)). To try and reduce the cost, consideration has also been given to the use of series tapping of the additional terminal (as shown in Figure 8.9(d)). There have been many contributions on the series tapping concept to show that it is technically feasible, but none has been built so far.

Back-to-back interconnections

The power handling capacity of a thyristor increases with increasing current. For the largest sizes presently used in HVDC schemes the use of water cooling permits higher current ratings. However, in two-terminal transmission schemes, the rated voltage and current values are mainly determined by the transmission line, which means that it may often not be possible to utilise fully the capacity of modern large thyristors.

In a back-to-back link there are no such external requirements and it is therefore possible to utilise the thyristors optimally, which means a high current and a low voltage. This will minimise the number of thyristors, and thereby also the valve cost. Other voltage-dependent costs will also be reduced, for instance those of the building and the transformers. As an example, a 500 MW back-to-back unit will, for DC of 3000 A, require a direct voltage of only 167 kV.

In these systems, the optimisation of the whole station regarding voltage and current ratings for a given power in order to achieve the lowest life cycle cost is a straightforward procedure. Generally, the DC voltage is low and the thyristor valve current high in comparison with HVDC long-distance interconnections. The main reason is that, on the one hand, valve costs are much more voltage dependent as the higher voltage increases the number of thyristors and, on the other, the highest possible current adds very little extra cost to the price of an individual thyristor.

The circuitry of a back-to-back link does not differ very much from that of two separate terminals. The main difference is that the transmission line is eliminated and, consequently, some DC equipment can be avoided or shared by the rectifier and the inverter.

The quadri-valve 12-pulse group is suitable also for a back-to-back station. With the back-to-back circuitry the two valve halls can be combined into one, with the DC loop maintained inside the hall and with transformers on both sides of the building.

Considering the cost–benefit of the back-to-back solution, zero-distance interconnections are often preferred when planning HVDC transmission between two asynchronous systems.

North America and India have already made extensive use of the back-to-back solution, as shown in Figure 8.10.

For relatively small ratings (say 50–100 MW) the back-to-back links are of monopolar design and usually include a smoothing reactor. Bipolar configurations with or without smoothing reactors are used for large interconnections (of 500 MW and over).

The valves of both converters of the back-to-back scheme can be located in the same valve hall together with their controls, cooling and other auxiliary services.

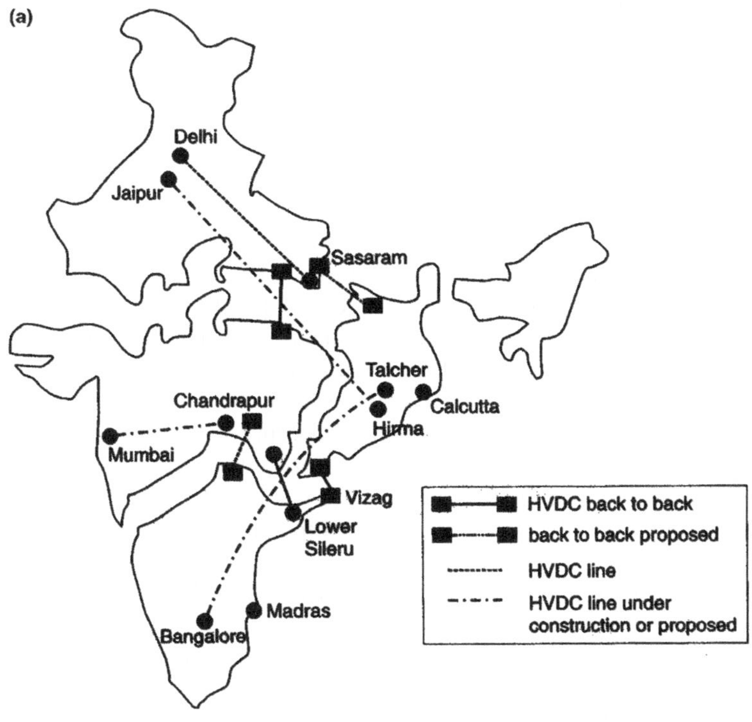

Figure 8.10 LCC HVDC link interconnections in India (a) and North America (b) (Arrillaga, J. (1998), *HVDC Transmission*, second edition, IEE Power and Energy series 29, reproduced by permission of the IET.)

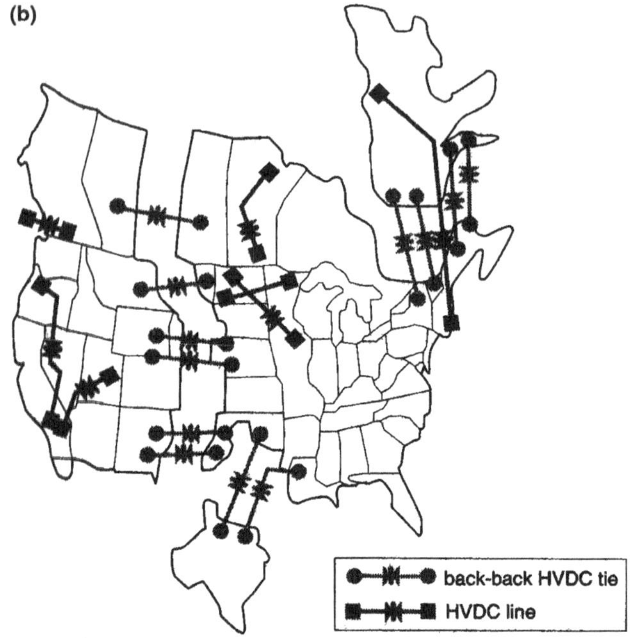

Figure 8.10 (Continued)

8.6 DC System Control and Operation

8.6.1 General Philosophy

From an operational point of view, the use of constant current control provides greater safeguard against disturbances. Therefore, if there is a choice and the economics are right, constant current should be used as the basic control philosophy. This, however, is not possible with conventional AC power systems which, being normally multi-ended, require reasonably constant voltages at points of common coupling.

Present DC schemes consist of point-to-point system interconnections and constancy of DC voltage levels is not a primary consideration because these are not directly available to consumers. There are, however, other considerations influencing the control philosophy, among them featuring prominently the overvoltages resulting from open-circuiting and load rejections and the high resistive losses resulting from constant current transmission at low power levels.

A hybrid voltage/current philosophy is possible with CSC-HVDC transmission to suit the needs of the particular operating conditions. This is achieved by adjusting the DC voltage levels on both sides of the link, by means of OLTC control on the steady state, and by thyristor control following large or small changes of operating conditions at either end of the link. The DC is limited only by the small resistance of the transmission line and is therefore very sensitive to such variations.

It will be shown in the following sections that the provision of current controllers at both ends, combined with transformer OLTC, offers a perfectly satisfactory solution to this problem; thus the use of current control is universally accepted in LCC HVDC transmission.

8.6.2 Different Control Levels

The control system of a DC converter station has a structured hierarchy with three layers designated as overall station, pole and converter unit controls:

1. *Overall station control:* This layer coordinates the power orders and distributes them to the pole control. These orders are generated in response to the required power transfer and other supplementary controls, such as system frequency control, system damping or combinations of them. The overall control also looks after the switching of harmonic filters and shunt capacitors as required by changing operating conditions.

2. *Pole controls:* The pole control layer derives the firing order of the pole converters following a power or a DC voltage order.

3. *Converter unit controls:* This layer is used to control the firing instants of the valves within a bridge and to define the γ_0 and α_{min} limits.

8.6.3 Overall Control Coordination

The overall control coordination of the Nelson River project [9], illustrated in the simplified diagram of Figure 8.11, is an ideal case to describe the degree of dynamic interaction which can be achieved in a DC transmission system:

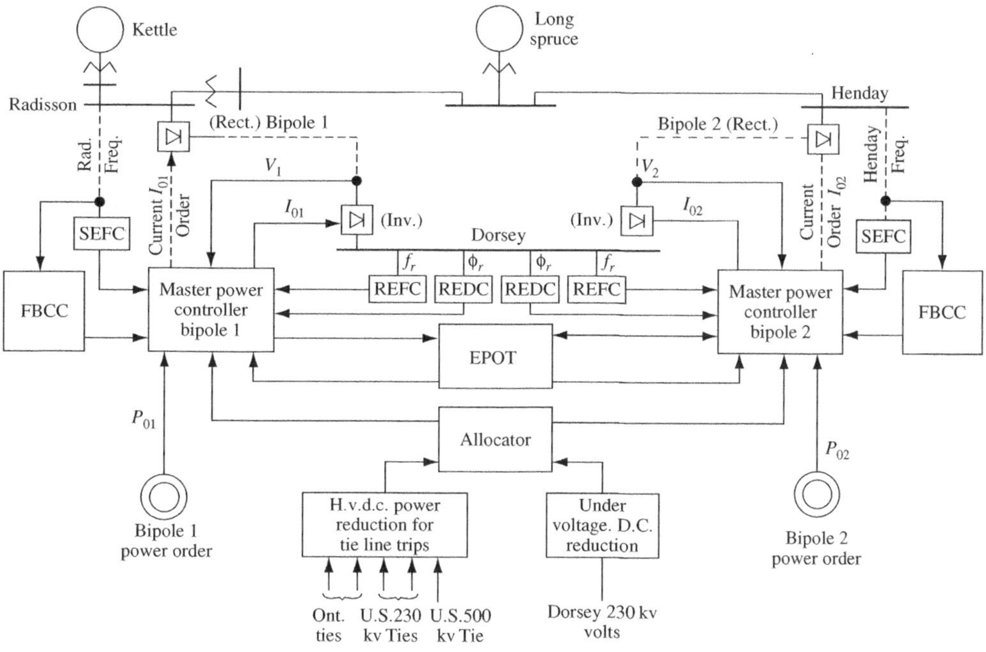

Figure 8.11 A simplified schematic of controls on the Nelson River HVDC link (Arrillaga, J. (1983), *HVDC Transmission*, IEE Power Engineering series 6, reproduced by permission of the IET.)

- The master power controller is designed to accept a power order (P_{01}) from the operator or the automatic generation control. This power order is divided by the measured bipole voltage (V_1) to determine the current order (I_{01}).

- A frequency-based capability control (FBCC) provides protection against HVDC collector system overloading. DC power is reduced if the collector system frequency drops below 59 Hz.

- An excess power order transfer (EPOT) controller is designed to utilise fully the combined available HVDC system capability. The EPOT controller transfers ordered power, which is in excess of the capability of one bipole, to the other operating bipole. This controller is activated only when a major outage occurs on the bipole, such as a valve group or pole block.

- The HVDC power reduction control for tie-line trips is designed to ensure system stability upon loss of interconnection lines. The DC power is reduced by an amount equal to tie line loading prior to tripping.

- An undervoltage DC reduction controller is used to reduce DC power whenever Dorsey voltage starts to collapse. A fixed amount of DC reduction releases MVARs both from the HVDC link and the AC system to restore the voltage. This control does not react to faults.

- The allocator accepts DC reductions from the tie-line reduction and undervoltage reduction controllers. The total DC reduction is then summed and allocated to each bipole according to a preset power order.

- A sending end frequency control (SEFC) is used to minimise oscillations in the collector system.

- The receiving end frequency control (REFC) minimises oscillations in the receiving system.

- Finally, the receiving end damping control (REDC) operates to prevent changes in the angle of the Dorsey 230 kV bus voltage.

The bipole control coordinates the power orders and distributes them to the pole controls. As described above, this is supplemented by inputs such as FC (Frequency Control) and PM (Power Modulation) to take into account the actual operating conditions.

Telecommunication requirements [10]

Without special arrangements to coordinate the operation at both ends of the link, a reduction of rectifier (or an increase of inverter) current order by more than the current margin, relative to the inverter (or rectifier) current order, can cause a complete DC voltage shutdown. This problem is normally avoided by the following procedure:

1. When the master controller operates from the sending (i.e. rectifying) end, an 'increase' in current order is implemented immediately at the local end, and as soon as possible (only subject to the unavoidable telecommunication delay) at the remote end. On the other hand, a 'decrease' in current order updates the remote current order first, while the local current order change waits for a 'true' check-back signal from the remote end, via a return telecommunication channel indicating the receipt of an error-free signal.

2. When the master controller is housed at the inverter end, the above procedure still applies but with the words 'increase' and 'decrease' interchanged.

3. If an error is detected in either telecommunication receiver then the local current order is not updated; the return check-back is sent as a zero.

The main purpose of the above system is to ensure that the sequence of current order updating is never such as to decrease the effective current margin, even in the presence of errors in the telecommunication systems. The telecommunication means used in earlier HVDC schemes include microwave radio, carrier on the power conductor's private wire, rented wire and the use of the public telephone system. All of them are liable to interference and even occasional failures. Carrier systems, using the HVDC line or cable, are affected by a continuous source of interference, i.e. the converters, which is difficult to filter because of the high impedance of the smoothing reactors; the use of the lowest possible bandwidth is thus recommended to reduce the effective noise. To avoid noise the analogue signals are converted to digital form and transmitted with an error checking signal.

Recent schemes make extensive use of fibre-optic cables, which are immune from interference. The New Zealand scheme, for instance, contains a high-capacity fully redundant communication link plus a physically independent back-up link. It is used for control, protection, data, status functions and speech facilities, between Haywards and Benmore.

8.6.4 Pole Controls

Characteristics and direction of DC power flow

Most DC schemes in existence are provided with bidirectional power flow capability. This property is inherent in the case of AC transmission, where the direction of power flow is determined by the sign of the phase angle difference of the voltages at the two ends of the line; the power flow direction is in fact independent of the actual voltage magnitudes.

On the other hand, in DC transmission the power flow direction is dictated by the relative voltage magnitudes at the converter terminals and the absolute or relative phase of the AC voltages plays no part in the process. However, if required, this condition could be altered by exercising a type of firing angle control which can make the power flow direction independent of the terminal AC voltage magnitudes and behave, instead, like the AC counterpart.

The basic characteristic of a converter from full rectification to full inversion, illustrated in Figure 8.12, is provided with constant current and constant extinction angle controls.

The level of the natural voltage characteristic can be adjusted by the transformer OLTC. This part of the characteristic takes place when $\alpha = 0$ (i.e. with diode operation); that is, the

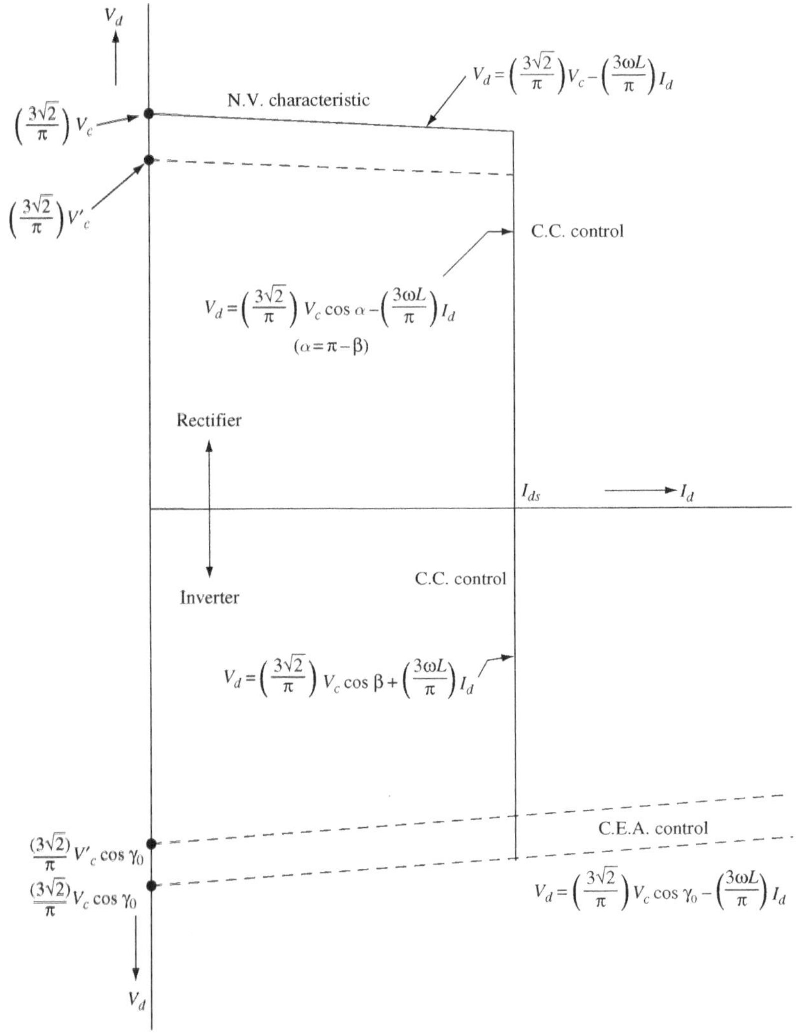

Figure 8.12 Complete control of a converter, from inversion to rectification N.V., natural voltage; C.C., constant current; C.E.A. constant extinction angle (Arrillaga, J. (1983), *HVDC Transmission*, IEE Power Engineering series 6, reproduced by permission of the IET.)

converter has no controllability in this region and the DC voltage (from Equation (3.7) with $\alpha = 0$) is

$$V_d = V_{c0} - \frac{3X_c}{\pi} I_d \tag{8.1}$$

For delay angles larger than $\alpha = 0$ the converter exercises constant current control, i.e. to maintain a current setting (I_{ds}), that has a practically vertical characteristic. This region is governed by Equation (3.7) and is limited by the need to maintain a certain minimum extinction angle required for safe commutation, as explained in Section 3.5. When this limit is reached, the converter again loses controllability and is governed by Equation (3.14).

Let us now consider the specific control parameter which implements the direction of power flow. We have already explained that each converter station is normally provided with current and extinction angle control facilities. The assignment of current control, either to the rectifier or to the inverter station, is made on consideration of the investment cost for reactive power compensation, the availability of reactive power, the minimisation of the losses and the total running cost. Normally the total reactive power compensation is least, and the utilisation of the line best, if the rectifier is assigned the current control task while the inverter operates on minimum extinction angle control.

This combination is achieved in a two-terminal DC link by providing the power sending station with a slightly higher current setting than the power receiving station. The difference between the two settings is termed the current margin (I_{dm}). Its effect can be better understood with reference to Figure 8.13(a), where the operating current is set by the constant current control at the rectifier end. The inverter end current controller then detects an operating current which is greater than its setting and tries to reduce it by raising its own voltage, until it hits the ceiling determined by the minimum extinction angle control at point A. This is the normal steady-state operating point, which presumes a higher natural voltage characteristic at the rectifier end, a condition which may require OLTC action at the converter transformer.

The no-load and direct voltage regulation along the transmission link are illustrated by the solid line in the diagram of Figure 8.14, with power flowing from left to right and with the operating current maintained by the rectifier controller.

Starting from the rectifier end, point a represents the maximum average direct voltage, $(V_{c0})_R$ with the rectifier unloaded. This is reduced by firing angle control to $(V_{c0})_R \cos(\alpha)$ (point b). The slope of lines bcA and $-d$A is determined by the DC current I_d, and their horizontal projections relate to the commutation reactances X_{CR} (for bc), $X_{C\mathrm{I}}$ (for dA) and line resistance R_d (for cA). Therefore point c represents the output voltage at the rectifier station and point A the voltage at the remote end of the line. This point is also reached from the inverter end open-circuit voltage $(V_{c0})_\mathrm{I}$ reduced by the extinction angle $(V_{c0})_\mathrm{I} \cos(\gamma)$ and by the commutation reactance. It should be obvious that point A represents the same operating condition as indicated by the crossing point in the characteristics of Figure 8.13(a).

Let us now assume that there has been a substantial AC voltage reduction at the rectifier end, such that the DC voltage ceiling (the natural voltage) of the rectifier becomes lower than that of the inverter. In the absence of a current controller at the inverter, the voltage

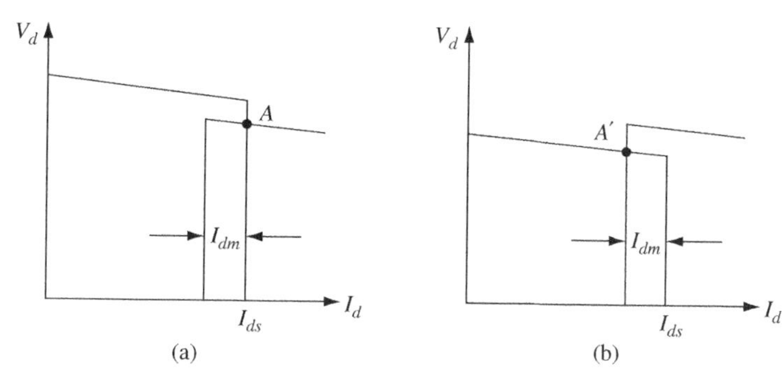

Figure 8.13 Steady-state characteristic and operating point: (a) under rectifier constant current control; (b) under inverter constant current control

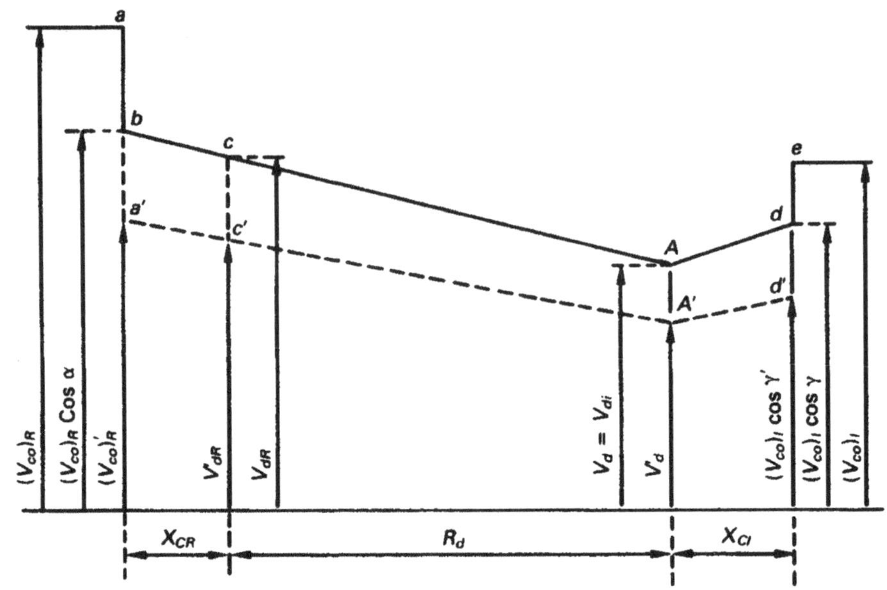

Figure 8.14 DC voltage profile (Arrillaga, J. (1983), *HVDC Transmission*, IEE Power Engineering series 6, reproduced by permission of the IET.)

across the line is reversed and the current reduces to zero (current through the valves cannot reverse). However, an inverter current controller will prevent a current reduction below its setting by advancing its firing (i.e. reducing α and hence inverter DC voltage), thus changing from extinction angle to constant current control. A new operating point A′ (Figure 8.13(b)) results at a current reduced by the current margin. In Figure 8.14 this condition is represented by the dashed lines and clearly shows that power flow will continue in the same direction, in spite of the lower AC voltage at the sending end.

The block diagram in Figure 8.15 shows the main components of a typical pole control system

Tap-changer control

Referring to Figure 8.13(a), the task of the tap-changer on the rectifier side is to place the voltage ceiling (i.e. the $\alpha = 0$ characteristic) relative to the operating point (A), so as to minimise the reactive power consumption subject to a minimum α limit. This limit is used to ensure that all the anodes (in the case of the mercury arc valve) or the series thyristors have a sufficient positive voltage across them to avoid the risk of misfiring.

On the inverter side the reactive power is minimised at the $\gamma = \gamma_0$ limit, which is the normal operating condition. It is therefore possible to use its tap-changer to keep the direct voltage within any desired limits.

When the current control is transferred to the inverter (as in Figure 8.13(b)), the rectifier tap-changer tries to raise the voltage ceiling and the tap-changer action on the inverter side must stop, since the latter no longer determines the voltage level. Otherwise the tap-changer would try and raise the inverter side AC voltage and thus reduce the power factor.

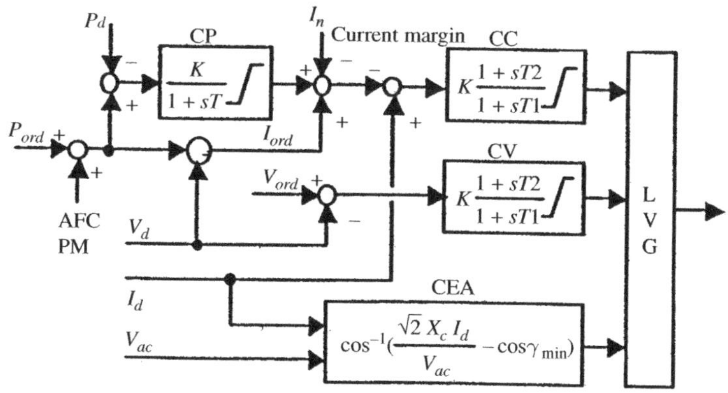

Pord: Power order
Iord: Current order
Vord: Voltage order
CP: Constant power
CV: Constant voltage
PM: Power modulation
CC: Constant current
CEA: Constant extinction angle

Figure 8.15 Block diagram of a typical pole control

It is interesting to note that the DC link, within its normal power rating, operates as a constant voltage circuit, i.e. the voltage is at its highest value under no-load conditions and power is raised by increasing the current settings. However, in order to avoid overloads, when the current limit is reached, the transmission link acts as a constant current system, i.e. a power change can then only be achieved by means of a voltage change and, because of the voltage ceilings, this is only possible to a limited extent by means of tap-changer action.

Reversal of power flow

The last section has explained that the direction of power flow is decided purely by control action. It can be expected, therefore, that a change in power direction cannot happen naturally as a result of a change in the operating conditions, but rather following a control order resulting from the overall requirements of the power system. The characteristics of Figure 8.13(a) can be extended below the voltage zero line so that the rectifier and inverter ends exchange their function. The result of this action, illustrated in Figure 8.16(a), shows that the two characteristics do not meet again. Station I increases the delay angle well into the inverting region and hits the limiting extinction angle voltage (γ_0). Station II advances its firing into the rectifying region and finally hits the rectifier voltage ceiling ($\alpha = 0$). In the process, the line current reduces to zero and the complete system is blocked.

Since, as indicated in the previous section, the rectifying station requires the higher current setting, it is thus necessary to subtract the current margin from the reference value of station I. This results in the characteristics of Figure 8.16(b), which displays one operating point of different voltage polarity and with the roles of the two stations interchanged. Thus the direction of power flow has been reversed without altering the direction of current flow, which of course is fixed by the converter valves.

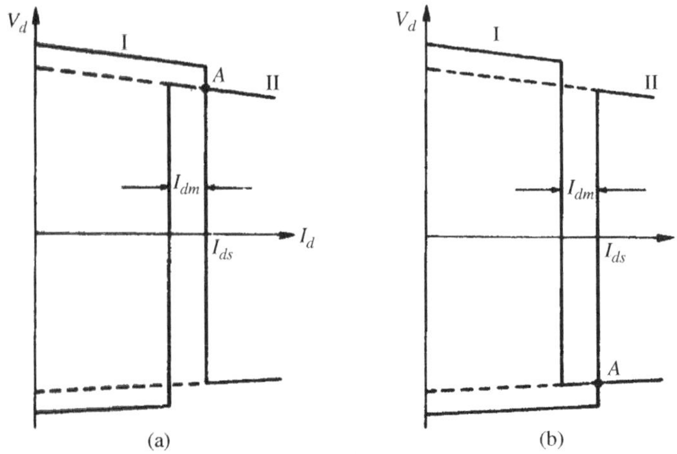

Figure 8.16 (a) Operating point with power flowing from station I to station II. (b) Operating point after power reversal

Modifications to the basic characteristics [11]

During AC system faults at the receiving end there is a big risk of commutation failure and, if the fault is electrically close, the inverter may not be capable of recovering by itself. In such cases it is important to reduce the stress on the inverter valves, and this is achieved by providing a low-voltage-dependent current limit to the rectifier control characteristic. The modified characteristics illustrated in Figure 8.17 consist of a branch CD′ at the rectifier and another EF′ at the inverter.

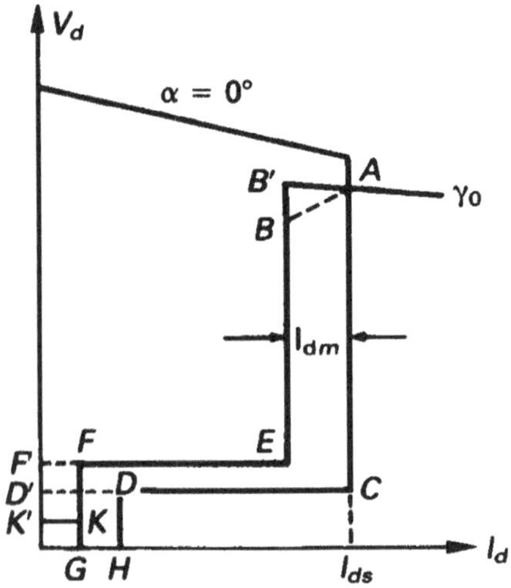

Figure 8.17 Rectifier and inverter modified characteristics

The break points C and E in Figure 8.17 are typically between 70% and 30% of DC voltage, and in special cases even higher depending on AC system requirements. The higher voltage is applied when the receiving AC network is sensitive to disturbances which could cause large voltage fluctuations.

The terms VDCOL (Voltage-Dependent Current Order Limit) and LVCL (Low-Voltage Current Limit) are commonly used for the function that reduces the current setting when the voltage decreases.

In the literature the added branches CD′ and EF′ are normally given finite slopes on the basis that the finite slope is expected to provide ‘smooth’ control; in fact this argument is not valid, since the slope constitutes a strong negative resistance which produces the opposite effect. A plain horizontal line, as shown in Figure 8.17, is better since the system will then operate stably down to a lower DC voltage during faults in the rectifier AC system; it will also recover faster from a total collapse of an inverter because the AC line/cable will charge more rapidly.

Also, an unwanted voltage reversal of the inverter is generally prevented by limiting α to some minimal value larger than 90° (shown by KK′ of the characteristic). Branches DH and FG represent maximum current limits at low voltage. A minimum current limit (not shown in the figure) is also normally provided to prevent operation with discontinuous current.

Another problem with the basic characteristics occurs when the rectifier voltage ceiling gets very close to that of the inverter, such that the characteristics intersect in the region between I_{ds} and $I_{ds} - I_{dm}$. In this region both current controllers are out of action. In practice the current will not settle at an intermediate point; instead, the inverter will be periodically entering the current controlling mode. A proven countermeasure against this type of oscillation is a small shape alteration of the inverter characteristic, as shown in Figure 8.17. The limiting $\gamma = \gamma_0$ characteristic cannot be changed without sacrificing heavily the reactive power; it is, however, sufficient to break the characteristic to a positive resistance slope in the transfer region from γ-control to current control of the inverter (AB in Figure 8.17 instead of AB′).

Operation at non-minimum margin angle [12]

In this case the margin angle γ is varied around an average value in order to stabilise the AC voltage. This can be achieved either by direct control of the AC voltage or indirectly by controlling the DC voltage. Another way of stabilising the receiving system AC voltage is to use the inverter, rather than the rectifier, as the current controlling station. These modes of control are normally used for operation in the ‘unstable’ region of the AC voltage/DC power characteristic. An alternative solution is shown in Figure 8.18, in which the operating point is moved down along the part of the characteristic with positive slope. The inverter is then able to increase the voltage at increased DC. This method is very effective when the inverter is connected to a weak AC network. The normal operating point A corresponds to a value of γ higher than the minimum.

Active power control

As the primary object of HVDC transmission, active power control should be the main consideration. However, for simplicity the early schemes still used the current control loop

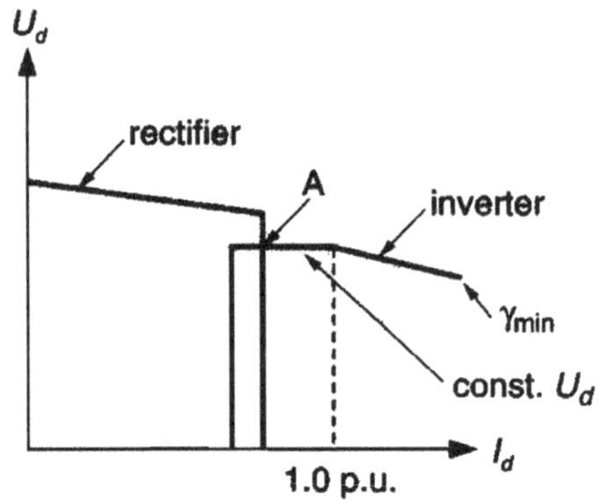

Figure 8.18 Alternative inverter characteristic with higher than minimum margin angle

as a basis. In such a case each station is provided with a dividing circuit, consisting of a power calculator and a high-gain operational amplifier, which receives the power order, voltage and current signals and produces the current order. Other input components of the station power control unit, illustrated in Figure 8.15, include the constant power order setting and any additional power modulation signals. A control error is formed from comparison of the calculated current order (minus the current margin in the case of the inverter station) and the measured current. Figure 8.15 also shows the control functions used in DC voltage and constant extinction angle at the inverter.

In most cases there is only one master controller (at one of the stations) which sends a current order to the pole controls of the two ends of the link. The power is monitored by multiplying voltage and current (summed from both poles) and fed back directly to the controller. As in previous control methods, to prevent unacceptable current orders (e.g. during startup) limits are normally built in.

Frequency control

The frequency of a network interconnected to a larger one by an HVDC transmission link can be controlled by means of a frequency feedback loop acting on the DC link controls, such that the small network draws the required power change from the larger one.

A typical case of frequency control is the Gotland link, as originally designed, which included a synchronous compensator. This network, once started, could be operated with no other power feed than the DC link. Similarly, when the power rating of the DC line is comparable with or greater than the rating of the running generators in the AC system to which the line is connected, the line terminal can share in the frequency regulation or even perform it unaided.

Power/frequency control

Often a combination of control modes is used, such that the control signal applied to the current controller is normally power, and it does so as long as the frequency remains within the predetermined limits. Outside these limits, frequency control takes over to assist the

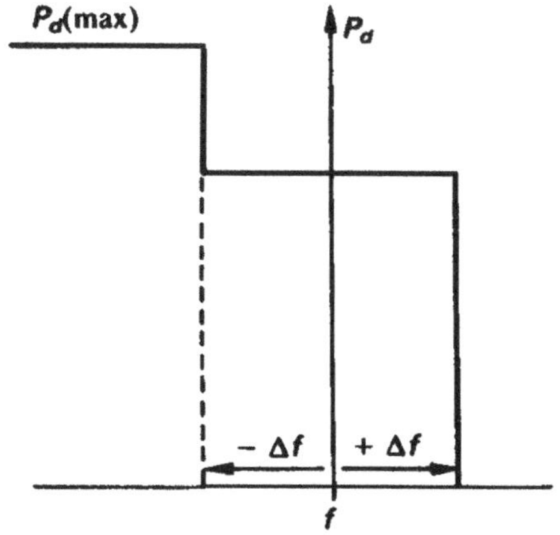

Figure 8.19 Power/frequency control characteristic

system in difficulty. However, when the maximum rated transmission power is reached, the frequency controller becomes inoperative. This combination is illustrated in Figure 8.19.

It should be remembered that the DC link is insensitive to frequency variations, unless some sensitivity is deliberately added to the control system. Without it, a constant power flow can speed up a receiving system which has lost part of its load and a sending system can eventually collapse if the required DC link power is more than the connected generation can produce. In general, therefore, the incorporation of an element of frequency control should be included.

Response to changing operating conditions

The link normally operates under constant power control as explained above. The response to power setting changes, or unforeseen variations in the terminal AC voltages, occurs in two stages. The first stage is very fast (limited only by the size of the smoothing reactors) since it is performed purely by firing angle control, according to the characteristics shown in previous sections. However, this process will increase the converter firing angles and with them the consumption of reactive power, which will in turn upset the terminal voltages.

In the second stage, the firing angles will be returned to their minimum settings by means of coordinated transformer OLTC control at both ends of the link; this process, requiring mechanical switchings, is very slow (in seconds) and therefore some voltage oscillations are unavoidable. The latter will also occur during large power transfer variations, which will require ON and OFF switchings of some filters and shunt capacitor compensation.

8.6.5 Converter Unit Controls

For the implementation of the firing signal of the bridge, most existing HVDC converters use the 'phase-locked oscillator' [13] control principle, illustrated in Figure 8.20. It consists of a voltage-controlled oscillator which delivers a train of pulses at a frequency directly proportional to a DC control voltage V_c. Various versions of the original phase-locked oscillator control system have subsequently appeared.

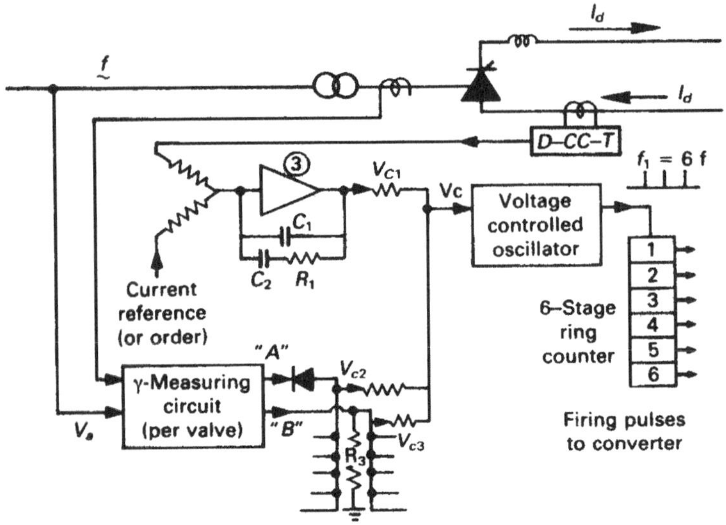

Figure 8.20 The phase-locked oscillator control system (Arrillaga, J. (1983), *HVDC Transmission*, IEE Power Engineering series 6, reproduced by permission of the IET.)

The train of pulses is fed to a six-stage ring counter in which only one stage is on at a time; the on stage is stepped cyclically from positions 1 to 6 by the oscillator pulses. As each ring-counter stage turns on, it produces a short pulse at the output (once per cycle). Therefore the complete set of six output pulses normally occur as successive intervals of 60°. The STOP pulses are also obtained from the ring counter but two stages later (e.g. the START pulse for valve 1 is from stage 1 and the STOP pulse for valve 1 is from stage 3, normally 120° later). One oscillator and one ring counter per bridge constitute the basic control hardware.

The various control modes only differ in the types of control loop which provides the oscillator control voltage V_c.

The phase of each firing pulse will have some arbitrary value relative to the AC line voltage, i.e. an arbitrary value of converter firing angle α. However, when the three-phase AC line voltages are symmetrical fundamental sine waves, α is the same for each valve.

For the normal working range, equidistant firing control is preferable to the individual phase control used in early schemes because of the reduction of abnormal harmonics. As with any control system, its firing pulse spacing may be modulated by any non-characteristic harmonics entering via feedbacks, e.g. of DC current. Thus its description as 'equidistant firing control' is somewhat anomalous; however the phase-locked oscillator control system is very stable when suitably designed, and this is not usually an important effect.

It is often claimed that, whenever the DC power contributes considerably to maintaining the network stability, such a system may benefit from a changeover to individual phase control under asymmetrical network fault conditions. However, under asymmetries the power transmitted is limited by the current (with individual control) or the voltage (with equidistant control) and therefore the above conclusion must be restricted to asymmetries which require no current limiting. For large asymmetries, with the power decreasing to very small levels (with either control), the reduction of direct voltage during the fault is less important; in such cases the use of equidistant firing control provides more reliable commutations and facilitates the rapid return of power flow when the asymmetry ends.

In practice the simple independent oscillator would drift in frequency and phase relative to the AC system; hence some method of phase locking the oscillator to the AC system is required. This is normally achieved by connecting V_c in a conventional negative feedback loop for constant current or constant extinction angle.

Constant current loop

With reference to Figure 8.20, let us first consider the constant current loop, i.e. only signal V_{c1} being effective. This voltage is obtained from the amplified difference (error) between the current reference and the measured DC line current; this forms a simple negative feedback control loop, tending to hold current constant at a value very close to the reference.

To visualise the operation of this loop, imagine that the current is nearly equal to the reference, such that the amplified error (V_c) happens to be precisely that value required to give an oscillator frequency of six times the supply frequency. The ring-counter outputs, and thus the valve-gate pulses, will have a certain phase with respect to the AC system voltage. Suppose further that this phase, which is identical to firing angle α, happens to be such as to give the correct converter DC output voltage, which, with the particular back emf of the DC link, results in the correct DC line current. This is steady-state operation.

The loop is self-correcting against disturbances of any source. For instance, a drop of back emf in the DC system causes a temporary current increase, which reduces V_c and hence slows down the oscillator, thus retarding its phase and finally increasing the firing angle α. This tends to decrease the current again, and the system settles down to the same current, with the same V_c and oscillator frequency but a different phase, i.e. different α.

The control system will also follow system frequency variations, in which case the oscillator has to change its frequency; this results in different V_c and hence current, but the current error is made small by using high-gain amplification.

This constant current scheme is the main control mode during rectification; it is also used during inversion whenever the inverter has to take over the current control.

The control system response is fast but, in practice, its effect will be slowed down by the relative slower response of the DC line which includes capacitance, inductance and smoothing reactance.

Inverter extinction angle

This control mode is implemented by a negative feedback loop very similar to the current loop and is also shown in Figure 8.20. The difference between the measured γ and the γ-setting is amplified and provides V_c as before. However, it differs in that γ is a sampled quantity rather than a continuous quantity. For each valve the extinction angle is defined as the time difference between the instant of current zero and the instant when the anode voltage next crosses zero, going positive. Typical waveforms of the γ-measuring technique are shown in Figure 8.21.

For each bridge there are six values of γ to be measured, which under symmetrical steady-state operation are identical. Under unbalanced conditions, however, the valve with greatest risk of commutation failure is the one having the smallest γ; this smallest measured

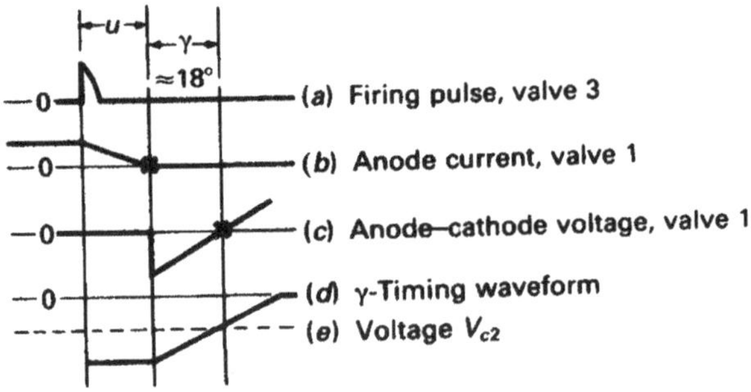

Figure 8.21 Waveforms of γ-control circuit (Arrillaga, J. (1983), *HVDC Transmission*, IEE Power Engineering series 6, reproduced by permission of the IET.)

γ produces the most negative output and thus causes its diode (signal 'A' in Figure 8.20) to conduct and produce the negative feedback voltage V_{c2}. During steady state operation and full inversion, V_{c2} controls the oscillator holding the smallest γ at a predetermined value by closed-loop control.

Under these conditions V_{c1} is zero because the inverter constant current (CC) setting is less than the DC line current (determined by the rectifier CC control); hence the inverter CC loop is trying to decrease α by making V_{c1} as low as possible. The minimum V_{c1} is clamped to zero volts and thus during normal inverter operation the CC loop is ineffective.

Component 'B' is an additional feedback voltage (V_{c3}) applied during the transient conditions when $\gamma < \gamma_{min}$. A sudden impulse is then applied to the voltage-controlled oscillator, which has an integrating characteristic and thus can suddenly shift the phase (i.e. angle α) by an appropriate amount.

Transition from extinction angle to current control

The current setting for inverter operation is below that of the current in the DC line, which is normally kept at a higher constant level by the sending end terminal. Under such conditions V_{c1} in Figure 8.20 is zero, the current control amplifier is saturated and the converter operates on constant extinction angle γ at full inversion.

In the event of a sudden AC system voltage rise at the inverter end, or a DC line voltage reduction, the direct current will decrease; the current amplifier then comes out of saturation and V_{c1} becomes positive. This additional input to the oscillator causes operation of a larger γ, i.e. advances the firings and the converter takes over current control.

Application to 12-pulse converter groups

When bridges are connected in series, the common DC is determined by one controller per pole and the control pulses are directed to the individual bridges from a common ring counter. With 12-pulse converter groups it is thus necessary to double the number of control impulses per cycle. In this case the ring counter provides the control pulses to the 12 different valves of the two bridges, keeping four valves activated simultaneously.

8.7 AC–DC System Interaction

A DC link can be operated according to the basic control modes described in the last section and thus remain passive to any special needs of the interconnected AC systems. Alternatively the link can be provided with more adaptive dynamic controls that respond to critical deviations from the expected operating conditions in the AC or DC systems.

The exclusive use of the basic controls often gives rise to unwanted interaction between the AC and DC systems, which is manifested in a variety of voltage, harmonic and power instabilities. When full advantage is taken of the fast and adaptable converter controllability a more useful interaction can be achieved which manifests itself in stable AC and DC system operation.

AC – DC system interactions are concerned with voltage stability, overvoltages, resonances and recovery from disturbances:

- Voltage stability criteria are used to determine the type of voltage control and of reactive power supply.

- The level of temporary overvoltage (TOV) influences station design, including thyristor valve and surge arrester ratings. The TOV levels will increase with decreasing values of the short-circuit ratio (SCR), the ratio of the AC system MVA fault at the converter terminals to the nominal DC power.

- Shunt capacitors are used in converter stations as part of the AC filters and VAR compensation. The larger the ratio of shunt capacitor MVAR to AC system short-circuit MVA, the lower will be the resonant frequency.

- Recoveries from AC and DC faults are slower with weak systems (i.e. high-impedance source), although modern controls are less affected by the system impedance than those used in earlier schemes.

8.7.1 Voltage Interaction

The AC terminal voltages at the converter stations depend on the active and reactive power characteristics of the converter. To minimise voltage variations it is essential to control the reactive power supply to match the converter reactive power demand.

It has already been explained that LCC consumes reactive power, which can be typically 60 % of the power transmitted at full load. Moreover, during transients the reactive power demand may vary widely, the duration of such variation depending to a large extent on the characteristics of the DC link control system.

If the presence of local generators (i.e. close to a rectifier) can be guaranteed, it is always more economical to supply most reactive power from these, with minimum size filters to reduce harmonics.

In general, however, it is necessary to provide full reactive power compensation to the converter and, sometimes, extra compensation for the AC system loads as well. A large proportion of the reactive power is supplied by the shunt harmonic filters, and additional compensation has so far been provided in the form of shunt capacitor banks or synchronous compensators. The filters and capacitors are normally connected to the high-voltage terminal, whereas the synchronous compensators are connected through a transformer tertiary winding. Other permutations have also been used, e.g. filters on a tertiary winding and synchronous compensators on separate transformers.

The size of the individual filters is the result of a compromise between economy (which demands the larger size) and the ability of the AC system to accept the step changes in system voltage caused by filter switchings. Switching is often needed to control reactive power at DC loads lower than the nominal. There is a further difficulty in discrete filter switching, which relates to the non-linear relationship between the increase of reactive power and harmonic current with load.

If the extra cost can be justified, the use of dynamic compensation, in the form of either synchronous compensators or controllable static compensators, can reduce or eliminate the step switching of filter branches.

8.7.2 Dynamic Voltage Regulation

During HVDC link disturbances the voltage control requirements depend on the nature and location of the disturbance. The reactive power consumption, although possibly higher initially, is partially or totally eliminated following the disturbance, with the result of considerable dynamic overvoltage regulation.

The regulation (dynamic) overvoltages are more significant at the rectifier end of the link. At both terminals the effective impedance angle is as important in determining the overvoltages as is the magnitude of the impedance. For links from hydro sources, the increase of frequency following load rejection will produce even higher dynamic overvoltages. This is an unacceptable situation for local consumers and must be allowed for in the insulation coordination of the converter station.

In practice, transformers start to saturate at typically 1.2 to 1.25 pu AC voltage and the fundamental frequency overvoltage will therefore be a little lower, with some distortion.

Single line-to-ground faults are also a source of dynamic overvoltages on the other phases or pole, due to mutual coupling between phases.

Following a voltage drop in the AC network, the initial effect is a fall in power. The power controller of the DC link then increases the current reference to try and restore the ordered power; the extra current increases the reactive demand and tends to reduce the AC system voltage further. With very weak AC systems this could lead to voltage collapse; however, power controllers always have limits built in to avoid excessive action.

By far the most important case is that of a nearby three-phase short circuit, assuming that the converters are blocked during the fault, with all the capacitors left on. This condition produces full magnetising inrush current on all transformers after fault removal, which results in substantial fundamental and harmonic overvoltages. Such overvoltages constitute in practice the determining condition for most valve, surge arresters and insulation voltage ratings.

Dynamic compensation equipment is used to reduce the dynamic voltage regulation, to help in the recovery of the AC system from faults, and to reduce the disturbances resulting from DC load variation or from the switching of filter banks. Ideally, to meet such a comprehensive range of duties, the compensation will generally require both reactive power absorption and generation capabilities. In each particular application the system SCR and the stability of the DC link and compensator controls must be considered, when trying to decide the type and dynamic range to be used.

A preliminary comparison of the technical characteristics of reactive power compensators for use with HVDC transmission, with an extensive bibliography on the subject, has been presented in a CIGRE document [14].

The traditional types of dynamic compensation for HVDC schemes are: synchronous compensator, AC self-saturated reactors, thyristor-controlled reactors and thyristor-switched capacitors. A more flexible type, gradually finding its way into the market, is the STATCOM (an electronic version of the rotating synchronous compensator).

The slow response of the synchronous compensators can be a problem, particularly in the absence of local generation. However, the synchronous compensator reduces the sensitivity to transients by increasing the SCR. It also increases the resonance frequency of the system, since it reduces the need for extra shunt capacitance.

It is often claimed that the static alternatives improve the voltage stability of the AC network and thus help HVDC control stability and speed of response. In practice the static VAR compensators rarely improve voltage stability, except that they act as limiters if the voltage tries to rise too much. In steady state the thyristor-controlled reactors may cause some destabilisation, though this can be made acceptable; saturated reactors appear to be better in this respect.

8.7.3 Dynamic Stabilisation of AC Systems

A power system is stable if after a disturbance it returns to a condition of equilibrium. This is manifested not by the constancy of absolute rotational speed of the various machines involved, but rather by these machines swinging together until a new common speed is reached. The power exchanged between them is determined by their relative angular position and therefore, when the equilibrium is disturbed, their rotor positions must give rise to corrective power flow leading to the new state of equilibrium.

If the angle between the machines increases steadily the system is transiently unstable. If the machines fall out of step after a period of increasing oscillations around the equilibrium point the system is dynamically unstable. Dynamic instability is rare in tightly connected systems, which are usually well damped for their characteristic frequencies of the electromechanical swing (between 1 and 2 Hz).

However, when large systems are connected by long, relatively weak interties, low-frequency swing modes result. The response of the power system controls to the synchronising swings associated with these low-frequency modes can produce sufficient negative damping to cancel the natural positive damping of the system. When this happens, oscillations of increasing amplitude occur.

An example of dynamic instability [15] is the northern and southern parts of the Western US power system, which are connected by the parallel Pacific AC and DC Interties with ratings of 2500 and 1400 MW, respectively. The AC Intertie has a long history of negatively damped 1/3 Hz oscillations resulting from interactions between generators with automatic voltage regulators and system loads. As a result of these oscillations, and because the oscillatory tendency imposed a constraint on the amount of surplus northwest hydro power which could be transmitted to the southwest, a control system to modulate the Pacific DC Intertie was developed.

Damping in the Pacific Intertie was produced by small signal modulation of the DC power in proportion to the frequency difference across the AC Intertie. This was accomplished by processing the AC Intertie power measured at the northern end, to obtain a filtered signal proportional to the derivative of AC power at frequencies near 1/3 Hz. This signal is applied, through a $\pm 3\,\%$ (± 40 MW) limiter, to the current regulator at the northern terminal of the DC Intertie; thus the current setting changes are well within the current margin.

While the small signal modulation described above is suitable to maintain the state of equilibrium, it is inadequate for the damping of large disturbances. Large signal modulation is thus needed to regain the equilibrium state following large disturbances. For instance, a large signal modulation scheme [15] has been added to the Square Butte HVDC system in the form of a frequency-sensitive power control (FSPC).

8.7.4 Controlled Damping of DC-Interconnected Systems

With an AC tie line, if one of the interconnected systems is in difficulty following a disturbance, the line is normally tripped to prevent the disturbance affecting the other system, and thus the system in difficulty loses an essential infeed.

An HVDC link, on the other hand, even with the basic controls, shields one system from disturbances on the other. Although the specified power flow can continue, the option is available to vary the power setting to help the system in difficulty to the extent which the healthy system can allow, without putting itself in difficulty, and subject to the rating on the link.

Although the policy of providing controls which enable the HVDC link to assist actively in the damping of disturbances should be encouraged, it must be considered that the DC link contains negligible energy storage and therefore any action to damp a disturbance at one end must naturally produce some disturbance at the other end. In some cases such assistance is readily acceptable, for example, when the local system has no directly connected consumers and it can be designed for greater than normal frequency variations. In effect, this allows the inertia of the system to be used to provide the energy for damping the distant system. Another example is where one system is very small compared with the other, such as the case of an offshore system where the total load is insignificant compared with the size of the mainland network.

With appropriate control, a disturbance originating in either system can be shared in a predetermined manner, and the resulting system oscillations can be damped simultaneously. Unlike transient stability, where the DC link must have the necessary overload capability to get through the first swing, dynamic stability can be achieved without overloading the DC link. Moreover, if the DC link is already operating close to its full capacity, substantial damping can be achieved exclusively by DC power reductions at the appropriate instants [16].

8.7.5 Damping of Sub-Synchronous Resonances [17]

The torsional oscillation modes of turbine–generators can interact with the system oscillation modes of the electric power transmission system. With a pure AC transmission system, the interaction is dominated by a synchronising component which gives rise to the system modes of oscillation. A relatively small negative damping component is also present, due to the resistance of the AC transmission line.

The torsional modes of vibration of the turbine–generator shaft are normally stable when connected to an AC transmission system, because of the relatively large positive damping contributed by the damping windings and mechanical damping resulting from steam flow, friction, etc. With the addition of series capacitor compensation in the AC network, however, the negative damping contribution of the AC system is dramatically increased when the electrical and mechanical resonant frequencies are close. Under these conditions the torsional modes of vibration can become unstable, a phenomenon which is commonly referred to as sub-synchronous resonance (SSR).

The potential destabilisation of torsional oscillations due to HVDC systems is similar to that caused by series-compensated AC transmission lines. However, the interaction with DC systems can be solved relatively simply by providing power modulation control to cancel the negative damping impact of the basic constant power control loop.

8.7.6 Active and Reactive Power Coordination

The degree of DC power modulation which can be achieved is restricted by terminal reactive power constraint. With only current or power modulation, an increase in active power transfer will be accompanied by a larger increase in terminal reactive power requirements and this effect is particularly noticeable during severe system disturbances. The reactive power variations can cause current control mode transitions between the rectifier and inverter ends, and hence DC changes equal to the DC system margin current.

Coordination between the active and reactive power modulation can be achieved by DC system voltage modulation. An increase in DC voltage will increase the DC power transfer as well as the power factor at both terminals, and hence decrease the reactive consumption as a percentage of active power transmitted.

8.7.7 Transient Stabilisation of AC Systems

Where system disturbances result in the reduction of transmission capability, the generating source will usually accelerate. Remote sources may decelerate as load exceeds generation as a result of the fault which decreases power into that area. When the fault is cleared, the generation and the remaining transmission experience a transient swing which may lead to instability. In particular, long fault clearance times can cause a loss of synchronism.

If the loss of synchronism is irrelevant, as in the case of an HVDC link connecting generation to load areas, it is advantageous to increase the sending end DC link power in the post-fault period in response to the increase of generator speed. This action will remove energy from the generator, reduce its speed, and thus reduce the angular displacement between the generator and the AC receiving system. An appropriate magnitude of the modulation applied for this purpose is in the range of 20 to 40 % of the DC link rating. Some systems have been designed with temporary overload limits as high as 65 %. In other cases even higher modulation limits have been utilised after taking into account the AC system power transfer need, the AC voltage support (VAR) capability and the DC system design ratings. Similarly, for receiving end phase angle or speed changes, DC link power can be controlled to correct this condition within the limits imposed by the controllability of the receiving end phase angle, the DC link capability, and the energy that may be taken from the generation source.

The thyristor valves used in HVDC transmission are rated to withstand considerable overloads without adverse effects to avoid unnecessary protective action. This capability provides the basis for first-peak transient stability improvement. The particular strategy, i.e. current increase or decrease, temporary power reversal, etc., will vary from scheme to scheme and in each case it can be assessed with the help of a multi-machine transient stability programme combined with a small-step transient converter simulation package, such as the EMTDC. The two solutions are carried out separately and compared periodically to ensure agreement on interface quantities. The time steps required for the DC solution are far shorter than those that would be used for the normal transient stability solution and therefore during this calculation a Thévenin equivalent AC network is used [18].

8.8 AC–DC–AC Frequency Interactions

8.8.1 Harmonic Cross-Modulation Across the DC Link

Under perfectly symmetrical operation and in the absence of waveform distortion in the commutating voltage and of DC ripple there are only characteristic harmonics on both sides of the link. However, any deviation from the ideal conditions will cause uncharacteristic harmonics via the mechanism discussed in Chapter 3 (Section 3.10). The latter situation is more critical in the case of an AC–DC–AC interconnection, considering the lack of synchronism between the two ends, where problems of frequency cross-modulation are frequent.

The numerous problems of frequency cross-modulation encountered by early HVDC schemes motivated a report on the subject by a Working Group of CIGRE SC-14 and the material used in this section comes from that report [19]. The presence of low-order uncharacteristic harmonics has been an important issue in early LCC schemes with relatively low SCRs, by exciting converter transformer asymmetrical saturation, which often leads to harmonic instability.

Figure 8.22 shows the modulated output current on the AC side of the converter in response to a DC that contains a ripple frequency.

In the absence of commutation overlap, the transfer functions are rectangular. The simplified modulation process can be extended to the more realistic case where the commutation process and the control system are included.

The frequency transfer relationships in the cross-modulation process of a line-commutated 12-pulse AC–DC–AC link have been collected together in Figure 8.23. The link interconnects two AC systems of frequencies f_1 and f_2.

The 'exciting' harmonic sources are multiples, integers or non-integer, of the frequency in system 1. k_1 is a current harmonic source, whereas $k_1 - 1$ and $k_1 + 1$ are voltage harmonic sources. The resulting harmonic orders in system 2 are related to the frequency of system 2.

The DC column refers to the DC side of the link, the AC_1^+ and AC_1^- columns represent the positive and negative sequences of system 1, and AC_2^+ and AC_2^- represent the positive and negative sequences of system 2, respectively.

When the DC link interconnects separate power systems, either of the same nominal (but in practice slightly different) frequency or different nominal frequencies, there will be a wide range of harmonic and non-harmonic frequency transfers. These can be divided into two groups:

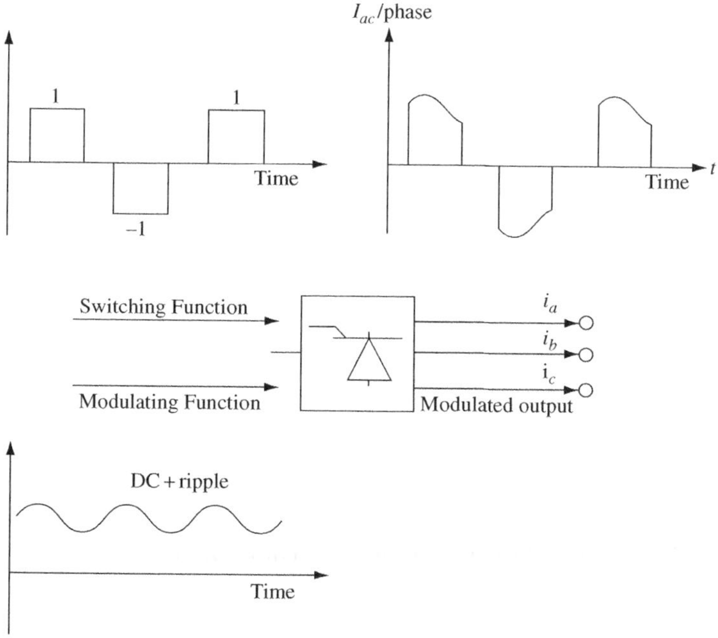

Figure 8.22 Idealised switching function, modulating function and modulated AC output waveform

AC_2^-	AC_2^+	DC	AC_1^+	AC_2^+
$k_1 f_1/f_2 - 1$ $(12n-1) \pm k_1 f_1/f_2$	$k_1 f_1/f_2 + 1$ $(12n+1) \pm k_1 f_1/f_2$	(k_1)	$k_1 + 1$ $(12n+1) \pm k_1$	$k_1 - 1$ $(12n-1) \pm k_1$
$k_1 f_1/f_2 - 1$ $(12n \pm k_1) \pm f_1/f_2 - 1$ $(12m-1) \pm k_1 f_1/f_2$ $(12m-1) \pm (12n \pm k_1) f_1/f_2$	$k_1 f_1/f_2 + 1$ $(12n \pm k_1) \pm f_1/f_2 + 1$ $(12m+1) \pm k_1 f_1/f_2$ $(12m+1) \pm (12n \pm k_1) f_1/f_2$	k_1 $12n \pm k_1$	($k_1 + 1$) $k_1 + 1$ $(12n \pm k_1) \pm 1$ $(12m+1) \pm k_1$ $(12m+1) \pm (12n \pm k_1)$	$k_1 - 1$ $(12n \pm k_1) - 1$ $(12m-1) \pm k_1$ $(12m-1) \pm (12n \pm k_1)$ ($k_1 - 1$)

Figure 8.23 Harmonic transfers across a 12-pulse HVDC link. The encircled elements indicate harmonic sources and $m,\ n \in (1, 2, 3, \ldots)$

1. Frequencies at terminal 1 caused by the characteristic DC voltage harmonics ($12nf_2$) and their consequential currents from terminal 2. These are represented in the expression

$$f_{AC1} = (12m \pm 1)f_1 \pm 12nf_2 \tag{8.2}$$

where m, n $(0, 1, 2, 3, \ldots)$ which can have any frequency including frequencies below the fundamental.

The back-to-back frequency conversion schemes represent the worst condition for non-integer harmonic frequencies. In this case, with small smoothing reactors the DC side coupling is likely to be strong, which means that the flow of harmonically unrelated currents on the DC side can be large. Such schemes can produce considerable sub-harmonic content even under perfect AC system conditions. However, 12-pulse converters do not produce sub-harmonic content under symmetrical and undistorted AC system conditions. These will produce interharmonic currents as defined by Equation (8.2).

When the link interconnects two isolated AC systems of the same nominal frequency but they differ by a small increment Δf_0, then the characteristic harmonics are different by $12n\Delta f_0$. A DC side voltage at frequency $12n(f_0 + \Delta f_0)$ is generated by one converter, and this will be modulated down again at the other converter by a characteristic frequency in the thyristor switching pattern as per Equation (8.2), i.e.

$$(12m \pm 1)f_0 \pm 12n(f_0 + \Delta f_0)$$

which on the AC side, among other frequencies, includes (for $m = n$)

$$f_0 \pm 12n\Delta f_0$$

The latter will beat with the fundamental component at a frequency $12n\Delta f_0$, which at some values of n will allow flicker inducing currents to flow.

2. Frequencies caused in system 1 by imbalance or distortion in the supply voltage of system 2. Negative sequence voltages at frequencies $(k-1)f_2$ produce the following non-characteristic frequencies on the DC side:

$$f_{DC} = (12n \pm k)f_2 \quad (n = 0, 1, 2, \ldots) \tag{8.3}$$

Cross-modulation of these current components produces the following frequencies in system 1:

$$f_{AC1} = (12m \pm 1)f_1 \pm (12n \pm k)f_2 \tag{8.4}$$

Let us first consider a frequency conversion scheme with a sinusoidal but negative sequence unbalanced voltage in system 2, i.e. $(k-1) = 1$ (and therefore $k = 2$). Substituting $m = n = 0$ and $k = 2$ in Equation (8.2) yields currents (and therefore voltage) at frequencies

$$f_{AC1} = \pm f_1 \pm 2f_2 \tag{8.5}$$

One of these frequencies $(f_1 - 2f_2)$ will beat with the fundamental frequency voltage of system 1 at a frequency

$$f_1 + (f_1 - 2f_2) = 2(f_1 - f_2) \tag{8.6}$$

which for a 50–60 Hz conversion scheme becomes 20 Hz. This is a flicker producing frequency. This same frequency will be referred to generator rotor shaft torque at 20 Hz, which may excite mechanical resonances.

Again, this type of cross-modulation effect is most likely to happen in back-to-back schemes due to the stronger coupling between the two converters, although it is also possible with any HVDC scheme in the presence of a suitable resonance.

Now consider two AC systems of the same nominal (but slightly different) frequency. Substituting $m = n = 0$ and $k = 2$ for fundamental frequency f_0 into Equation (8.4), a current and resultant voltage (through the AC system impedance) of frequency

$$f_{AC1} = \pm f_0 \pm 2(f_0 + \Delta f_0) \tag{8.7}$$

which leads to $f_0 \pm 2\Delta f_0$ is induced on the AC side. This will either beat with the fundamental frequency f_0 or produce generator/motor shaft torques at $2\Delta f_0$. This frequency is generally too low to produce flicker but may induce mechanical oscillations.

Substituting $m = n = 1$ and $k = 2$ in Equation (8.4) gives, among others, a current (and thus voltage) at the frequency

$$(12+1)f_0 - (12+2)(f_0 - \Delta f)$$

and for $f_0 = 50\,\text{Hz}$ and $\Delta f = 1\,\text{Hz}$, the resulting AC (and thus voltage) in system 1 is

$$13 \times 50 - 14 \times 49 = 36\,\text{Hz}$$

This distorting voltage will, therefore, beat with the fundamental producing 14 Hz flicker. However, the sub-harmonic levels expected from this second-order effect will normally be too small to be of consequence.

8.8.2 Complementary and Composite Resonances

The traditional definition of resonance is still commonly used with reference to either the AC or DC sides of the converter independently from each other. This sort of resonance is well defined, being the frequency at which the capacitive and inductive reactances of the circuit impedance are equal. At the resonant frequency, a parallel resonance has a high impedance and a series resonance a low impedance.

This approach has led to the concept of 'complementary resonance', i.e. a high-impedance parallel resonance at a harmonic on the AC side closely coupled to a low-impedance series resonance at an associated frequency on the DC side. The first-order associated AC and DC side frequencies are shown in the three-port model of Figure 3.25. A reported experience of this condition, and used in the commonly used CIGRE benchmark HVDC test system [18], involves a second-harmonic parallel resonance on the AC side and a fundamental frequency series resonance on the DC side.

Moreover, when the AC and DC systems are interconnected by a static converter, the system impedances interact via the converter characteristics to create entirely different resonant frequencies. The term ***composite resonance*** has been proposed [13] to describe this sort of resonance, emphasising its dependence on all the components of the system. A composite resonance may be excited by a relatively small distortion source in the system, or by an imbalance in the converter components or control. The resulting amplification of the small source by the resonant characteristics of the system can compromise the normal operation of the converter and even lead to instability.

Further to this, the converter impedance comprises several contributions. First, there is the AC side and converter transformer impedance, which usually sums to be largely inductive. Second, there is the end of commutation period dynamics, which is such that if DC current out of the converter increases, the DC voltage reduces. This impedance looks mainly resistive. Finally, the constant current control modifies the converter DC terminal voltage according to the DC current. This can also be described as an impedance, although over a range of frequencies, the resistive component of this impedance will be negative.

A true instability results when, at the composite resonant frequency, the resistance of the overall circuit is negative. This can occur at non-integer frequencies, and is driven by conversion from the fundamental frequency and DC components to the composite resonance frequency via the converter control. Light damping, or ringing, during fault recovery indicates that the negative resistance offered by the current controller is close to the natural resistance of the circuit.

8.9 DC Link Response to External Disturbances

8.9.1 Response to AC System Faults

The severity of a three-phase short circuit is greatly reduced compared with an alternative AC interconnection, because the DC link, due to its fast current controller, feeds virtually no additional current into the fault.

If the fault occurs on the rectifier side no special control action is needed; provided there is some commutating voltage, the rectifier will continue operating with the highest possible direct voltage, and when the fault is cleared, the rectifier will again recover without special control action. On the other hand, a fault close to the inverter end causes commutation failures and produces higher current peaks, but these can be minimised by quickly reducing the firing angles.

The behaviour of line-commutated DC transmission during AC system faults is best illustrated by EMTDC simulation. The CIGRE benchmark model [20] is used as the test system, a model specifically designed to assess the dynamic performance of DC transmission controllers under a difficult combination of AC and DC system components, as shown in Figure 8.24. The simulation circuit is shown in Figure 8.25.

The controller, modelled in Figure 8.26, is of the proportional–integral type in both current and minimum extinction angle control. The circuit is first simulated for 1 second to achieve the steady state, whereupon a snapshot is taken of the system state. Figure 8.27 illustrates selected oscillograms of the response to a five-cycle, three-phase fault applied to the inverter commutating bus. The simulation starts from the snapshot taken at the 1 second point.

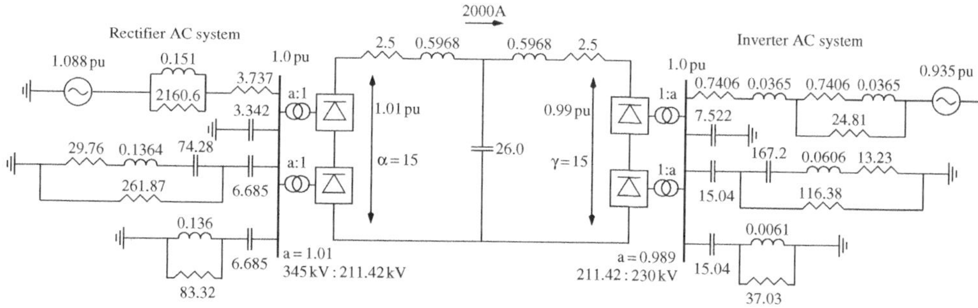

Figure 8.24 CIGRE HVDC benchmark model

A clear advantage of starting from snapshots is that many transient simulations, for the purpose of control design, can be initiated from the same starting state.

These results clearly indicate that the link controllers respond satisfactorily to the disturbance and provide very fast recovery.

8.9.2 Response to DC Line Faults

A DC line short circuit, once started, will not be extinguished until the current is brought down to zero and the arc deionised. When the fault occurs the line voltage collapses, the rectifier current tends to rise and the inverter current to fall. The inverter will then advance its firing angle to provide sufficient voltage to maintain the inverter set current. The rectifier will increase its firing delay to maintain is own current setting. In practice, to reduce the current to zero the two terminals are driven temporarily into inversion to clear the energy stored in the DC circuit. A typical simulated response, using the New Zealand original link parameters, is shown in Figure 8.28. This figure shows that the initial fault current is kept within twice the rated current (due to the presence of the smoothing reactor) and that the total interruption time is of the order of 10 cycles.

8.10 Reliability of LCC Transmission

CIGRE Advisory Group B4.04 collects data annually on the reliability performance of HVDC systems in operation throughout the world and produces a biennial report. The report contains data on energy utilisation, which includes forced energy unavailability (FEU) and scheduled energy unavailability (SEU), statistics on the frequency and duration of forced outages, data on thyristor failure rates and commutation failure rates.

The data on forced outages gives a good indication of the system reliability over the years. This information is classified into six categories as follows:

- AC and auxiliary equipment (AC-E)
- Valves (V)

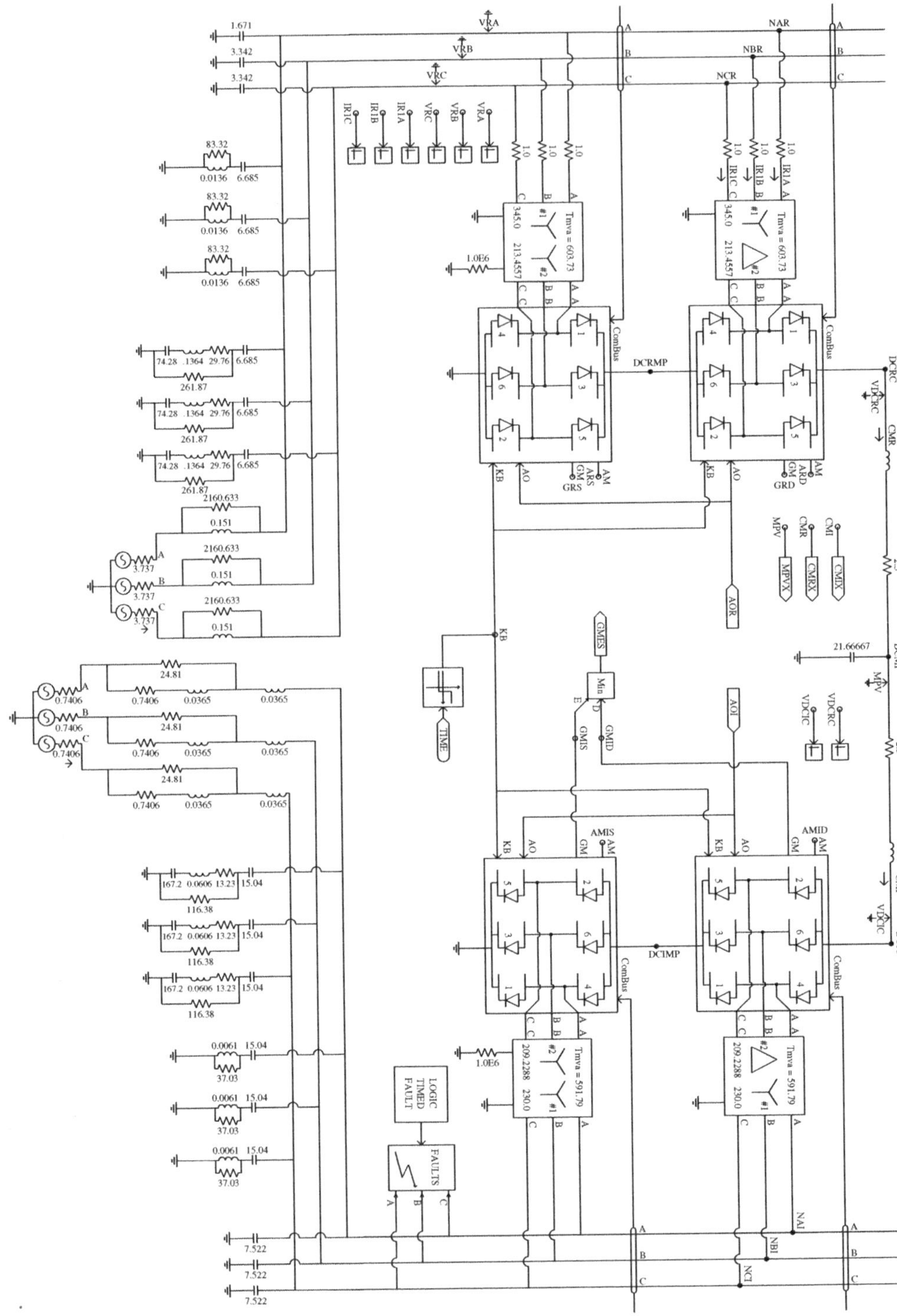

Figure 8.25 CIGRE benchmark model as entered into the PSCAD software

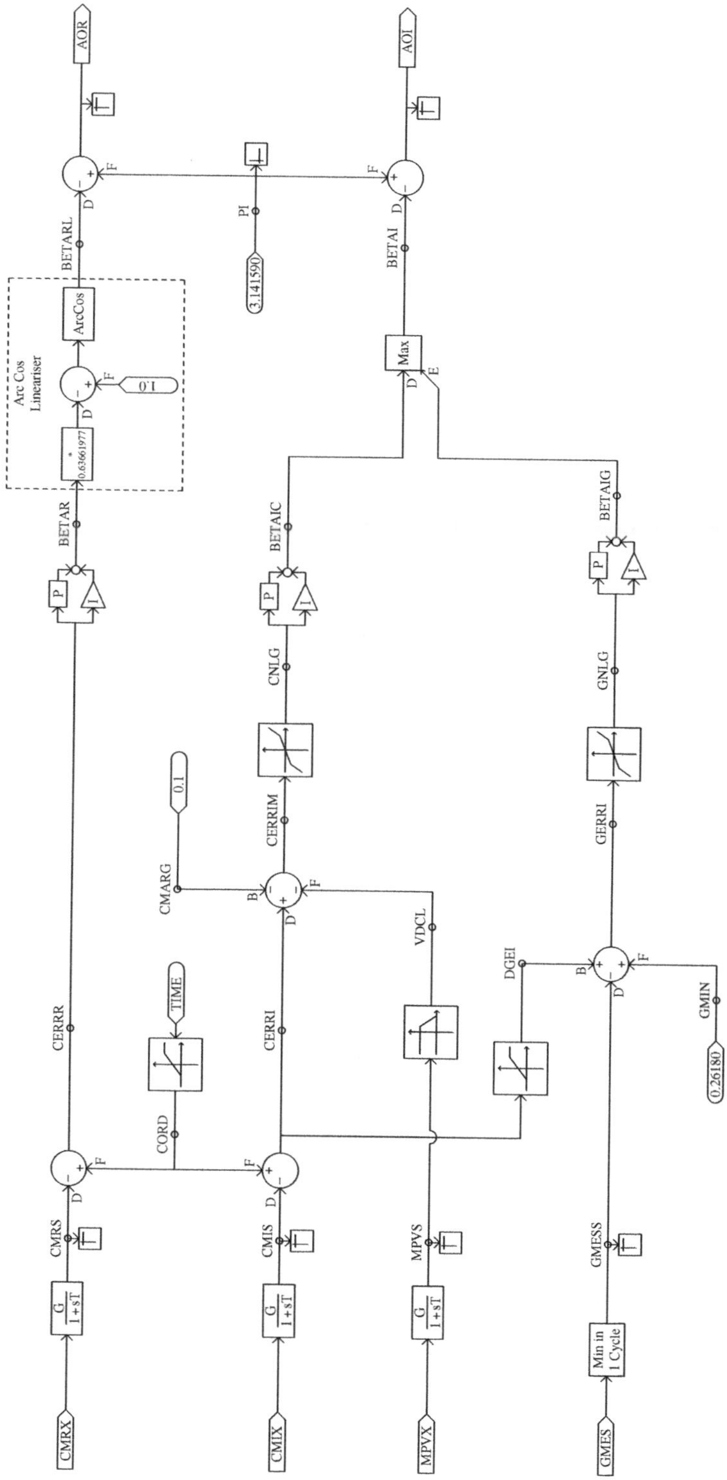

Figure 8.26 Controller for the PSCAD/EMTDC simulation of the CIGRE benchmark

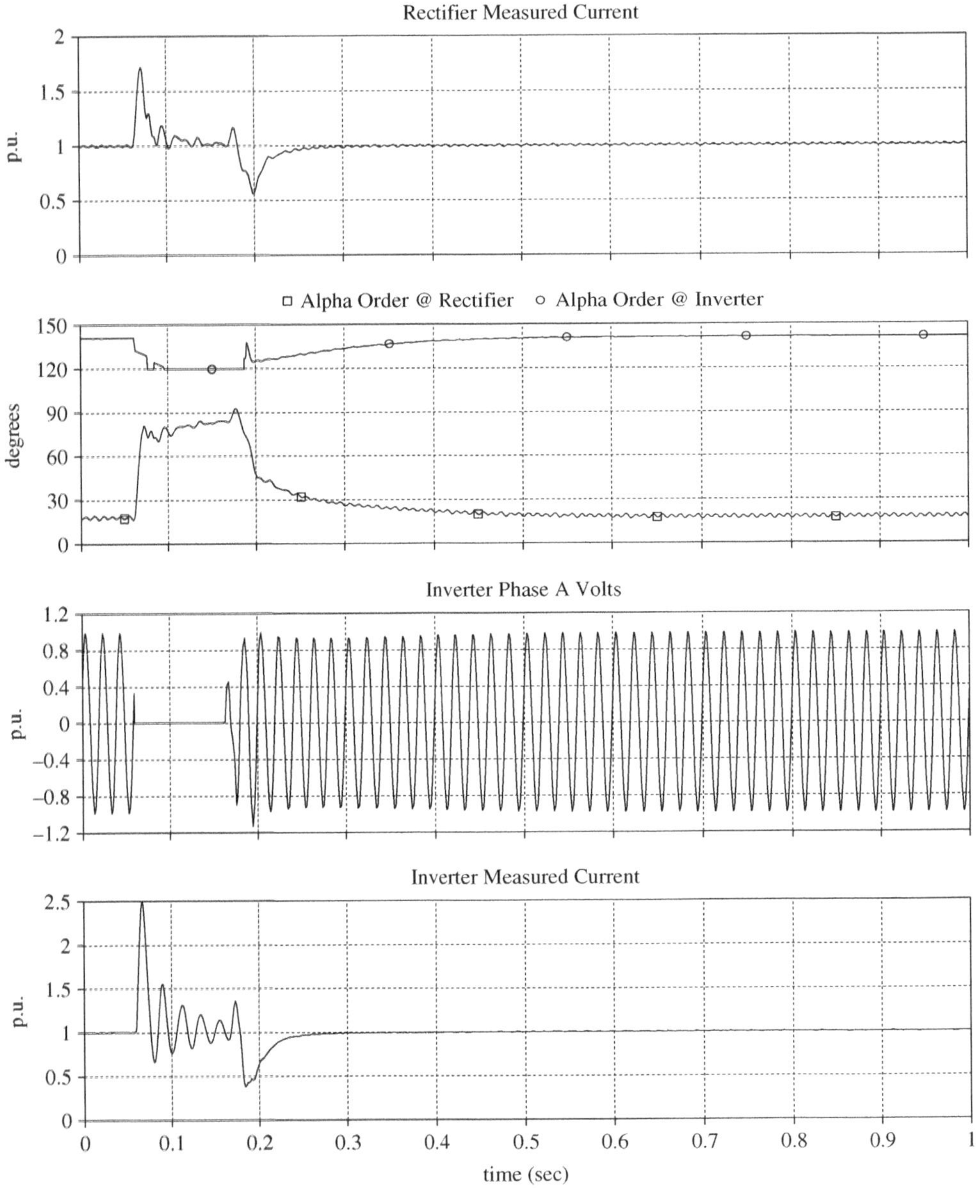

Figure 8.27 Response of the CIGRE benchmark model to a five-cycle, three-phase fault at the inverter bus

- Control and protection (C&P)
- DC equipment (DC-E)
- Other (O)
- Transmission line or cable (TL).

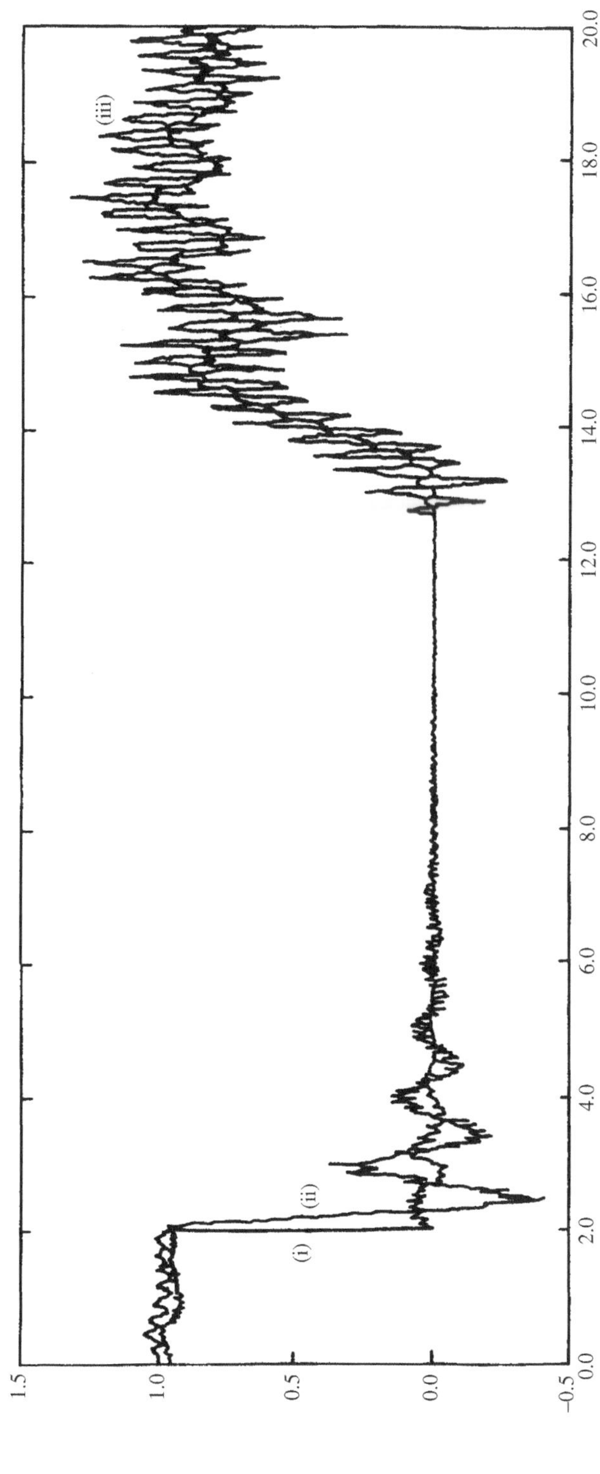

Figure 8.28 DC response to a DC line fault: (i) rectifier end; (ii) inverter end

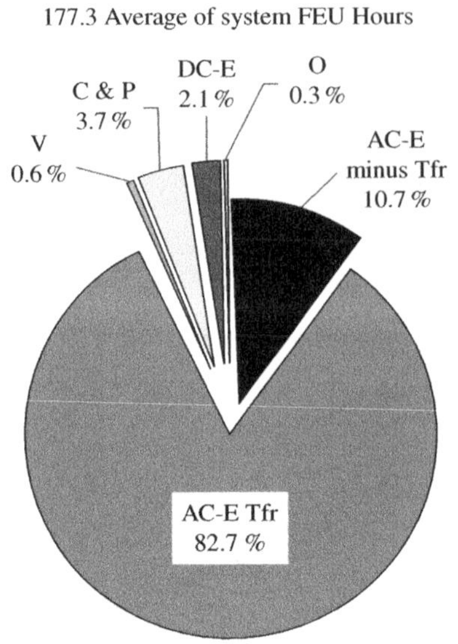

Figure 8.29 FEU breakdown in the period 2003–2004 (Reproduced by permission of CIGRE.)

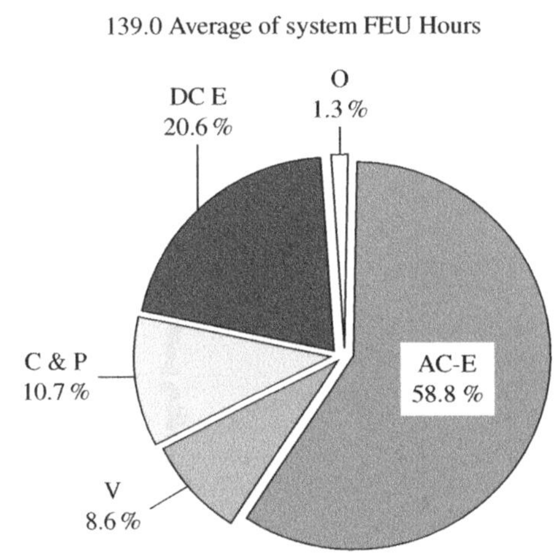

Figure 8.30 FEU breakdown in the period 1983–2002 (Reproduced by permission of CIGRE.)

Figures 8.29 and 8.30 display the breakdown of average FEU, by categories, of all reporting systems for the 2003–2004 and 1983–2002 periods respectively [21]. These results clearly show a great improvement in the reliability of the converter components over the years, as compared with the more conventional AC components. To a large extent the poor performance of the latter is attributable to converter transformer outages that occurred at four systems in 2003.

The initial performance reliability of the recently commissioned Three Gorges scheme, not available in time to be included in the latest CIGRE report [22], has been very satisfactory considering its size and complexity. No bipolar outages have occurred in the first two years of operation. In 2004 and 2005 there have been four and eight monopolar outages, half of the latter caused by icing on the pole lines that led to repeated flashovers. None of the 8352 thyristors contained in the four converter stations of the 3GC and 3GG links has failed so far and no other disturbance directly related to the thyristors has been recorded.

8.11 Concluding Statements

The chapter has described the components and operating characteristics of the present LCC-based HVDC technology. Particular emphasis has been given to reactive power control and harmonic-related issues, the two main problems of the thyristor converter.

After 50 years of experience, regularly monitored by CIGRE Study Committee B4 (earlier SC-14), the reliability of LCC-based HVDC transmission has surpassed all expectations. Unexpected outages are rarely due to the power electronics components, but rather to the conventional plant components of the converter stations and, thus, the utilisation levels are very high. About 83 % of the forced outages are due to transformer faults and only 2.7 % are attributed to major DC equipment [21]. An indication of the link reliability is the small number of thyristors replaced every year in the English terminal of the Cross-Channel link between England and France, namely only 5 thyristors out of the total of 12 000.

Long-distance transmission schemes in thousands of megawatts and voltages of up to ±600 kV are already in existence and the possible use of up to ±800 kV is currently under consideration. Moreover, a large proportion of the HVDC interconnections commissioned in recent years have been of the back-to-back type, where, due to the synchronous constraints of the interconnected systems, there is no alternative to DC transmission. On the other hand, the earlier enthusiasm for the development of a multi-terminal HVDC technology based on CSC seems to have faded. No schemes beyond three terminals are being given consideration. Instead, the multi-terminal concept is again discussed with reference to distribution systems, based on the recently developed VSC technology described in Chapter 10.

Further enhancements to the line-commutated HVDC technology, both for the switching devices and for the converter configuration, are under active consideration to improve further the DC transmission reliability and increase the control flexibility. These are discussed in Chapter 9.

It is reasonable to expect that, despite its relatively low flexibility, the thyristor-based HVDC technology will continue to dominate the market for bulk power and long-distance transmission. It is up to the new alternatives (discussed in the following chapters) to show that they can achieve the required power and voltage ratings with similar levels of cost and operational reliability.

It will be shown in Chapter 11 that the addition of multi-level current reinjection (MLCR), described in Chapter 7, allows the converter valves to commutate naturally; in the future, developments such us this are likely to make the self-commutating thyristor-based HVDC technology a practical and more flexible alternative.

References

1. Arrillaga, J. (1998), *High Voltage Direct Current Transmission*, (Second Edition), IEE Power and Energy Series 29, London.

2. Arrillaga, J. and Smith, B.S. (1998), *AC-DC power system analysis*, IEE Power and Energy Series 27, London.
3. Flisberg, G. and Funke, B. (1981), 'The CU h.v.d.c. transmission', *ASEA Journal*, **54**(3), 59–67.
4. Eriksson, G. and Haglof, N. (1975), 'HVDC station design', *ASEA Journal*, **48**(3), 61–65.
5. CIGRE (1999), 'Guide for the specification and design evaluation of ac filters for hvdc systems', Document WG-14.30, Paris.
6. Holmgren, T. *et al.* (1995), 'A test installation of a self-tuned ac filter in the Konti-Skan 2 HVDC link', IEEE/KTH Stockholm Conference on Power Technology, paper, Sweden, SPT PE 02-06-0367.
7. Kimbark, E.W. (1971), *Direct current transmission*, Wiley Interscience, New York.
8. Uhlmann, E. (1975), *Power Transmission by Direct Current*, Springer-Verlag, Berlin.
9. Chand, J., Rashwan, M.M. and Tishinski, W.K. (1981), 'Nelson River HVDC system-operating experience', *Thyristor and Variable Static Equipment for A.C. and D.C. Transmission*, IEE Conference Publication 205, London, pp. 223–26.
10. Ainsworth, J.D. (1981), 'Telecommunication for h.v.d.c.', *Thyristor and Variable Static Equipment for A.C. and D.C. Transmission*, IEE Conference Publication 205, London, pp. 190–193.
11. Jotten, R. Bowles, J.P., Liss, G., Martin, C.J.B. and Rumpf, E. (1978), 'Control of h.v.d.c systems—the state of the art', CIGRE paper 14-10, Paris.
12. CIGRE WG 14-07 (1992), 'Guide for planning DC links terminating at ac system locations having low short-circuit capacities', Part 1: AC-DC interaction phenomena.
13. Ainsworth, J.D. (1968), 'The phase-locked oscillator - a new control system for controlled static convertors', *IEEE Transactions on Power Apparatus and Systems*, **PAS-87** (3), 859–865.
14. Le Du, A. (1981), 'Use of static or synchronous compensators in h.v.d.c. systems', CIGRE Study Committee 14, Rio de Janeiro.
15. Grund, C.E., Pohl, E.R.V., Cresap, R.L. and Bahrman, M.P. (1980), 'Increasing power transfer capabilities of a.c./d.c. transmission systems by coordinated dynamic control', Symposium sponsored by the Division of Electric Energy Systems, US Dept of Energy, Phoenix, Arizona, 371–387.
16. Uhlmann, E. (1975), *Power Transmission by Direct Current*, Springer-Verlag, Berlin, p. 169.
17. Hingorani, N.G., Nilsson, S., Bahrman, M., Reeve, J., Larsen, E.V. and Piwko, R.J. (1980), 'Subsynchronous frequency stability studies of energy systems which include h.v.d.c. transmission', Symposium sponsored by the Division of Electric Energy Systems, US Dept of Energy, Phoenix, Arizona, 389–398.
18. Arrillaga, J. and Watson, N.R. (2001), *Computer Modelling of Electrical Power Systems*, Second Edition, John Wiley & Sons, Ltd, Chichester.
19. CIGRE SC-14 WG 14.25 (1997), 'Harmonic cross-modulation in HVDC transmission', HVDC Colloquium, Johannesburg.
20. CIGRE WG 14-02 (1991), 'First benchmark model for hvdc control studies', *ELECTRA*, No. 135, 55–75.
21. Vancers, I., Christofersen, D.J., Leirbukt, A. and Bennettt, M.G. (2006), 'A survey of the reliability of HVDC systems throughout the world during 2003–2004', CIGRE paper B4-202, Paris.
22. Rebati, D., Borje, L., Rinaldo, S. and Cheung, S.P. (2006), 'Operation experience from bulk power HVDC links from Three Gorges complex', CIGRE paper B4-208, Paris.

9

Developments in Line-Commutated HVDC Schemes

9.1 Introduction

Line-commutated conversion (LCC) is still the preferred practical option for large-power and long-distance HVDC transmission. However, the conventional technology is not at a standstill.

The main development influencing the continuation of this technology is the thyristor switch itself. Research in this area is far from over, with the power and voltage ratings of the thyristor module continuing to increase. In particular, the prospective application of new materials such as silicon carbide (SiC) appears to be distancing the thyristor further from its more controllable competitors. This topic has already been covered in Chapter 2.

Other promising developments likely to influence the design and operation of future LCC HVDC transmission schemes are:.

- capacitor-commutated conversion (CCC);
- continually tuned AC filters;
- active DC side filters;
- STATCOM-aided conversion;
- conversion of existing AC lines for use by HVDC transmission;
- use of DC voltages higher than 600 kV.

Flexible Power Transmission: The HVDC Options J. Arrillaga, Y.H. Liu and N.R. Watson

9.2 Capacitor Commutated Conversion [1–3]

CCC includes series capacitors between the converter transformer and the valves, as shown in Figure 9.1. Although in principle the capacitors could be connected on the AC side of the converter transformers, it is not possible to avoid ferro-resonance problems completely with such a configuration.

The series capacitors contribute to the valve commutation voltage and, thus, make it possible to operate CCC with much lower reactive power consumption compared with a conventional converter. This permits reducing the power rating of the filter banks; also the latter are permanently connected throughout the operating range, i.e. from minimum to maximum load, while still providing the required reactive power. Thus, the filter banks' subdivision and switchgear required by the conventional converter are eliminated. Further, CCC gives a more robust and stable dynamic performance of the inverter station, especially when it is connected to a weak AC system or a long DC cable.

With weak AC systems the AC load rejection voltages are reduced, due to the low reactive power consumption of the converter.

A long DC cable has large capacitive energy storage. In the event of a temporary AC voltage reduction in the inverter AC system, e.g. caused by a remote single phase-to-ground fault, the DC cable will partially discharge into the AC system. Thus the transient current increase at the inverter end will not be immediately detected at the rectifier end, resulting in a delay before the rectifier reduces the DC current. Therefore AC voltage collapse may occur unless the inverter is able to counteract the current increase by raising the terminal voltage. However, this is not possible as a conventional inverter operating on minimum commutation margin has a negative impedance characteristic. On the other hand, a CCC inverter operated at minimum commutation margin has an almost constant or slightly positive impedance, which improves the response to a transient current increase compared with a conventional converter.

If required, larger commutation margins can be achieved, without increasing the reactive power consumption of the converter station, by reducing the capacitance of the commutating capacitors in order to increase their contribution to the commutation voltage. The firing angle (when referred to the AC system voltage) can even be increased beyond 180°, thus permitting the converter to supply reactive power.

A study carried out by the Manitoba HVDC Research Centre has indicated that series capacitor compensation is one of the most economical methods of AC voltage control for

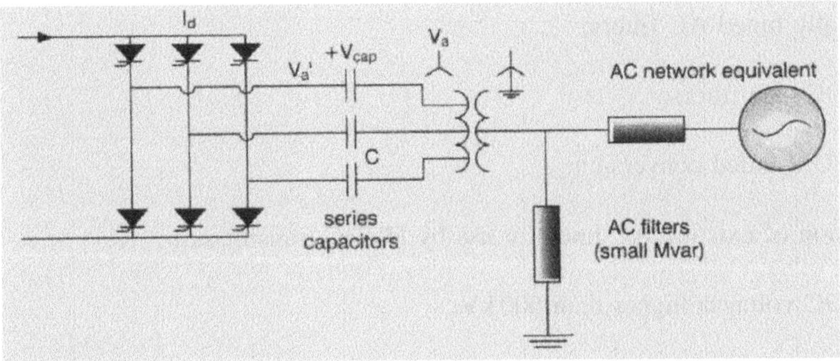

Figure 9.1 Capacitor commutated converter CCC (Gole, A.M. and Meisingset, M. (2002), 'Capacitor commutated converters for long distance HVDC Transmission', *Power Engineering Journal*, June, reproduced by permission of the IET.)

Table 9.1 Cost comparison of voltage control devices

Voltage control device	Economic evaluation (% of station capital cost)		
	Capital cost	Capitalised operational cost	Total lifetime cost
Synchronous condenser (1 unit)	21	11	32
Synchronous condenser (2 units)	40	15	55
Thyristor-controlled reactor	17	6	23
Thyristor-switched capacitor	18	5	23
Metal–oxide varistor	14	Negligible	14
Series capacitor	13	1	14

HVDC converter stations [4]. Table 9.1 summarises the centre's findings with reference to a 810 MW inverter station.

The first commercial application of the CCC configuration is the Garabi 1100 MW back-to-back connection between Argentina and Brazil [5], which has been in operation since the year 2000.

9.2.1 Basic CCC Operation

Typical waveforms for the AC, capacitor voltages and AC side voltages for two different values of the DC are illustrated in Figure 9.2. These refer to a link operating at 250 kV, 2 kA (on the DC side) and 207 kV phase-to-phase voltage on the secondary side of the converter transformer. As the commutated current (Figure 9.2(a)) flows through the series capacitor, it develops the voltage waveform shown in Figure 9.2(b) with a magnitude proportional to the DC current. The phase voltage on the valve side of the converter is then increased by the series capacitor voltage. The resulting line voltage waveform on the valve side of the series capacitor and the AC bus voltage are shown in Figure 9.2(c).

The valve side voltage is seen to be lagging the AC bus voltage by an angle γ in proportion to the DC current. Therefore the off-going valve has a longer interval of reverse bias than in the conventional converter. Figure 9.2 also shows the real and apparent extinction angles (γ and γ_{app}), where γ is measured from the turn-off instant of the off-going valve to the instant when forward biasing voltage is applied. The apparent extinction angle γ_{app}, on the other hand, is the value corresponding to the conventional converter.

For the case of $I_d = 2\,\text{kA}$, $\gamma_{app} = 1°$ and $\gamma = 39°$, the small apparent extinction angle corresponds to a near unity power factor in this case. Thus CCC reduces the need for shunt capacitive compensation at the converter bus and the commutating voltage is self-adjusting, i.e. the larger the current, the larger the lag γ of the valve side voltage. This makes CCC attractive for long cable applications, where large currents often occur as the cable capacitance discharges following an AC voltage reduction.

9.2.2 Simulated Performance

The test system chosen to compare the performance of CCC with the conventional 12-pulse conversion consists of a 600 km DC cable transmission link rated at 1200 MW and 500 kV

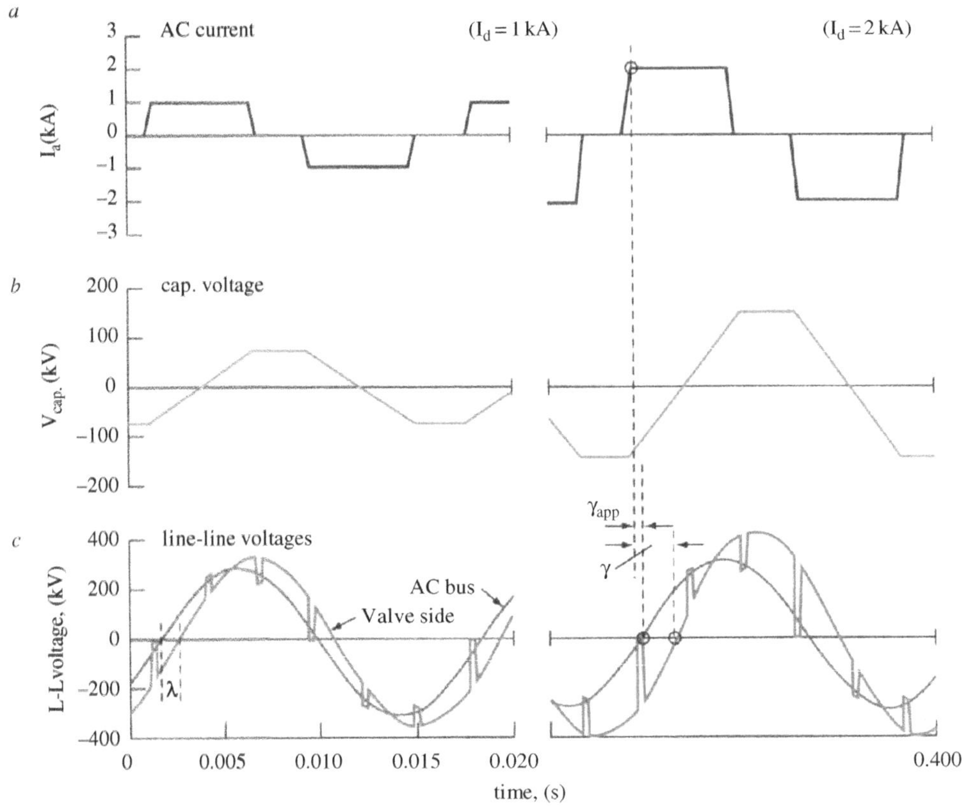

Figure 9.2 CCC operating waveforms for two values of the DC (Gole, A.M. and Meisingset, M. (2002), 'Capacitor commutated converters for long distance HVDC Transmission', *Power Engineering Journal*, June, reproduced by permission of the IET.)

DC voltage; the inverter is connected to a 300 kV weak AC system (with a short-circuit ratio of 1.82) and maintains a constant extinction angle of 20°. All the reactive power required by the converter is assumed to be provided locally. The results are derived with the help of electromagnetic transient simulation.

Steady-state characteristics

The superior voltage regulation of the CCC configuration is clearly illustrated by the characteristics shown in Figure 9.3. Graph (b) shows that the load rejection (i.e. zero-current) overvoltage is reduced by about 25 %. Graph (c) shows that CCC increases the maximum available power, thus providing a higher stability margin when operating in the power control mode. Graph (d) shows that the real extinction angle (set up at 20° for the conventional case) increases with DC loading. This feature of CCC is particularly useful in the case of long-cable transmission, as the probability of a commutation failure (caused by an undervoltage condition that increases the DC due to the discharge of the cable capacitance) is substantially reduced.

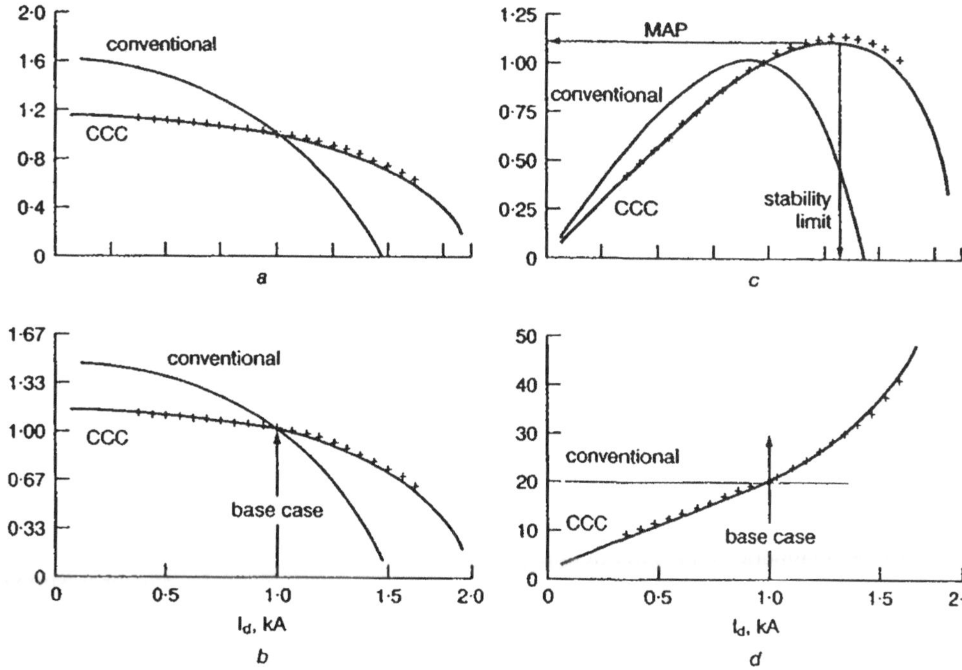

Figure 9.3 Steady-state characteristics for the CCC and conventional configurations: (a) DC voltage in per unit; (b) inverter AC voltage in per unit; (c) power in per unit; (d) real extinction angle in degrees (Gole, A.M. and Meisingset, M. (2002), 'Capacitor commutated converters for long distance HVDC Transmission', *Power Engineering Journal*, June, reproduced by permission of the IET.)

Dynamic performance

The results of extensive EMTDC simulation [2], carried out to compare the performance of the conventional and CCC configurations, have shown that the latter:

- reduces considerably the AC overvoltage following load rejection;
- allows larger set-point control changes;
- improves the recovery from three-phase faults;
- slows down the recovery from unbalanced faults. The reason for this behaviour is the asymmetrical charge of the series capacitors caused by the unbalanced fault; the stored energy from this charge must be dissipated to restore the capacitors' balanced operating mode.

Moreover, the possibility of transformer core saturation instability, a common occurrence in the conventional schemes, does not exist in CCC because the series capacitors block any DC entering from the valve side.

EMTDC simulation has also shown that the current control mode of operation is more effective when implemented at the inverter end. This is so because, for AC side faults, the cable discharges into the converter and it is more difficult to control the discharge current from the remote rectifier end of the link.

9.3 Continuously Tuned AC Filters [6]

Filters are characterised by their impedance and bandwidth. A low impedance is required to achieve a sufficiently low harmonic voltage and a reasonable bandwidth is needed to reduce the effect of filter detuning.

Detuning of conventional filters is caused by network frequency excursions and component variations, e.g. capacitance changes due to temperature differences.

A filter in which tuning can be adjusted to follow frequency variations and component variations can be designed with a high-**Q** factor to provide a low impedance for the harmonics. Automatic tuning will also ensure that all risks of resonances and current amplification phenomena are eliminated, implying that the ratings of the AC filter components can be reduced.

In the proposed filter, tuning is achieved by varying the inductance by means of a transverse DC magnetic field. This field is perpendicular to the main flux direction and does not affect the linearity of the magnetising process. A cross-section of the reactor is shown in Figure 9.4. The control winding, placed at ground potential, is insulated from the main winding by an epoxy filament cylinder.

With the continuously tuned filter, separation of filtering and reactive power compensation is possible, since filters with low reactive power generation can be built.

In combination with the CCC concept (described in Section 9.2), where only limited reactive power compensation by means of shunt filters is needed, the continuously tuned filter

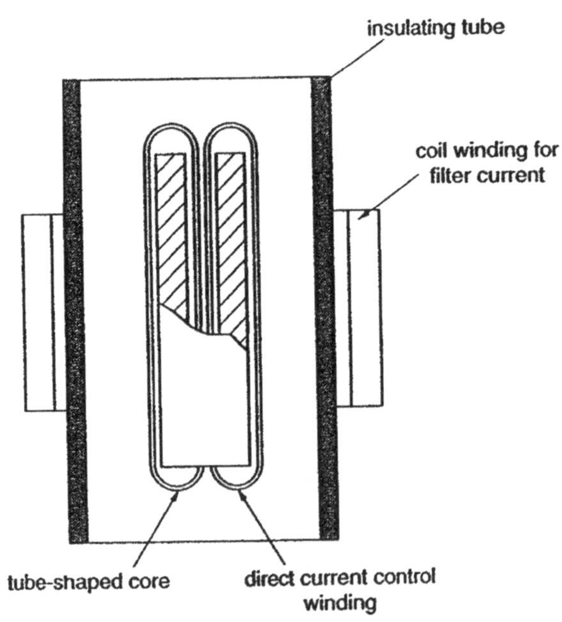

Figure 9.4 Cross-section of the self-tuned filter reactor (Arrillaga, J.(1998), *HVDC Transmission*, second edition, IEE Power and Energy series 29, reproduced by permission of the IET.)

is an attractive solution. With automatically tuned filters for the 11th and 13th harmonics and conventional high-pass filters, only between 0.12 and 0.15pu total reactive power is needed. With such a small AC filter, the harmonic resonance, formed by the AC system inductance in parallel with the shunt filter capacitance, will be sufficiently high, i.e. above the third harmonic, and as a consequence the use of low-order filters will not be required.

9.4 Active DC Side Filters

Considering the part played by the AC filters in the provision of the LCC reactive power requirement, the replacement of present passive filters by an active alternative is not envisaged. However, this is not the case on the DC side, where an interesting hybrid solution has already been implemented [7, 8].

The principle of the passive/active filter concept is that the harmonics in the DC line current are measured; a controller reproduces a current in counter phase with the disturbance; this signal is then amplified in a high-power amplifier and finally fed into the neutral bus end of the DC filter by a transformer

The main components of the active filter are a harmonic current transformer (HCT), a computerised controller, a PWM power amplifier, and a high-frequency transformer together with a transient overvoltage protection. These are shown in Figure 9.5.

The harmonic current transformers consist of a Rogowski coil and electronic circuits that convert the measurement to optical information which is fed to ground potential by an optical fibre.

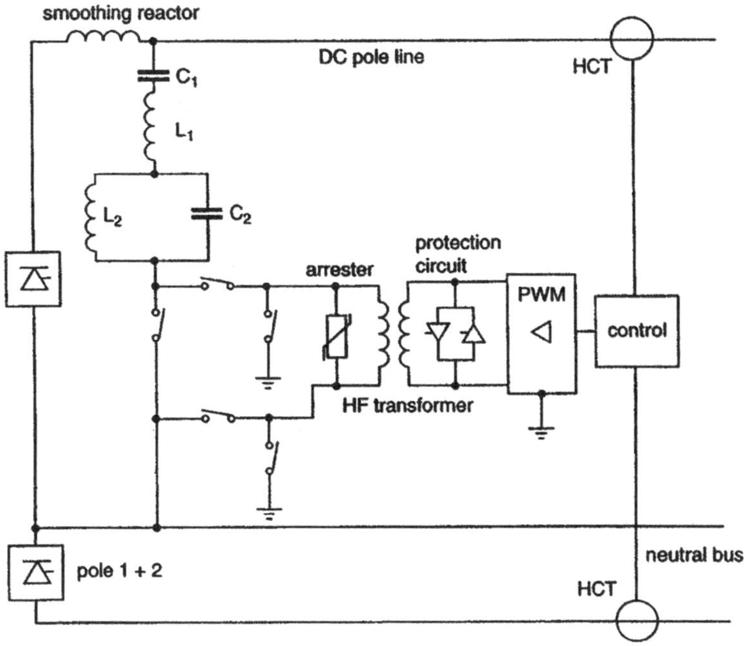

Figure 9.5 Active DC filter (Arrillaga, J. (1998), *HVDC Transmission*, second edition, IEE Power and Energy series 29, reproduced by permission of the IET.)

The controller is realised in the same hardware and software environment as the control equipment for the HVDC pole. A digital signal processor handles the fast controller mathematics in conjunction with a single board computer.

The PWM power amplifier consists of a large number of digitally modulated transistors working in parallel in switched bridges. The output ranges up to 4 kHz at a peak voltage of about 300 V and a power of 100 kVA.

The high-frequency transformer is of a dry isolated type and has a ratio of 1:20 with the low-voltage side connected to the amplifier.

The transient overvoltage protection is realised by two anti-parallel thyristors of the same type as the thyristors used in the HVDC valve. The protection is triggered by overvoltage or overcurrent.

9.5 STATCOM-Aided DC Transmission

LCC needs an AC voltage source to operate; the source can be in the form of synchronous generators or synchronous compensators. The availability of VSC-based STATCOM provides an interesting alternative that combines the robustness and cost-effectiveness of LCC and the flexibility of the PWM-VSC technologies. This hybrid solution has been proposed as an effective alternative for the connection of large-scale wind and oil platform applications.

Large-scale wind power application

Offshore wind power generation schemes in hundreds of megawatts are being planned, located at relatively long distances from the main load centres. The integration of such schemes with the network requires the use of cable transmission. The present capacity of a three-core AC submarine cable is typically 200 MW at 145 kV and, therefore, several cables would be needed for larger powers. Using HVDC instead, considerably larger powers can be transmitted with only one cable. Also, feasibility studies are being carried out for inland wind farms in remote areas, such as the USA and Sahara Desert. In such cases the number of conductors required to transmit say 1850 MW at 400 kV would be 12 (of 282 mm^2), whereas only 4 (of 644 mm^2) would be needed for a bipolar DC overhead line operating at $\pm$500 kV.

Apart from transmission simplicity, the DC option offers various advantages compared with AC, such as full control of power flow, less disturbance following network faults, reduced capital costs and lower power loss.

The use of doubly fed induction generators (DFIGs) seems the preferred option for large wind farms due to their advantages in terms of speed control, reduced flicker and four-quadrant (i.e. active and reactive) power capability, achieved via the VSC-controlled rotor side. However, most of the AC transmission limitations listed above still remain.

VSC transmission with self-commutating devices is already used in several schemes (see Chapter 10). However, at present, this technology does not offer an economic solution for very large power ratings due to the high power losses and capital costs of the multiple converters and cables required. The line-commutated HVDC technology provides high reliability, requires little maintenance and can be operated at a distance, as well as being more cost effective in terms of equipment and energy loss. The main problem of LCC, besides the need for an AC voltage supply, is its inability to provide independent control of the active and reactive powers.

In line with conventional technology, the use of LCC complemented by a synchronous compensator has been proposed for the integration of very large offshore wind farms into the network [9].

A more advanced solution, still benefiting from the robustness of LCC [10], consists of a thyristor-based converter and a STATCOM at the generating end of the scheme. The STATCOM provides the necessary commutation voltage to the HVDC converter, continuous AC voltage control, fast reactive power compensation to the network under transient conditions and removal of possible non-characteristic harmonic interactions. It also provides limited active power support to the network during transient conditions, such as active power changes of the wind farm output. Of course, the ability to provide active power support depends on the energy storage on the DC side, which in the case of a conventional DC capacitor will be very limited; larger energy storage could be provided by means of batteries, SMES (Superconductive Magnetic Energy Storage), etc. In the absence of passive reactive power compensation the AC harmonic filters are considerably reduced, an important consideration as space is at a premium in offshore locations. It is a robust, cost-effective and flexible solution for large-power and long-distance transmission.

The proposal made in [10] relates to the transmission of power from a large wind farm based on DFIG over a long distance to the load centre. Diagrams of the scheme and its control strategy are shown in Figures 9.6 and 9.7 respectively and the comprehensive simulation studies reported show that the dynamic performance of the scheme is equivalent to that of VSC transmission

Supply of power to offshore oil platforms

The power requirement of offshore oil platforms is of the order of tens of megawatts, much of that in the form of direct on-line-started induction machines used for pumps and compressors. The platforms are often located hundreds of kilometres away from the onshore grid and a number of platforms are usually placed in close proximity to each other. They usually contain their own power generation in the form of gas turbines, as an alternative AC supply from the shore is impractical for other than short distances. However, private generation occupies valuable space, is expensive to run and requires maintenance outages which reduce the platform production capability.

The main advantages of a network connection by means of a STATCOM-aided LCC HVDC link have already been outlined in the previous section. For this application the

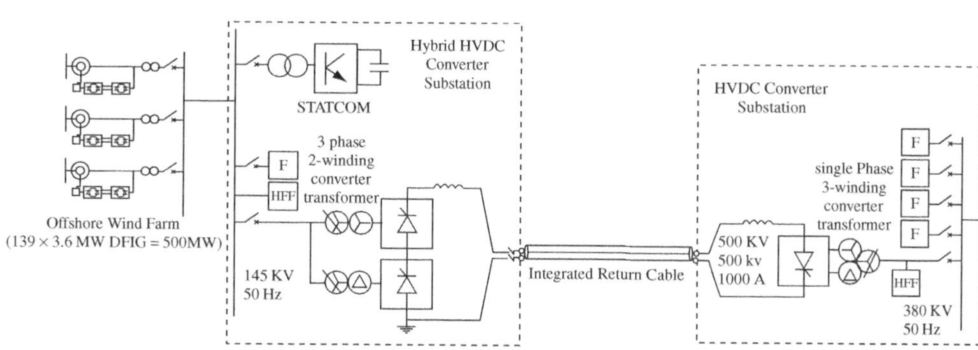

Figure 9.6 Diagram of a hybrid HVDC system connecting a large DFIG-based wind farm (Reproduced by permission of CIGRE.)

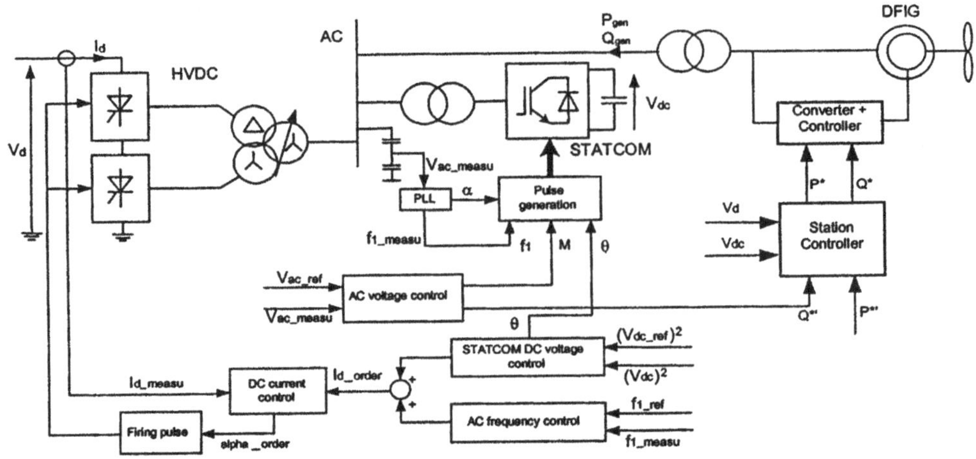

Figure 9.7 Control strategy of the system of Figure 9.6 (Reproduced by permission of CIGRE.)

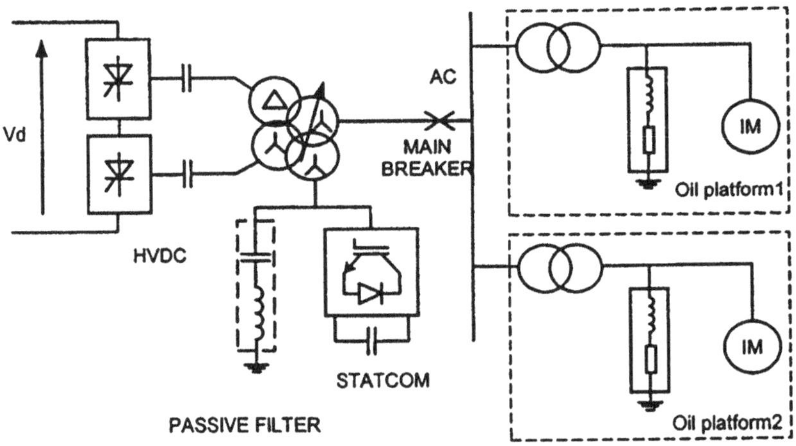

Figure 9.8 Diagram of a hybrid HVDC system supplying offshore oil platforms (Reproduced by permission of CIGRE.)

offshore terminal, shown in Figure 9.8, consists of a thyristor-based CCC, a STATCOM and a passive filter. Because of the use of CCC, the reactive power requirement of the converter is small and this reduces the size of the passive filter as well as the area of the site.

The STATCOM can be of the cascaded H-bridge type or any of the other topologies discussed in Chapter 6, and its power rating is much smaller than that of the overall scheme rating. Moreover, the overall power loss is substantially reduced with respect to that of an equivalent VSC transmission link.

Simulated performance

The dynamic performance of a hybrid scheme for the connection of an offshore wind farm has been carried out using EMTDC simulation. The wind farm and DC transmission power rating is 500 MW and that of the STATCOM ±100 MVAR. The wind generation is

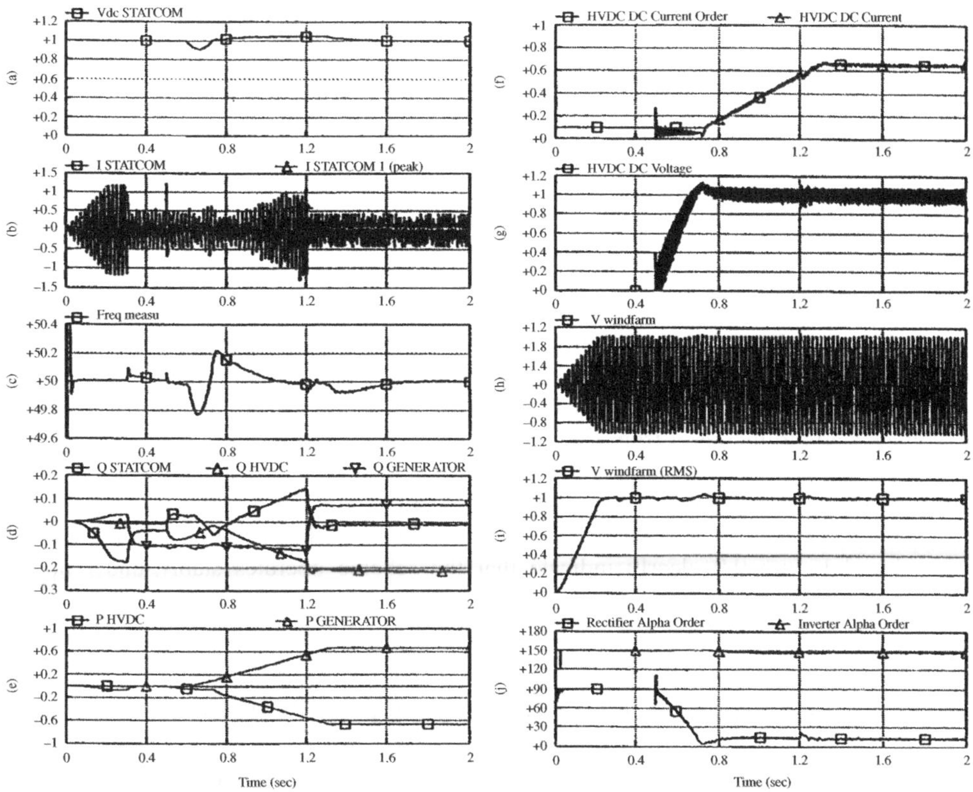

Figure 9.9 Simulation results of system starts and variation of wind generation: (a) STATCOM DC link voltage; (b) STATCOM AC; (c) offshore AC system frequency; (d) reactive powers; (e) active powers; (f) HVDC converter dcs; (g) HVDC rectifier DC voltage; (h) offshore AC system voltage waveform; (i) offshore AC system rms voltage; (j) rectifier and inverter firing angle orders (Reproduced by permission of CIGRE.)

represented as a lump DFIG model with an operating power factor range of ±0.95. The DC link is based on the CIGRE benchmark model [11] with an SCR of 3 at the inverter end. The STATCOM is a two-level VSC switching at 1350 Hz and its DC capacitor has a total energy storage of 4 MJ at the nominal (500 kV) voltage. There are two passive filters, one rated at 60 MVAR and tuned at the 12/24 harmonics, while the other is a high-pass filter rated at 12 MVAR. The following graphs are shown in Figure 9.9:

(a) STATCOM DC side voltage

(b) STATCOM AC current

(c) offshore AC system frequency

(d) reactive powers

(e) active powers

(f) HVDC converter dc currents

(g) HVDC rectifier DC voltage

(h) offshore AC system voltage waveform

(i) offshore AC system rms voltage

(j) rectifier and inverter firing angle orders.

The results show the dynamic response for the following operating sequence. Between 0 and 0.2 s the STATCOM ramps up the AC voltage to offset the filter capacitors; at 0.3 s the DFIG is enabled (absorbing reactive power); at 0.5 s the DC link is deblocked and its DC voltage starts to increase; from 0.6 s on the DFIG active power ramps up and at 1.2 s a reactive power step is applied to the DFIG to make it operate with leading power factor.

The reactive power changes can be followed in graph (d). The inverter controls the DC voltage, which therefore stays practically constant throughout the operating conditions. Thus the results of Figure 9.9 clearly indicate that the scheme operates stably under normal operating conditions.

9.6 AC Transmission Lines Converted for Use with HVDC

Normally AC transmission lines are not loaded to their maximum thermal rating and the firm power capability is lower in any AC transmission line arrangement than it would be if the line were used to transmit power by DC. If an AC link is used to transmit DC power, the conductors can form poles of a DC system which may be operated independently, if necessary, up to the thermal rating.

The insulation, chosen to satisfy AC system requirements, could generally sustain a DC voltage to earth equalling the peak value of the AC voltage to earth, or exceeding it if heavy atmospheric pollution is not a limiting factor. The possibility of using earth return under outage conditions offers an added attraction.

Similar arrangements could be made for both land and submarine-type cables. This principle has already been applied in the case of DC transmission to Vancouver Island. The conversion capability is not restricted to overhead lines and cables. Capacitor units for series or shunt compensation, if suitably specified initially, could be reused at converter stations to constitute harmonic filters. Existing AC switchgear could also be reapplied at converter stations to provide isolation facilities and, in the event of schemes with earth return, AC switchgear could be used for high-speed changeover duty. An additional benefit will be a reduction in the fault level.

The major problems expected in the implementation of this method of increasing power transmission capability are:

- The withdrawal from service during the period of changeover.

- The need, in many schemes, to provide loads at intermediate points.

This method is a practical alternative to the addition of further parallel AC circuits or the introduction of a higher AC voltage. It thereby provides an economic means of overcoming potential amenity problems which may eventually emerge as the limiting factor of public acceptance of very high-voltage overhead transmission.

9.6.1 Modulated (Tripole) DC Transmission

The possible conversion of existing AC lines to HVDC transmission is a subject regularly under consideration, but so far ignored by the power industry due to the cost of the converter plant required and, in the case of single circuit lines, to the bipolar character of DC which makes one of the AC phases idle. However, a document presented at a fairly recent CIGRE meeting [12] proposes an effective way of achieving the AC to DC conversion by adding a third pole to the bipolar station. One pole is positive, while the other two share the negative pole role. To avoid the positive pole carrying more current than the two negative poles, the scheme suggests rotating the overload among the three poles. Each pole carries the overload current one-third of the time, thus causing temporary excursions in conductor temperature; a typical time of 4 minutes is suggested to reduce the temperature excursions to about 2 °C. The pole rotation is achieved by reversing the polarity of one pole, which, in the case of LCC, requires the use of double valves in anti-parallel, as shown in Figure 9.10. The scheme claims to achieve about 40 % extra power rating, while reducing power losses in the range of 20 to 25 % compared with a bipolar system transmitting the same power.

The modulation ratio, defined as $r = I_{max}/I_{min}$, affects the power distribution, the pole powers (1, 2 and 3), total power (Tot.) and the losses as shown in Table 9.2.

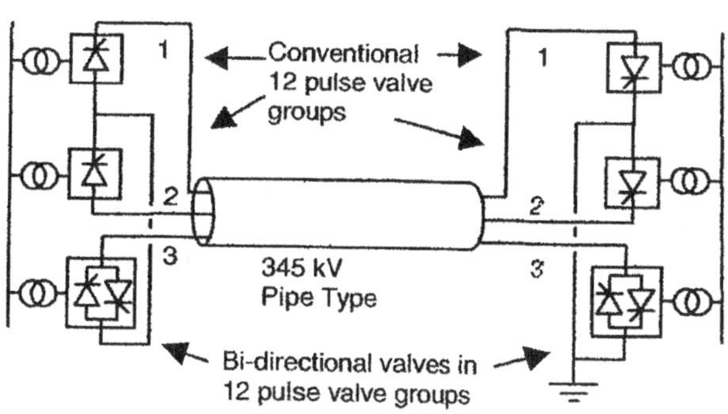

Figure 9.10 Tripolar HVDC transmission configuration

Table 9.2 Power distribution and losses for different modulation ratios

Criterion		Power				
	r	1	2	3	Tot.	Loss
Minimum losses	2.00	0.47	0.47	0.32	1.27	0.75
Equal power/pole	3.00	0.45	0.45	0.45	1.34	0.78
Maximum power	3.73	0.43	0.43	0.50	1.37	0.80

At maximum rating the tripole drops only 27 % of its capacity with either a pole or conductor outage. Its internal redundancy is greater than a double bipole with no permanent ground return. Moreover, in the presence of a metallic ground return path, the power transfer of the tripole could be operated at 1.5 times the bipole rating for as long as the emergency ground return is permitted (typically 30 minutes).

9.7 HVDC Transmission at Voltages above 600 kV

As indicated in the Introduction, ±600 kV is the highest DC voltage level implemented so far. There is, however, increasing interest in the application of higher transmission voltages, possibly up to ±800 kV. The main reason is that most of the hydro-power resources that are close to the consumer centres have already been exploited. Thus, to meet the increasing demand for renewable energy, remote hydro generation plants with very large power ratings are being built (or seriously considered) in various regions of the world, especially China, India, South Africa, Congo and Brazil. A recent comparison [13] carried out for the transmission of 6400 MW over 1800 km indicates that the total cost for the 800 kV alternative is about 25% lower than for 600 kV and the total losses 50 % lower.

The transmission of very long-distance power of such capacity to important load centres requires high reliability, as the probability of line faults increases with the length. Concentrating the power transfer on a smaller number of overhead lines increases the risks involved when some of these lines are taken out of operation in an unplanned manner due to weather-related incidents, technical failures or sabotage (problems that also apply to AC transmission links). The risks can be mitigated by a variety of measures. Normally the terminal stations will have more than one converter group per pole, so as to minimise the disturbances during faults. Another reason for increasing the number of groups is the transport restriction on converter transformers. Each 12-pulse group should have a separate valve hall, with six double valves and six single phase two-winding transformers. The converter stations should be designed with some redundancy to reduce the likelihood of line trips. The two poles of the bipolar HVDC line must be designed with as much independence as possible; when one pole is tripped, monopolar operation should be possible with the remaining pole. A majority of line faults are likely to affect only one pole and these can be cleared in a few cycles, during which period the healthy pole compensates for the loss of power.

If a number of HVDC lines are available when one of the lines is taken out of operation, the other lines can be provided with temporary overload capability. Finally, the complete power system should be designed in such a way that the overall system stability is maintained even if the power from one of the 800 kV DC links is totally lost.

In some existing schemes the two converters in the bipole can be paralleled and the power transmitted on one pole line, although with higher losses. Switching stations can also be placed along the line to cope with simultaneous line faults on different segments along the line.

The operational experience from existing HVDC schemes, from 250 to 600 kV has shown that the flashover rate has no direct correlations to the DC voltage level. The specific creepage distance needed is dictated by the site pollution severity [14].

A critical design issue is the insulation coordination of the stations. Considering the larger expected stresses at 800 kV, the converter stations in this case should be provided with higher insulation margins. The study of external insulation is a key topic in the 800 kV HVDC programme development [15].

Although the use of 800 kV is considered achievable in the near future, at the top of the power and distance range considered (5000 MW and 4000 km) the optimum transmission voltage is more likely to be around 1000–1100 kV.

The main competition to HVDC for the power ratings and distances considered appears to be from bulk gas transmission, with the electricity generation needed for consumption in locations far from the gas field.

Considering that the costs of DC terminals and of combined-cycle plants are not affected by the transmission length, the transmission of electricity is more attractive than gas transmission for:

- longer distances;
- higher cost areas for transmission lines and gas ducts;
- lower prices for gas at the gas field;
- smaller powers.

9.8 Concluding Statements

The reliability and transmission efficiency of conventional CSC HVDC technology will be difficult to be matched by the emerging self-commutating VSC transmission technology. While the power rating and characteristics of the more controllable switching devices continue to improve, the thyristor technology is not at a standstill. Apart from the continuing research in the thyristor itself, the main advances in CSC technology relate to the development of converter configurations with lower reactive power and filter requirements; recent additions in this respect are CCC, hybrid STATCOM/conventional conversion, self-tuned AC filters and active DC filters.

New ideas to reduce the cost of possible conversions of existing AC lines for use by DC continue to emerge, the tripole current-modulated concept being the most recent proposal. Finally, the development of a higher voltage (beyond 600 kV) HVDC technology currently under active investigation is likely to provide further incentive for the continuing used of CSC thyristor-based transmission.

References

1. Reeve, J., Baron, J.A. and Hanley, G.A. (1968), 'A technical assessment of artificial commutation of hvdc converters', *IEEE Transactions on Power Apparatus and Systems*, **PAS-87**(10), 1830–40.
2. Gole, A.M. and Meisingset, M. (2002), 'Capacitor commutated converters for long-cable hvdc transmission', *IEE Power Engineering Journal*, 129–34.
3. Jonsson, T. and Björklund, P.E. (1995), 'Capacitor commutated converters for HVDC', IEEE/KTH Stockholm Power Technology Conference, Sweden, paper SPT PE 02-3-0366.
4. Woodford, D.A. and Zheng, F. (1995), 'Series compensation of DC links', CIGRE Symposium on Power Electronics in Electric Power Systems, Tokyo, paper 420–04.

5. Menzies, D., Graham, J. and Ribeiro, F.U. (2001), 'Garabi, the Argentina-Brazil 1000 MW interconnection commissioning and early operating experience', BERLAC Conference, Foz do Iguacu, Brazil.
6. Holmgren, T. *et al.* (1995), 'A test installation of a self-tuned ac filter in the Konti-Skan 2 HVDC link', IEEE/KTH Stockholm Power Technology Conference, Sweden, paper SPT PE 02-06-0367.
7. Jansson, B. *et al.* (1993), 'New features of the Skagerrak 3 HVdc intertie', International Colloquium on HVdc and FACTS, Wellington (NZ), paper 6.3.
8. Nyman, A., Ekenstierna, B. and Waldhauer, H. (1996), 'The Baltic cable HVdc project', CIGRE 14–105, Paris.
9. Cartwright, P., Xu, L. and Ekanayake, J.B. (2004), 'The integration of large scale wind power generation into transmission networks using power electronics', CIGRE SC B4-305, Paris.
10. Andersen, B.R., Xu, L. and Cartwright, P. (2004), 'Hybrid HVDC for the supply of power to offshore oil platforms', CIGRE SC B4-304, Paris.
11. CIGRE WG 14-02 (1991), 'First benchmark model for hvdc control studies', *ELECTRA*, No. 135, 55–75.
12. Barthold, L.O., I-MOD Inc., (2004), US patent 6714427 B1.
13. Asplund, G., Astrom, U. and Wu, D. (2005), 'Advantage of HVDC transmission at 800 kV', 14th ISH keynote lecture, Beijing, China.
14. Lescale, V.F., Astrom, U., Nunes, J., Weimers, L. and Wu, D. (2006), 'Power transmission with HVDC at 800 kV', CIGRE paper B4-106, Paris.
15. Krishnayya, P.C., Lambeth, P.L., Maruvada, P.S., Tring, N.G., Desilets, G., and Nilson, S.L. (1988), 'An evaluation of the R&D requirements for developing HVDC converter stations for voltages above ±600 kV', CIGRE 14-01, Paris.

10

VSC Transmission

10.1 Introduction

The great advances made in the past two decades on power semiconductors, described in Chapter 2, encouraged the manufacturers to investigate their possible use in HVDC transmission. As a result, a more flexible VSC transmission technology has been developed that does not rely on the AC system voltage for the valves' commutations.

The structural and operating characteristics of self-commutating VSC have already been described in Chapter 4. With reference to power transmission, self-commutating VSC offers the following advantages over conventional LCC-CSC:

- Each end of the link can be controlled to absorb or generate reactive power independently of the active power transfer.
- The DC link can be connected to a weak, and even passive, AC network.
- VSC transmission has no minimum DC limits.
- VSC transmission can be designed to provide a variety of ancillary services to the interconnected AC systems, such as reactive power compensation, harmonic and unbalanced voltage compensation, flicker elimination, etc.

At the time of writing, self-commutating HVDC is only available in the form of PWM-VSC, a technology developed by ABB under the code name of HVDC Light. Therefore, much of the chapter is devoted to this subject. HVDC Light provides the most flexible power transmission alternative in terms of power controllability. However, the high-frequency switching required by the PWM process results in substantially higher losses than those of line-commutated CSC transmission, a factor that has stimulated the development of the multi-level self-commutating VSC transmission options discussed in Chapter 11.

Flexible Power Transmission: The HVDC Options J. Arrillaga, Y.H. Liu and N.R. Watson

10.2 Power Transfer Characteristics

The basic configuration of a point-to-point VSC transmission link consists of two VSC units and a DC line as shown in Figure 10.1. Each end of the link may be connected to a separate AC system or to a different bus of a common grid. Moreover, the receiving end of the link may supply power to a totally passive system.

To simplify the description, the interface transformers are represented as ideal transformers in series with their leakage reactance (X_C). V_C is the converter output voltage (on the primary side of the ideal transformer) and V_T the converter terminal voltage (behind the transformer reactance). The AC system is represented by a series reactance (X_L) and an ideal source (V_S). The three voltages (V_S, V_C, V_T) indicated in Figure 10.1 are the line (i.e. phase-to-phase) voltages.

On the DC side, V_{dc} is the converter DC voltage and the line is modelled as a series inductive (L_{dc}) resistive (R_{dc}) circuit.

The AC system voltage and current relationships at the sending and receiving ends are illustrated in the phasor diagrams of Figures 10.2 and 10.3. In these diagrams δ is the phase angle difference between the AC system source and converter voltage and δ' the phase angle difference between the AC system source and the converter terminal voltage. The following

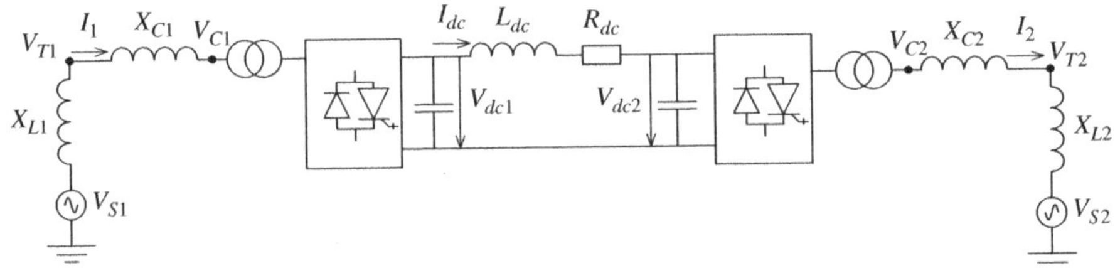

Figure 10.1 Basic VSC transmission link

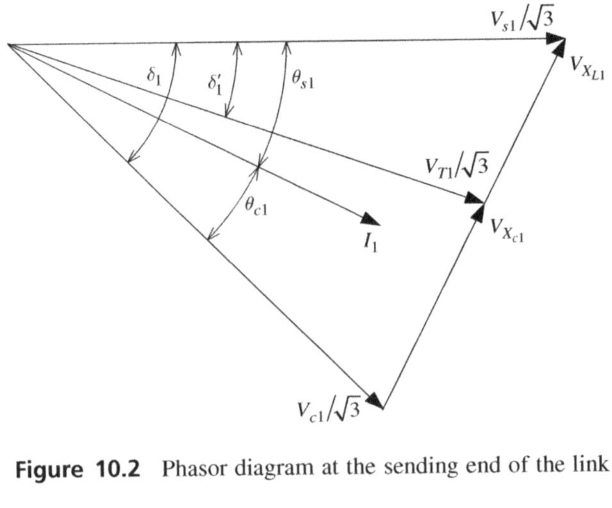

Figure 10.2 Phasor diagram at the sending end of the link

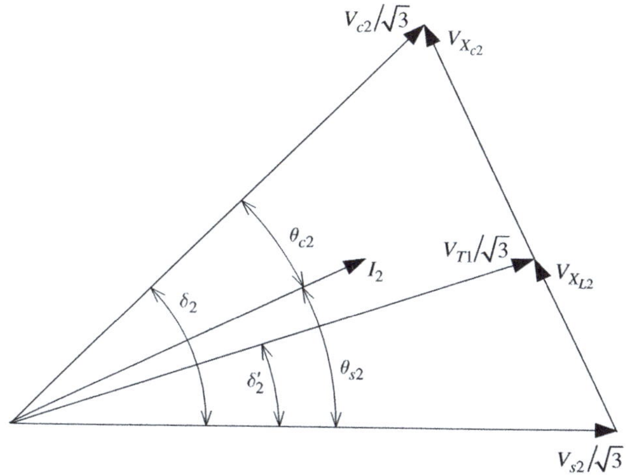

Figure 10.3 Phasor diagram at the receiving end of the link

expressions apply to the (fundamental frequency) active and reactive powers at the sending and receiving ends of the link:

$$P_1 = \frac{V_{S1}V_{C1}}{X_{L1}+X_{C1}}\sin(\delta_1) = \frac{V_{S1}V_{T1}}{X_{L1}}\sin(\delta_1') = \frac{V_{T1}V_{C1}}{X_{C1}}\sin(\delta_1-\delta_1') \tag{10.1}$$

$$P_2 = \frac{V_{S2}V_{C2}}{X_{L2}+X_{C2}}\sin(\delta_2) = \frac{V_{S2}V_{T2}}{X_{L2}}\sin(\delta_2') = \frac{V_{T2}V_{C2}}{X_{C2}}\sin(\delta_2-\delta_2') \tag{10.2}$$

$$Q_{SC1} = \frac{V_{S1}(V_{S1}-V_{C1}\cos(\delta_1))}{X_{L1}+X_{C1}} \tag{10.3}$$

$$Q_{SC2} = \frac{V_{S2}(V_{S2}-V_{C2}\cos(\delta_2))}{X_{L2}+X_{C2}} \tag{10.4}$$

The reactive powers between the AC sources and their respective interface transformer terminals are expressed by

$$Q_{ST1} = \frac{V_{S1}(V_{S1}-V_{T1}\cos(\delta_1'))}{X_{L1}} \tag{10.5}$$

$$Q_{ST2} = \frac{V_{S2}(V_{S2}-V_{T2}\cos(\delta_2'))}{X_{L2}} \tag{10.6}$$

Ideally, the transfer of active power should take place with minimum loss (i.e. with minimum AC at each end), a condition that in the case of PWM-VSC transmission is achieved by independently controlling the reactive power injected at each end of the link.

10.2.1 Current Relationships

The following relations apply to the phasor diagram of Figure 10.2:

$$(V_{S1}/\sqrt{3}) - (V_{C1}/\sqrt{3})\cos(\delta_1) = I_1(X_{L1} + X_{C1})\sin(\theta_{S1}) \tag{10.7a}$$

$$I_1(X_{L1} + X_{C1})\cos(\theta_{S1}) = (V_{C1}/\sqrt{3})\sin(\delta_1) \tag{10.7b}$$

$$(V_{S1}/\sqrt{3})\sin(\delta_1) = (X_{L1} + X_{C1})I_1\cos(\theta_{C1}) \tag{10.7c}$$

Eliminating θ_{S1} between Equations (10.7a) and (10.7b) gives

$$I_1 = \frac{\sqrt{V_{S1}^2 + V_{C1}^2 - 2V_{C1}V_{S1}\cos(\delta_1)}}{\sqrt{3}(X_{L1} + X_{C1})} = \frac{\sqrt{V_{S1}^2 + V_{T1}^2 - 2V_{T1}V_{S1}\cos(\delta_1')}}{\sqrt{3}X_{L1}} \tag{10.8}$$

Then combining Equations (10.7c) and (10.8),

$$\theta_{C1} = \cos^{-1}\left[\frac{V_{S1}}{\sqrt{V_{S1}^2 + V_{C1}^2 - 2V_{C1}V_{S1}\cos(\delta_1)}}\sin(\delta_1)\right] \quad \text{for} \quad V_{C1} > V_{S1}\cos(\delta_1)$$

or

$$\theta_{C1} = -\cos^{-1}\left[\frac{V_{S1}}{\sqrt{V_{S1}^2 + V_{C1}^2 - 2V_{C1}V_{S1}\cos(\delta_1)}}\sin(\delta_1)\right] \quad \text{for} \quad V_{C1} < V_{S1}\cos(\delta_1) \tag{10.9}$$

and

$$\theta_{S1} = \delta_1 - \theta_{C1}$$

Similarly from Figure 10.3,

$$I_2 = \frac{\sqrt{V_{S2}^2 + V_{C2}^2 - 2V_{C2}V_{S2}\cos(\delta_2)}}{\sqrt{3}(X_{L2} + X_{C2})} = \frac{\sqrt{V_{S2}^2 + V_{T2}^2 - 2V_{T2}V_{S2}\cos(\delta_2')}}{\sqrt{3}X_{L2}} \tag{10.10}$$

$$\theta_{S2} = \cos^{-1}\left[\frac{V_{C2}}{\sqrt{V_{S2}^2 + V_{C2}^2 - 2V_{C2}V_{S2}\cos(\delta_2)}}\sin(\delta_2)\right] \quad \text{for} \quad V_{S2} > V_{C2}\cos(\delta_2)$$

or

$$\theta_{S2} = -\cos^{-1}\left[\frac{V_{C2}}{\sqrt{V_{S2}^2 + V_{C2}^2 - 2V_{C2}V_{S2}\cos(\delta_2)}}\sin(\delta_2)\right] \quad \text{for} \quad V_{S2} \leq V_{C2}\cos(\delta_2) \tag{10.11}$$

and

$$\theta_{C2} = \delta_2 - \theta_{S2}$$

The AC real and imaginary components (using their respective source voltages as a reference) are

$$\begin{aligned} I_{\mathrm{Re}1} &= I_1 \cos(\theta_{S1}) \qquad I_{\mathrm{Im}1} = I_1 \sin(\theta_{S1}) \\ I_{\mathrm{Re}2} &= I_2 \cos(\theta_{S2}) \qquad I_{\mathrm{Im}2} = I_2 \sin(\theta_{S2}) \end{aligned} \tag{10.12}$$

AC current minimisation

The receiving end AC current is at its minimum value (for a specified value of P_2) when

$$\frac{V_{C2}}{V_{S2}} = \sqrt{1 + \frac{P_2^2 (X_{L2} + X_{C2})^2}{V_{S2}^4}} \tag{10.13}$$

Using instead the interface transformer terminal voltage (V_{T2}) as a reference, the minimum current condition becomes

$$\frac{V_{T2}}{V_{S2}} = \sqrt{1 + \frac{P_2^2 X_{L2}^2}{V_{S2}^4}} \tag{10.14}$$

The components $P_2^2(X_{L2} + X_{C2})^2/V_{S2}^4$ in Equation (10.13) and $P_2^2 X_{L2}^2/V_{S2}^4$ in Equation (10.14) can also be written in terms of the SCR (short-circuit ratio), which expressed in the per unit system become

$$\begin{aligned} \frac{P_2^2 (X_{L2} + X_{C2})^2}{V_{S2}^4} &= \left[\frac{P_2}{P_{2rated}}\right]^2 \left[\frac{V_{2rated}}{V_{S2}}\right]^4 \left[\frac{X_{L2} + X_{C2}}{X_{2rated}}\right]^2 \\ &= \left[\frac{P_2}{P_{2rated}}\right]^2 \left[\frac{V_{2rated}}{V_{S2}}\right]^4 \left[\frac{1}{\mathrm{SCR}_2}\right]^2 \end{aligned} \tag{10.15}$$

$$\begin{aligned} \frac{P_2^2 X_{L2}{}^2}{V_{S2}^4} &= \left[\frac{P_2}{P_{2rated}}\right]^2 \left[\frac{V_{2rated}}{V_{S2}}\right]^4 \left[\frac{X_{L2}}{X_{2rated}}\right]^2 \\ &= \left[\frac{P_2}{P_{2rated}}\right]^2 \left[\frac{V_{2rated}}{V_{S2}}\right]^4 \left[\frac{1}{\mathrm{SCR}_2'}\right]^2 \end{aligned} \tag{10.16}$$

where V_{2rated} and P_{2rated} are the rated values of voltage and power used as base parameters in the per unit system.

To achieve minimum AC current, Equations (10.13) and (10.14) indicate that the ratios V_{C2}/V_{S2} and V_{T2}/V_{S2} have to increase when the active power demand increases. Moreover, their maximum increments are strongly dependent on the AC system impedance, i.e. a strong AC system requires a low increment and a weak system a high increment. Also a slight variation in the AC source voltage (V_{S2}) can cause a significant change of the ratio due to the fourth exponent.

For example, if the SCR of the receiving end is 2.5, to transfer the rated active power with a source voltage V_{S2} equal to $0.96V_{2rated}$ the required ratio is $V_{C2}/V_{S2} = 1.091$ and the converter AC output voltage needs to be adjusted to $V_{C2} = 1.047V_{2rated}$.

A similar process will achieve the current minimisation at the sending end in the case of PWM-VSC transmission, where the reactive power injection required to get the right V_{C1}/V_{S1} ratio can be controlled independently from that at the receiving end. This is not the case with the multi-level HVDC alternatives, as will be discussed in Chapter 11.

10.3 Structure of the VSC Link

The structure of each terminal in a VSC transmission link is shown in Figure 10.4 [1] and the physical layout of a typical station is illustrated in Figure 10.5 (the total building site is 45 m by 18 m). Chapters 4–6 and 7 have described the characteristics of the various self-commutating converter topologies in existence or under consideration.

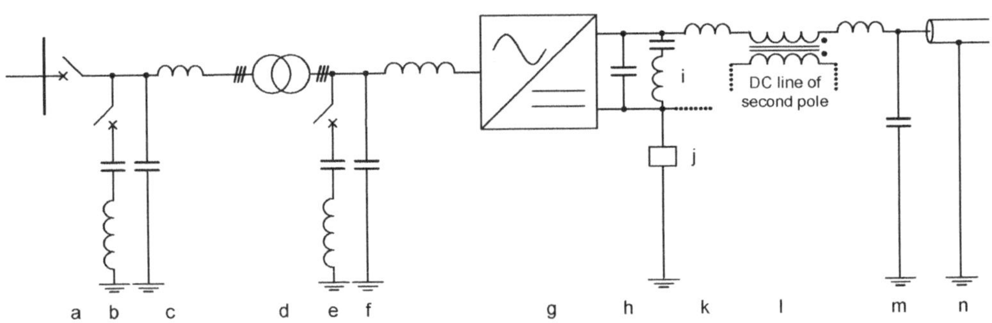

Figure 10.4 Main components of a VSC substation: (a) AC system circuit breaker; (b) AC side harmonic filter; (c) AC side radio-frequency interference filter; (d) interface transformer; (e) converter output harmonic filter; (f) high-frequency-blocking filter/phase reactor; (g) VSC converter unit; (h) DC side capacitor; (i) DC side harmonic filter; (j) neutral point grounding; (k) DC reactor; (l) common mode blocking reactor; (m) DC side radio-frequency interference filter; (n) DC cable or overhead line (Reproduced by permission of CIGRE.)

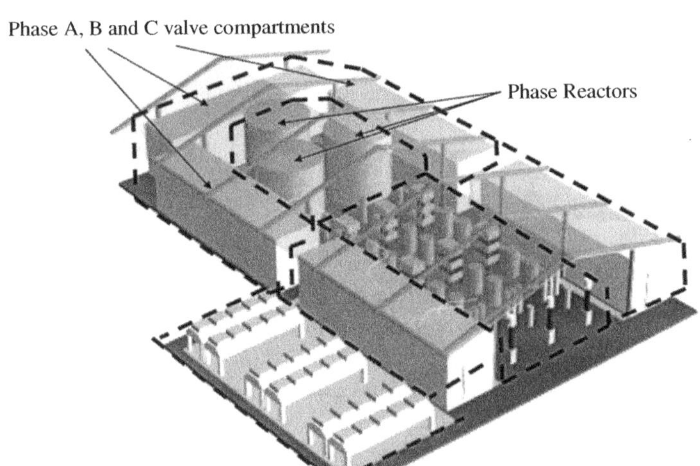

Figure 10.5 Physical layout of an HVDC converter station (Reproduced by permission of ABB.)

The main function of the station circuit breaker, shown on the far left of Figure 10.4 (component a), is to disconnect the AC system from the DC link during faults. The breaker is needed because, unlike the CSC-based option, VSC transmission has no inherent capability to clear DC line faults. The latter will cause the DC capacitor to be discharged and the fault current will flow through the diodes until the station circuit breaker opens. Also, when the DC capacitor is charged from the AC system by switching the station breaker, the latter requires the assistance of a closing resistor to reduce the charging current. This resistor also helps to reduce the inrush currents of the transformers and filters during station energisation.

The next component (b in Figure 10.4) represents the AC side harmonic filters, their need depending on the particular design concept being used. Generally this requirement will be very limited in the case of VSC, as compared with conventional CSC transmission technology. In both cases the filter design principles are the same, but they will be cheaper and more compact in the case of VSC. The use of PWM only produces high-frequency harmonics and, therefore, a high-pass filter is sufficient (i.e. no tuned filters are needed); moreover, as the filters do not have to contribute to the reactive power demand, there is no need for filter subdivision switchgear. A radio-frequency interference (RFI) filter (component c) may be needed to reduce the penetration of high-frequency harmonics into the AC system.

An interface transformer (component d) is required to adapt the converter voltage and provide reactance between the AC and converter unit in order to control the AC output current. Instead of the transformer, an air-core reactor can be used for this purpose in a VSC terminal station. The transformer also fulfils the other tasks indicated in conventional LCC transmission, except that in this case the only purpose of the on-load tap-changer (if installed) is to optimise the VSC operation and, thus, reduce the power losses.

A harmonic filter (component e) and a high-frequency blocking filter (component f) are shown on the converter side of the interface transformer, the latter consisting of a series reactor and a shunt capacitor.

Next is the DC storage shunt-connected capacitor (component h), which keeps the DC voltage within tight limits and controls the DC voltage ripple. A DC link will require an independent DC capacitor at each end, in order to maintain a stable DC voltage, i.e. the use of a common capacitor in the middle of the line is not recommended.

DC filters (component i) can be placed in parallel with the DC capacitor where the limits of voltage or current distortion in the line may be exceeded.

Grounding at one point in the DC side (component j) is needed to define the potential of the DC circuit. The grounding branch can consist of reactance, capacitance, resistance or even an electrode (if ground return operation is planned).

The remaining components in Figure 10.4 are only needed for long-distance transmission. These are a DC reactor (component k), a common mode blocking reactor (component l), a DC side RFI filter (component m) and, of course, the DC line (component n). The DC line can in theory be either overhead or cable. The latter is, however, the preferred option as explained in the following section.

10.3.1 VSC-HVDC Cable Technology

As the cable is not exposed to lightning strikes, storms, falling trees, etc., the probability of DC line short circuits is greatly reduced. This is an important consideration in VSC transmission because such faults require isolation of the link by circuit breakers at both ends to permit clearance of the reactive energy and the return to normal power flow is very slow.

There is no limit in DC cable length without the need for intermediate stations. Generally, cables have far less impact on the environment than overhead transmission. The magnetic fields are almost completely eliminated by adopting the bipolar system and they do not produce ground currents. Moreover, the lifetime of the cable insulating materials is better for DC than AC.

VSC transmission allows only one DC polarity and thus the cable does not require to be designed for polarity reversals. This greatly simplifies the cable design, permitting the use of polymeric insulation, instead of the conventional oil-impregnated paper insulation. The polymeric insulation material can withstand high forces and repeated flexing and is thus more suited for deep-water installation. This is because this type of cable can use galvanised steel wire armour, whereas AC cables need to use non-magnetic, less strong armour. Present standard voltages for the polymer-insulated cables used in recent HVDC schemes are 84 and 150 kV [2], but cable ratings of 300 kV are currently under development.

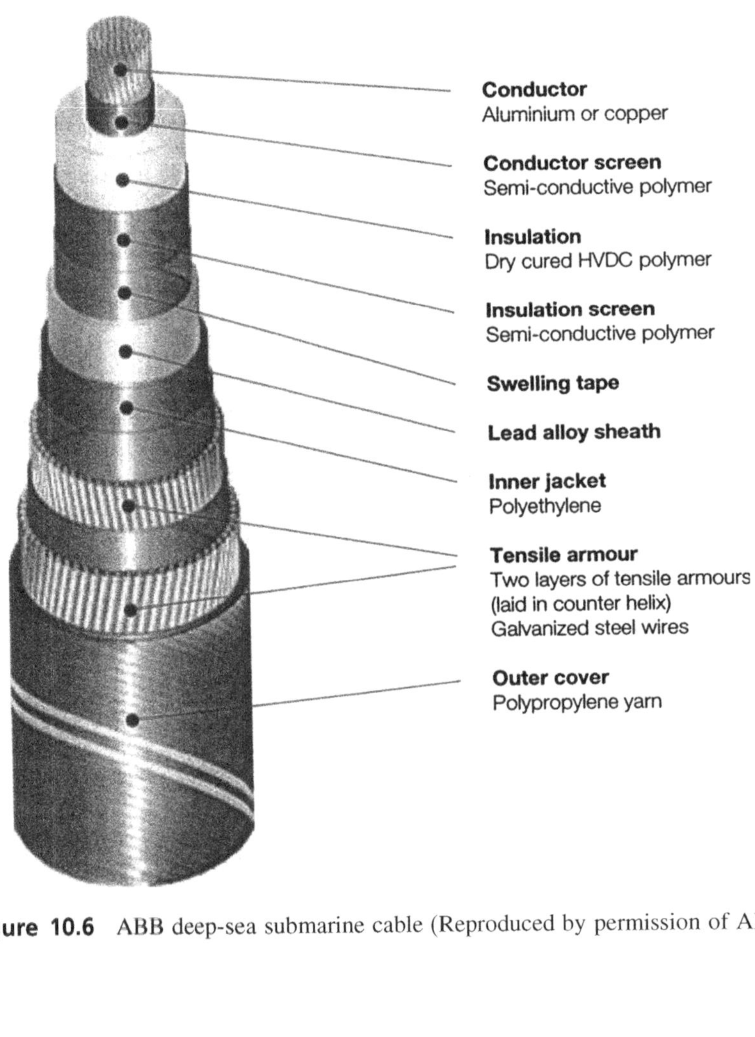

Figure 10.6 ABB deep-sea submarine cable (Reproduced by permission of ABB.)

In these cables the polymeric insulation is triple extruded together with the conductor screen and the insulation screen. This offers a very robust design, suited to land installation (using ploughing techniques), as well as submarine use under severe conditions.

Extruded HVDC single core cables are installed close to each other in bipolar pairs with anti-parallel currents that cancel the magnetic field.

A general cutaway drawing of the deep-sea submarine cable used in recent ABB HVDC schemes is shown in Figure 10.6.

The main problems of cable transmission for long distances over land are laying costs and reliability, due to the large number of joints required. These are caused by the limited length of cable that can be transported by road.

10.4 VSC DC System Control

10.4.1 General Philosophy

The common feature of all VSC configurations is the generation of a fundamental frequency AC voltage from a DC voltage; the control of this voltage, in both phase and magnitude, is the basic function of the VSC.

The phase angle (δ), and therefore the active power transfer, are controlled by shifting the fundamental frequency voltage produced by the converter. The power transfer can be either from the AC system to the converter or vice versa depending on the sign of the phase angle difference, i.e. the VSC can act as a rectifier or an inverter.

In multi-pulse and multi-level configurations with the valves switching at the fundamental frequency, the magnitude of the generated AC voltage will be directly proportional to the DC capacitor voltage. The latter can be varied by feeding power from the AC side into it or out of it, by means of small variations in the phase angle difference between the AC system voltage and the converter voltage. When power is fed into the capacitor, its charge increases and so does its voltage. When power is taken from the capacitor, its voltage decreases. A disadvantage of using the DC voltage level to control the AC voltage is the time taken to charge the large DC capacitor.

Present VSC-HVDC schemes do not effectively operate in this control mode. Instead, they are designed to keep the DC voltage nominally constant and the control of the converter AC voltage is achieved by means of PWM. In VSC-PWM conversion, as explained in Chapter 5, the AC voltage output is varied by means of a modulation index signal (λ) defined as the ratio of the required AC voltage magnitude to the maximum AC voltage that can be generated for a given DC size capacitor. When this modulation index is close to one the converter voltage is greater than the AC system voltage and reactive power is transferred to the AC system. When the index is low (i.e. the converter voltage is lower than the system voltage), the converter absorbs reactive power. There are two possible strategies to implement the modulation index, namely ***direct control*** and ***vector control*** [1].

In direct control the modulation index λ or the phase angle (δ) are adjusted directly from the parameters being controlled as shown in Figure 10.7. Vector control, the alternative strategy, decouples the adjusting action of the modulation index from the phase angle, as shown in Figure 10.8. The current control attached to this strategy protects the valves from overloading, but this extra current loop slows down the speed of response.

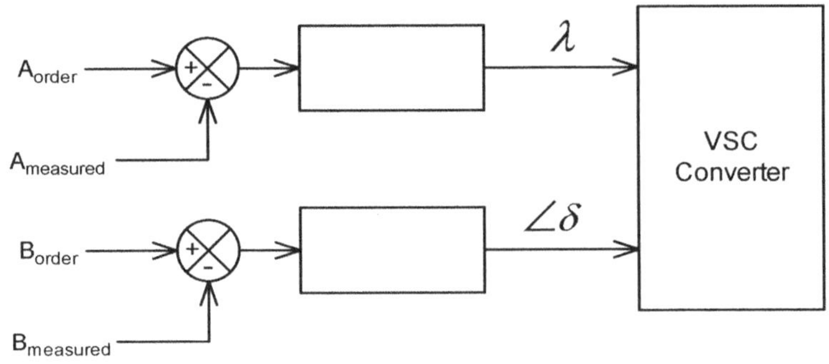

Figure 10.7 Direct control of the modulation index λ and phase angle (Reproduced by permission of CIGRE.)

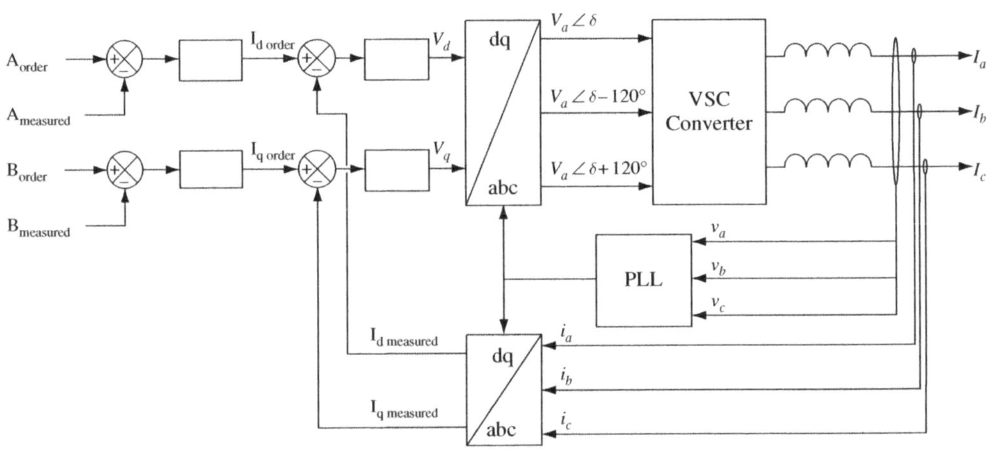

Figure 10.8 Vector control in terms of d and q axis current components (Reproduced by permission of CIGRE.)

In the vector control strategy, the three-phase currents are transformed to d and q axes, which are then synchronised with the AC system three-phase voltage via a phase-locked loop (PLL). The d and q voltages generated by vector control are transformed to three-phase quantities and converted into line voltages by the VSC.

A number of (up to three) degrees of freedom are provided by the PWM-VSC technology, as follows.

Frequency control

Controlling the frequency of the oscillator that determines the valve pulse firing sequence is essential when the VSC is the only source of power, i.e. the VSC link supplies power to an isolated load. When the VSC is connected to an active power system, the VSC can participate in the system frequency control by regulating the power delivered to or taken from the AC system.

AC voltage control

The AC voltage can be controlled by regulating the magnitude of the fundamental frequency component of the AC voltage produced by the VSC on the converter side of the interface transformer, either by altering the DC capacitor voltage (in the case of two-level multi-pulse converters) or by varying the modulation index (in the case of PWM conversion). If the VSC feeds into an isolated load, the AC voltage controller also provides automatic control of the power going into the load (this assumes that the sending end of the link is controlling the DC side voltage).

The converter transformer may be provided with OLTC (On-Load Tap-Changing) control with the purpose of keeping the bus voltage, and therefore the modulation index, within specified limits.

Active power control

The control of active power transfer is achieved by regulating the phase angle of the fundamental frequency component of the converter-generated AC voltage. Power is taken from or delivered to the AC system depending on the sign of this angle. The transfer of active power through the link requires simultaneous coordinated action at both ends of the link.

Fast power transfer control can be used to damp electromechanical oscillations and for the improvement of transient stability following disturbances.

Reactive power control

The reactive power generated or absorbed by the VSC is controlled by the magnitude of the converter AC voltage source, which in PWM conversion is determined by the modulation index. The use of this function is important when the other converters in the transmission system are operating to maintain their respective AC voltages.

DC voltage control

As the various converters in a DC link share a common DC voltage, at least one of them is required to control the DC capacitor voltage, a task achieved by regulating the small extra power required to charge or discharge the capacitor to maintain the specified DC voltage level. A proportional or proportional–integral controller will be used for this task.

In the case of multi-pulse conversion DC voltage regulation is the only way for the VSC to achieve AC voltage control.

AC control

Current control is often a desirable feature to ensure that the converter valves are not overloaded. As indicated earlier, this control can be achieved directly or through vector control (as shown in Figure 10.8), where the control of the current is an intermediate step in the control of other parameters such as the active and reactive powers.

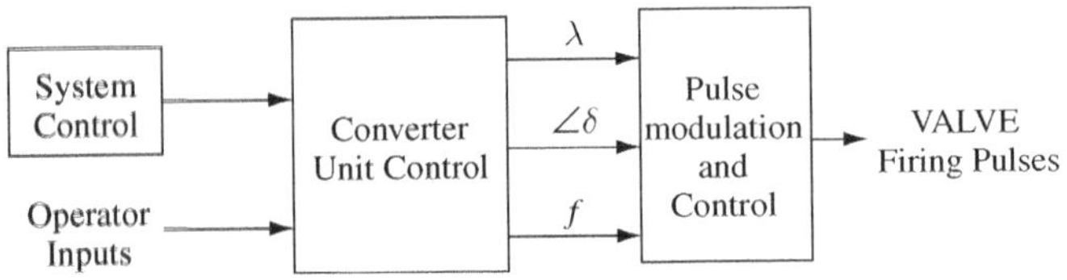

Figure 10.9 Control levels of VSC (Reproduced by permission of CIGRE.)

10.4.2 Different Control Levels

There are three control levels in a VSC scheme (shown in Figure 10.9); these are referred to as firing, converter unit and system controls.

Firing control

The type of firing control of the bridge valves depends on the switching device and VSC configuration.

In the PWM method, the valve firing sequence is regulated by the type of PWM applied. A number of selective harmonic elimination techniques have been discussed in Chapter 5 to achieve harmonic cancellation with the lowest possible switching frequency. In each case a compromise needs to be made between the degree of lower harmonic cancellation through PWM or multi-pulse configuration and the size of filters to be used.

The type of oscillator used in firing control is a PLL, if it is to be synchronised to an active AC system, or independently locked, if feeding an isolated system. The input of the PLL is the three-phase voltage set measured at the filter bus. The output of the PLL is a time-dependent phase angle, which should be equal to the phase angle of the filter bus voltage vector in steady state, and a phase angle error.

Converter unit control

When the VSC uses the multi-pulse two-level configuration, the converter control determines the DC voltage across the DC capacitor by adjusting the phase of the PLL so that the phase angle created across the interface reactor causes the necessary power flow. If, instead, the VSC uses the PWM principle, the converter control changes both the phase angle of the generated fundamental frequency component of the converter AC voltage as well as its magnitude. The use of a single chop within the fundamental frequency cycle can be used to control the fundamental component magnitude in multi-level schemes. However, this action results in higher harmonic content.

The VSC can also participate in the AC system frequency control by adjusting the frequency of the oscillator in the firing controls (if feeding an isolated system), or modulating the power (if feeding into an active AC system).

System control

By providing the appropriate signals as inputs and feedback quantities, the system control can achieve a variety of important functions, such as to control the real and reactive power flows, control the voltage magnitude and phase angle, enhance the system transient stability, help to damp system oscillations, provide frequency control, etc.

As in the case of LCC transmission, the specified power flow in VSC transmission is normally controlled at the sending end of the link and the DC voltage at the receiving end (at a nominally constant value). The difference between them is that, while in LCC the current is unidirectional (a change in power direction requiring DC voltage polarity reversal), the DC voltage polarity in VSC is always the same and power reversal is achieved by a change in the DC current flow direction.

10.4.3 DC Link Control Coordination

Normally one or more of the converters connected to the common DC bus will be used to regulate the active power needed to maintain the DC voltage across the capacitor within the prescribed limits.

The converters at the two ends of a point-to-point link must be controlled to work together in order to transfer active power across. This action involves setting the appropriate phase angles (with opposite signs) for the AC voltages at the converter sides of their respective interface reactors.

When reactive power as well as AC voltage control is used, interference between the different controllers can be avoided by selecting only those converters not used for AC voltage control to provide the reactive power control.

If all the converters working in close proximity are controlling the same quantities, it is possible for each to participate in AC voltage control through a carefully designed droop characteristic. However, if their controlling functions are different (such as acting as independent power controllers), the droop characteristic may be difficult to define. As explained earlier, current control is an inherent feature of the vector control strategy; this is important to ensure that the AC current through the VSC is within the permissible operating range to avoid overloading.

When the VSC feeds into an active AC system, where the frequency is fixed by synchronous generators and load frequency control, the VSC can participate in the frequency control by regulating the power delivered to, or taken from, the AC system. When the VSC feeds into a passive system, the voltage frequency is controlled by the frequency of the oscillator that determines the valve pulse firing sequence.

Under normal operation each station controls its reactive power independently of the other. On the other hand, the active power flow of the sending end station must equal that of the receiving station plus the losses in the DC system. Any difference will result in an increase or decrease in the DC voltage. The power balance is achieved by using one of the converter stations to control the DC voltage, while the other station controls the active power order. The DC voltage controlling station will adjust its power order to ensure power balance (which means maintaining the voltage constant); this is achieved without the need for telecommunications between the stations, purely based on measurements of the DC voltage.

However, power imbalances will occur during transient conditions. Assume for instance that a converter supplying power to the AC side is temporarily blocked. Then the energy stored in the inductances of the DC circuit will charge the DC capacitors and the DC voltage will increase; this will be counteracted by the DC voltage controlling station, which will decrease, and even reverse, the active power flow to maintain the DC voltage constant. At startup both stations are energised separately. When the AC breakers are closed, the DC buses are energised through the anti-parallel diodes in each bridge. Once the gate drive units are charged, the converters in the two stations can be connected by the switches on the DC side. The first converter which is deblocked will control the DC voltage.

During emergencies, power reversal through the link can be implemented very quickly. From the control viewpoint the converter can reverse full power in milliseconds, but the speed is limited by the network components. Reversal can be achieved without change of control mode and without any filter switching or converter blocking. It is implemented by changing the DC current direction, while the DC voltage remains constant. Under PWM control the reactive power controllers at each of the terminals continue to operate independently of the active power direction.

As the AC current can be controlled, the HVDC link contribution to the short-circuit power is expected to be small. However, simulation studies carried out by ABB have shown that, in contrast to conventional CSC transmission, HVDC Light can contribute some short-circuit current, which varies in inverse proportion to the SCR, and that the maximum contribution occurs when the converters operate at zero active power. The contribution depends on the control strategy used. Under reactive power control, the short-circuit current contribution will be small, because the current order limit reduces with the voltage. In the AC voltage control mode, the contribution increases with decreasing active power if the current limit is not changed. Of course the increased short-circuit contribution is associated with an improved performance in voltage stability, i.e. the voltage dips during distant faults are reduced. During AC faults the current control will rapidly lower the fundamental frequency voltage of the bridge to reduce the DC current down to the pre-fault level.

The gate units are provided with a primary valve/bridge protection system that acts in nanoseconds and a back-up protection system that acts in a few microseconds, the latter based on the current flowing in the DC capacitors and phase reactors.

10.4.4 Control Capability of VSC Transmission [3]

The capability of VSC transmission to help the stability of the interconnected power systems is limited, as shown in Figure 10.10, by the following three factors:

1. The maximum current through the IGBTs. This gives rise to a maximum MVA circle in the power plane where the maximum current and the actual AC voltage are multiplied. When the AC voltage drops, the MVA capability reduces accordingly.

2. The maximum DC voltage level. The reactive power is mainly dependent on the voltage difference between the AC voltage that the VSC can generate and the grid AC voltage. If the latter is high, the difference between the maximum DC voltage and the AC voltage will be low. The reactive power capability is then moderate, but increases with decreasing AC voltage.

3. The maximum DC current through the cable.

When the voltage level reduces, the capability is determined by the maximum current level. The small bias in the Q axis direction is due to the presence of a line reactor and filter

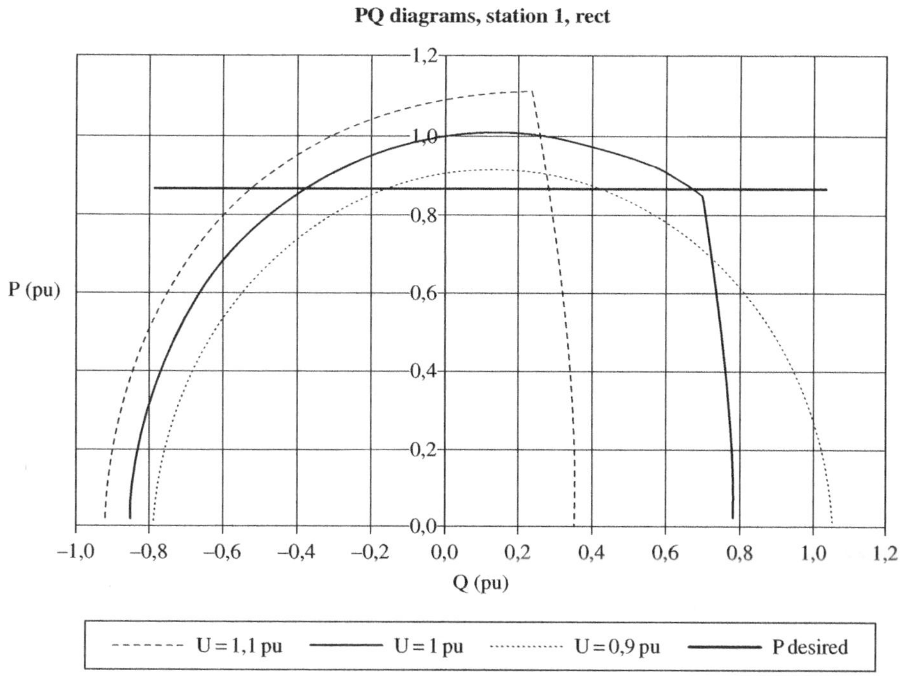

Figure 10.10 Active and reactive power converter capabilities

capacitance within the VSC transmission system. When available, the converter transformers' tap-changers also play an important part in the control of the reactive power transfer.

The VSC link can alter the operating point anywhere within the capability chart (shown in Figure 10.10) practically instantaneously. Fast power reversal is also possible, which requires reversing the DC current direction and not the voltage polarity. The XLPE-based cable technology used can handle such reversals without any problems.

The practically instantaneous variation of the active and reactive power within the capability curve can be used to support the grid with the best mixture of active and reactive power during stressed conditions. In general, active power modulation will provide the best damping, but a mix of active and reactive power control will in some cases be the most practical solution. VSC transmission is able to support the AC grid with a suitable power factor to improve system stability.

10.4.5 Assistance During Grid Restoration

Following a blackout on the AC system, the VSC link will be able to assist recovery almost immediately by its inherent fast voltage controllability, regardless of the short-circuit capacity of the AC system. The extent of the benefit will, of course, depend on whether one or both ends of the link are exposed to the blackout.

The DC link can then energise some of the AC transmission lines at a lower voltage in order to avoid the overvoltages caused by the Ferranti effect, thus permitting the remote connection of transformers at a safer voltage level. Following the transformer connection, the DC link can ramp the AC voltage up to its nominal value. The VSC link can also feed auxiliary power to the local plants at the remote end and thus provide them with a stable voltage and frequency to start on.

10.5 HVDC Light Technology

HVDC Light is a pioneering technology developed by ABB [4], which combines the high controllability of PWM conversion (based on high-frequency IGBT switching) with the more reliable and environmentally friendly underground cable transmission. Other attractive features of the technology are its compact and lightweight design (based on a modular concept), short installation and commissioning times, low operation and maintenance costs (as the stations can be operated remotely) and great control flexibility. It is suitable for a large spectrum of applications in the transmission and sub-transmission of electric power. It was initially designed as standard units between 5 and 150 MW, which are built in self-contained modules. However, as a result of the successful performance of the early schemes and the availability of ever increasing component ratings, the power capability of this technology is already in hundreds of megawatts (which questions the continued use of the term *Light*!).

With reference to the basic configuration of Figure 10.1, the three-phase bridge converters of an HVDC Light link are arranged to produce either two-level or three-level voltage waveforms. In the three-level case, IGBT valves may also be used, instead of diodes, for neutral point clamping. Each valve consists of a number of series-connected IGBTs with anti-parallel diodes and each IGBT position is individually controlled and monitored via fibre optics. The valves, DC buses and DC capacitors are designed with low inductance to reduce the overvoltage across the valve at turn-off. Auxiliary power for the gate drive unit is generated from the voltage across the IGBT. Turn-on/off of each IGBT is ordered via an optical link from the control equipment placed at ground potential.

The use of IGBT valves requires low energy to switch the individual devices, due to their high-impedance gate, and this in turn permits the use of high-frequency switching, as required by PWM.

The main purpose of PWM is to achieve fast control of the amplitude and phase of the fundamental frequency voltage, which, in turn, permits control of the active and reactive power independently. However, as the rating of the converter is based on maximum currents and voltages, the reactive power capabilities of a converter must be treated against the active power capability [5]. The combined active/reactive power capabilities are illustrated in Figure 10.10. The capacitive limit in the figure is needed to impose a voltage limitation, i.e. if the voltage is reduced, this limit increases.

Harmonic content of the converter output voltage

A variety of modulation alternatives have been described in Chapter 5 to reduce the low harmonic orders. The quality of the waveform needs to be further improved for power transmission and distribution applications and this is achieved by means of a series reactor and some AC filters.

The magnitude of the voltage harmonic components varies with the DC voltage, the switching frequency, the number of converter levels and the PWM strategy chosen. The use of an optimal pulse width modulation (OPWM) technique provides harmonic elimination and reduces the converter losses. OPWM consists of two functions, their specific purposes being to calculate the time to the next sample instant and modulate the reference voltage vector. This technique concentrates the harmonics in a narrow bandwidth, and thus reduces substantially the size of the filters. The converter losses are reduced by switching the valves less often when the current is high.

Its effect is illustrated in Figure 10.11 for the case of a converter using a PWM technique designed for optimum harmonic cancellation. The figure shows the converter terminal phase-to-ground voltage (with the fundamental component as the dotted line). The harmonic spectrum is given in terms of sequence components. The filter of a typical HVDC Light scheme contains two or three tuned branches and reduces the individual harmonics to 1 % and the THD to 2.5 % with reference to the fundamental voltage measured at the point of common coupling.

On the DC side, the DC capacitor and the line smoothing reactor will normally be sufficient to limit the harmonic content. Some additional DC side filtering may be required, in the form of a common mode reactor (to eliminate the zero-sequence harmonic content) and/or one or more single filters.

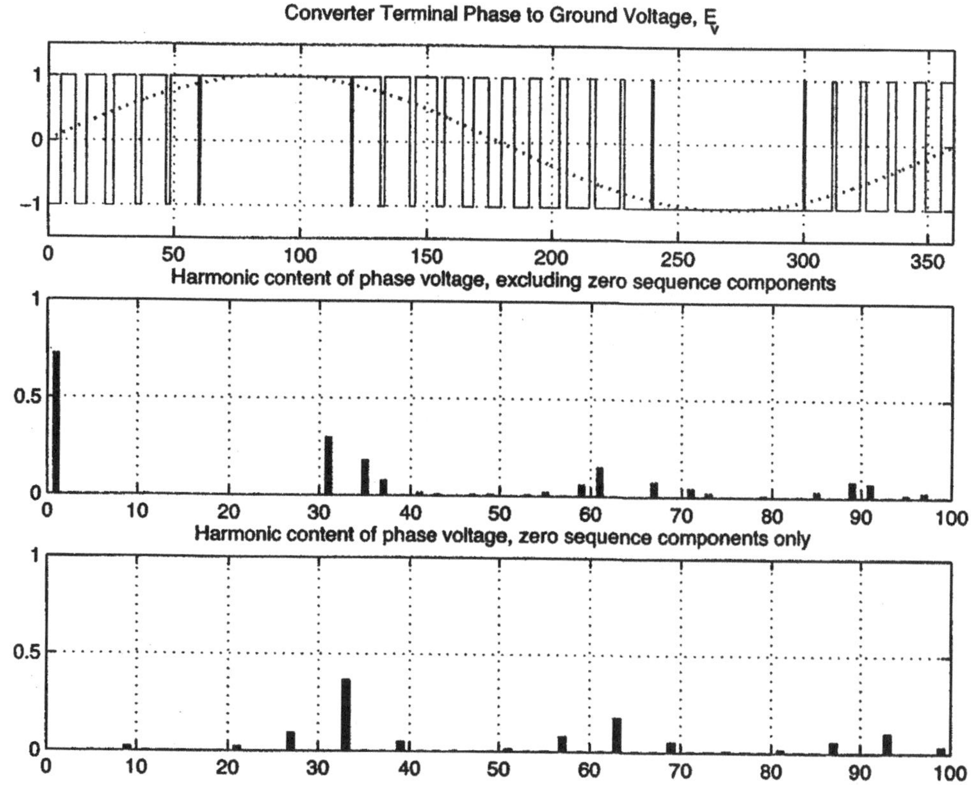

Figure 10.11 VSC-PWM with optimum harmonic cancellation

10.5.1 Two-Level PWM Schemes

This section describes the purpose and main characteristics of the two-level HVDC Light projects already in existence or under consideration at the time of writing.

The positive experience gained with the small Hellsjon project between Hellsjon and Grangesberg [4], a 10 km, 3 MW, 10 kV link (operating since 1997), in central Sweden gave confidence to extend the potential of this technology to larger power ratings and distances in a range of applications such as power supply to isolated remote loads (even passive ones), power supply to islands, infeed to cities, remote scale generation (such as low-head hydro power and wind power), offshore generation, etc.

Tjaereborg (commissioned in 2000)

Tjaereborg is an isolated wind farm consisting of four turbines with a total installed capacity of 6.5 MW. The purpose of this test project was to investigate how the PWM-VSC transmission scheme could be used to optimise the use of the wind energy.

The isolated wind farm is thus connected to the 10.5 kV power distribution network of Eltra (Denmark), via a 4.3 km underground cable of extruded polymer type. The power ratings are 8 MVA, 7.2 MW, $-3/+4$ MVAR and the nominal DC voltage ± 9 kV. It uses a single two-level IGBT-based converter unit under PWM control, the switching frequency being 1950 Hz. The control modes include the AC voltage and power frequency (between 30 and 65 Hz) in the converter connected to the wind farm. The ability to change the stator frequency of the induction generator permits optimisation of the power output from the turbine, by adjusting the frequency in relation to the wind velocity and thus operate at maximum power. The scheme is provided with automatic start when the wind is capable of supplying 700 kW.

Figures 10.12 and 10.13 illustrate the control structures at the sending and receiving ends [6]. The sending end station sets the voltage and frequency for the wind farm controller. Since each wind turbine may have a different wind speed at a particular moment (and therefore a different operating point), a droop control of voltage forms the basis of the system of Figure 10.12. This ensures that each turbine delivers active power and absorbs reactive power according to wind speed variations. Voltage stability is very dependent on the demand for reactive power, and thus the voltage control loop should ensure that the reactive currents circulating in the wind farm are not too large. The droop characteristic of the voltage controller minimises the stationary error with respect to the voltage reference and thus reduces the circulation of reactive currents. As the induction generator always draws reactive power from the PWM converter, the voltage reference should be increased when the generator demand for reactive power increases, i.e.

$$V^{ref} = V^{sp} + k_v Q^{meas}$$

where k_v is the voltage droop coefficient.

Normally the grid side PWM converter will control the DC link voltage and this is achieved by the droop control characteristic shown in Figure 10.13. A PLL is used to derive the angle for the transformation between axes. The actual power of the DC link is calculated from the measured DC voltage and current signals. The signals for the reactive power calculation are obtained from the receiving end AC system side.

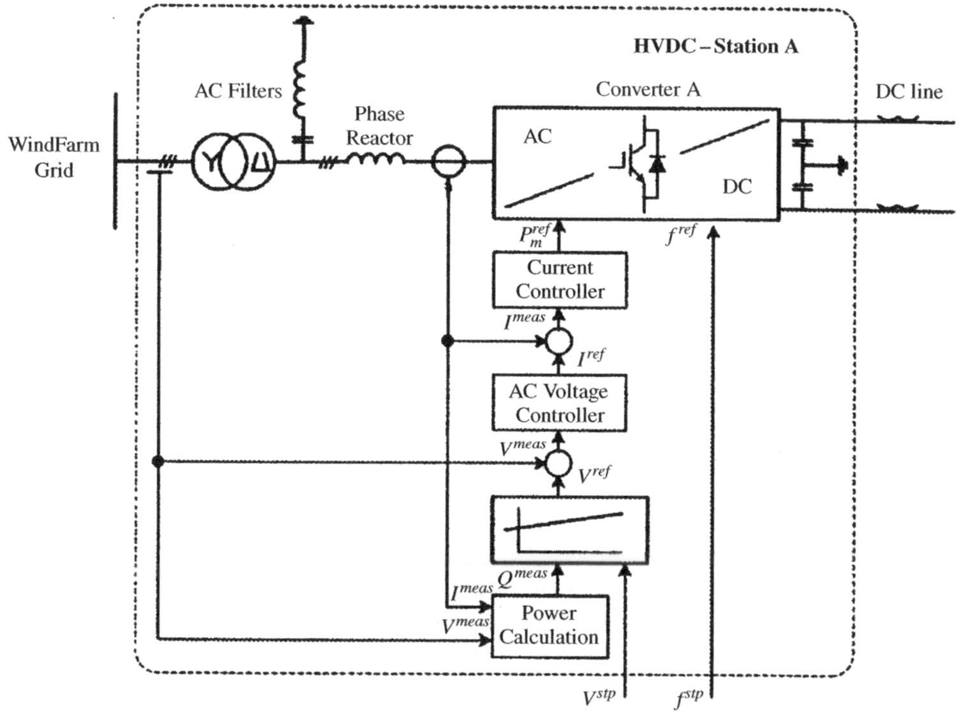

Figure 10.12 Control structure of the sending end station of a wind farm [6] (Reproduced by permission of ABB.)

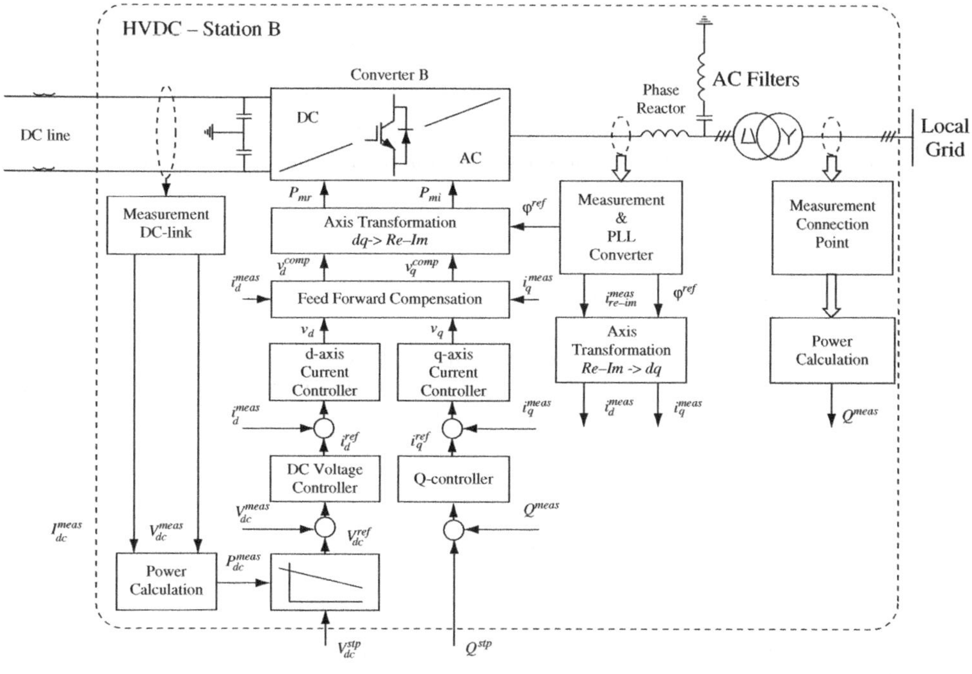

Figure 10.13 Control structure of the receiving end station of a wind farm [6] (Reproduced by permission of ABB.)

The DC link voltage setting is calculated from the measured DC power and the droop characteristic; the error signal between the voltage setting and the measured value is fed to a PI-type DC voltage controller.

The reference for the d axis current is derived by the DC link voltage controller, whereas that of the q axis current is obtained from the output of a PI controller that processes the difference between the set reference and actual value of the reactive power. An inner current control loop and a feed-forward compensation of the voltage references are provided for both axes. The feed-forward compensation requires the exact value of the total reactance that exists between the PWM converter and the point of grid connection, which involves the phase reactor and transformer (the influence of the AC filter impedance at 50 Hz can be neglected).

Computer simulation has shown that during AC grid faults the wind farm can be quickly isolated and controlled to recover to full wind power generation once the AC system fault has been cleared. Tests have also shown that the frequency at the wind site can be varied smoothly by the converters, while the receiving end grid frequency remains constant.

Direct Link (commissioned in 2000)

Direct Link is an ITP (Independent Transmission Project) developed by the Hydro-Quebec group and North Power. This scheme pioneered the use of HVDC Light for large-power, long-distance underground interconnection between two separate AC transmission systems, namely those of the Queensland 110 kV and New South Wales 132 kV transmission systems in Australia, its main justification being the price differential of electrical energy in the two states. The transmission distance is 65 km and the cable is buried following existing rights of way with no land acquisition required. It consists of three separate links, each rated at 65 MVA (180 MW, ±75 MVAR), with a DC voltage of ±80 kV. It uses two-level IGBT-based converters under PWM control, with a switching frequency of 1950 Hz. The specific control modes are active power, reactive power and AC voltage.

Direct Link is a non-regulated project, operating as a generator by delivering energy to the highest value regional market. The VSC terminals can act independently of each other to provide ancillary services (such as VAR support) to the relatively weak networks interconnected by the link.

Gotland (commissioned in 1999)

This is a 70 km link designed to connect the Nas wind farm to the city of Visby via the GEAB (a subsidiary of the Swedish National Energy Administration) power distribution system. The AC voltage at both ends is 77 kV and the DC voltage ±80 kV. The rated powers are 50 MW, ±30 MVAR. It uses a two-level IGBT-based converter under PWM control with a switching frequency of 1950 Hz.

The Gotland system has a peak load of about 160 MW and contains some 165 windmills with a total installed capacity of 90 MW. The fault level is very low at the connection point. HVDC Light was the only realistic way to overcome the technical problems caused by the large content of wind generation on such a low SCR. So far the experience with the scheme has shown that the flicker problem has been practically eliminated, the system stability improved, the voltage dips mitigated and the harmonic content reduced.

The transmission link uses bipolar DC extruded cables ploughed into the ground and, thus, eliminates the environmental problems as the scheme crosses bird and wildlife reserves. The specific control modes are active and reactive power control, AC voltage control and power quality control.

Motorformer DC Light (commissioned in 2005)

This is the first offshore HVDC Light installation to provide power from the mainland to the motor drives of the Troll A platform. This avoids the need to generate power at the platform. Gas-fired power plants onshore are three times more efficient than gas turbines on the platform and CO_2 emission is substantially reduced (by more than 50 %).

The Troll A offshore platform is used to delivered gas to Kollnes in Norway. The distance is 67 km. Two HVDC Light transmission links, each 40 MW at $\pm$60 kV, are used as feeders from the mainland to transmit power to the inverter placed on the Troll gas platform.

The platform-based converters were lifted onto the platform in 2004. The DC link directly feeds a high-voltage variable speed synchronous machine (used as the compressor drive) with variable frequency (0–63 Hz) and variable voltage (0–56 kV).

The inverter control software is designed to perform motor speed and torque control, while the control hardware is identical for the rectifier and motor converters. Over the entire motor operating range, unity power factor and low harmonic content are assured as well as a good dynamic response. No telecommunication link was initially installed between the rectifier control on land and the inverter control on the platform, the only detectable quantity at both ends being the DC link voltage. Provision is, however, made for a telecommunication link to permit land-based operation, fault finding and maintenance of the platform station. The high-voltage converter equipment needs to be protected against the salt-laden and humid air at sea. With this aim the high-voltage equipment has been installed inside a compact module offshore and in an indoor building onshore. Space and weight are also scarce resources on offshore installations and in this respect the low filter requirements make the HVDC Light concept very compact and lightweight.

ABB has announced an order for the second offshore HVDC Light station from BP Norway for connection from shore to the Valhall field in the North Sea over a distance of 292 km.

Powerformer DC Light

At the time of writing this is a proposed application, largely based on the unit connection concept [7], which involves the use of single generators directly connected to HVDC converter units, without converter transformers, filters and switchgear, thus providing, as well as higher efficiency, considerably reduced investment and operating costs.

Unlike the original unit connection concept (based on thyristor technology), the use of HVDC Light provides reactive power and waveform control. These properties reduce considerably the cost of the synchronous generator and permit the connection to the grid of asynchronous generation from small hydro, wind turbines, solar farms, etc. The VSC-based unit connection will permit the small hydro generators to be designed for higher frequency operation and thus optimise their weight and cost; this is particularly important in the case of variable speed wind turbines, as it can boost the energy delivery by some 2.5 to 5 %.

10.5.2 Three-Level PWM Schemes

This section describes the purpose and main characteristics of the three-level HVDC Light projects already in existence.

Eagle Pass (commissioned in 2000)

This project is a back-to-back link connected in series in the 138 kV AC line from Eagle Pass to Piedras Negras on the Mexico–USA border (the utilities involved are CSW Central Power and Light Co. and Comision Federal de Electricidad). Its main purpose is to stabilise the AC voltage and possibly import power from Mexico during emergencies. An alternative conventional back-to-back link would not have provided the necessary reliability because of the weakness of the AC system on the US side of the border. The power ratings are 36 MW and $\pm$ 36 MVAR, which means that the link can interchange active power, while controlling the reactive power or, alternatively, operate as a $\pm$72 MVAr STATCOM. It uses three-level IGBT converter units under PWM control, with a switching frequency of 1500 Hz. The nominal DC voltage is $\pm$15.9 kV. The reactive power is controlled independently at each end. In voltage control mode, the AC voltage control uses the full capability of the VSC regardless of the set power order and the magnitude of the AC voltage may be controlled as long as the converter valve current is below the permissible level.

Murray Link [8, 9] (commissioned in 2002)

This is a 177 km long submarine link of extruded polymer cable installed for the Victoria (Australia) transmission system. The rated powers are 200 MW, +140/ − 150 MVAR, DC voltage $\pm$150 kV and AC voltages at the connecting points 132 and 220 kV. As shown in Figure 10.14, the scheme uses the three-level neutral point connection, as well as IGBT-based converters under PWM control with a switching frequency of 1350 Hz. The figure also shows the tuning frequencies for the AC and DC side filters; as the limits for the DC side harmonics in this scheme are very tight, a zero-sequence reactor and a 9th/21st harmonic filter were added as shown in the figure.

The specific control modes are active power, reactive power and AC voltage. Although this link was initially unregulated, it has recently become regulated.

Cross Sound [10] (commissioned in 2002)

This is a submarine extruded polymer cable interconnection 40 km long between Long Island and Connecticut in the USA. The AC voltages at the connection points are 345 and 138 kV respectively, the DC voltage $\pm$150 kV and the power rating 330 MW, $\pm$75 MVAR.

A key feature of this scheme is to control the power transfers continuously from 0 to 330 MW, in accordance with scheduled transactions by those who have purchased rights to its capacity

The scheme uses a single three-phase VSC at each end. The converter is a three-level bridge using IGBT/diode packs, both for the main converter switches and for the neutral point clamping. The three-level concept reduces the rate of change of voltage, as the valves

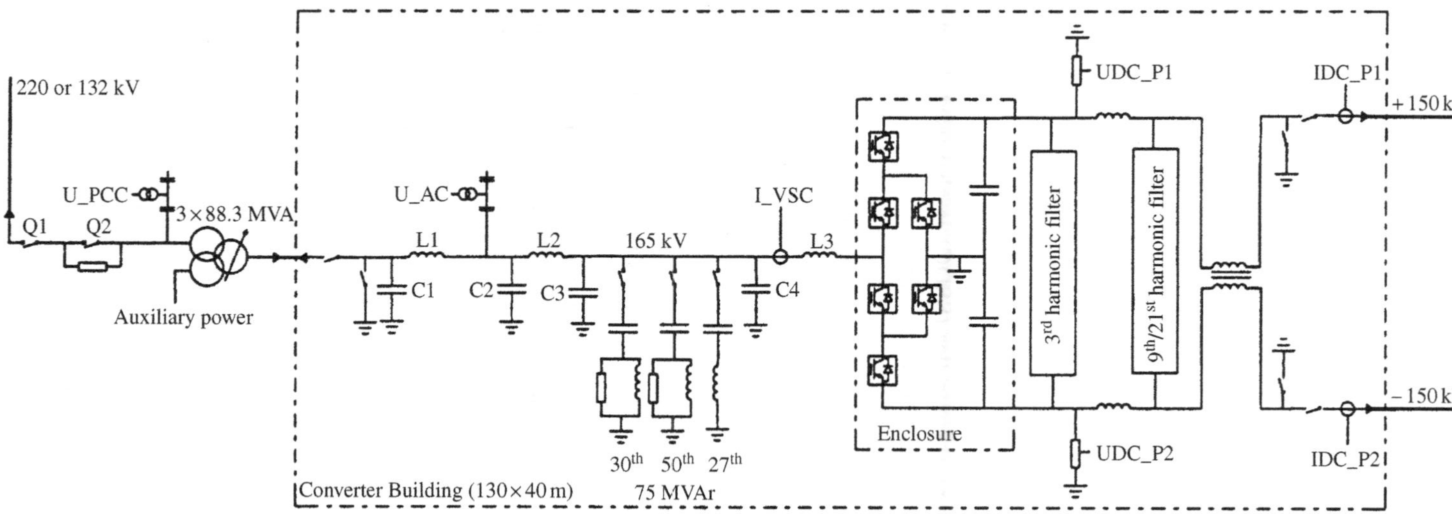

Figure 10.14 Single line diagram of one end of the Murray link (Reproduced by permission of CIGRE and ABB.)

switch on and off only between ±150 kV and 0 V. The switching frequency is 1260 Hz (corresponding to the 21st harmonic), which is half the frequency used in the two-level projects. Therefore the losses are correspondingly reduced, without increasing the harmonic content.

The active and reactive powers are controlled independently, although the latter is limited to one-half of the former, as the main purpose of the link is to transfer active power. The ability to control reactive power may also be used to keep the AC voltage constant at the point of common coupling (PCC). By using PWM, the resulting converter terminal voltage is achieved by switching the valves according to a modulation scheme determined by several factors, such as calculated reference voltage, available DC voltage, harmonic generation and valve currents.

The controller includes a zero-sequence third-harmonic component modulation (additional to the sinusoidal reference) to reduce the peak AC converter voltage and increase (by approximately 15 %) the fundamental frequency component produced by the VSC process.

Non-linearities in the valve switchings create some low-order harmonics (mostly the fifth and seventh) and thus a special controller is used to act on the PWM pattern to minimise the low-order harmonic currents at the PCC.

A sub-synchronous damping controller (SSDC) is also used, because both ends of the link contain power generating plant in the vicinity of the converters. The angular frequency deviation given by a PLL is band-pass filtered to extract the sensitive frequency range; this signal is then used in the SSDC and the limited output signal is added to the current orders

Following the US blackout of August 2003, this link played an essential part in the restoration of power across Long Island; this event has showed the importance of enhancing the transmission infrastructure.

Estonia to Finland connection (expected commission in 2007)

The largest project using prefabricated converter valves mounted in modular metal enclosures is the EstLink designed to supply 350 MW at ±150 kV from Estonia to Finland. HVDC Light was chosen over conventional HVDC due to the requirement for 31 km of underground cable along with 74 km of submarine cable using the specially developed PEX technology. The current rating of the IGBT switches has been increased from the 1050A (used in the Cross Sound scheme) to 1130 A. This increase has been achieved by improving the heat sink and optimising the switching pattern. The project time frame is only 20 months.

10.5.3 HVDC Light Performance

A test system of a three-level HVDC Light offshore wind farm scheme is shown in Figure 10.15 and the structure of the sending end converter is shown in Figure 10.16. At the sending end the nominal frequency is 50 Hz and the PWM switching frequency 1050 Hz. A three-phase 3×10 MVAR high-pass filter bank is used on the AC side of the converter tuned at 600 Hz and the series impedance between the converter and the filter bank is 2.178 Ω. The corresponding components and frequencies at the receiving end are a nominal frequency of 60 Hz and a PWM frequency of 1050 Hz, a 3×4 MVAR tuned filter

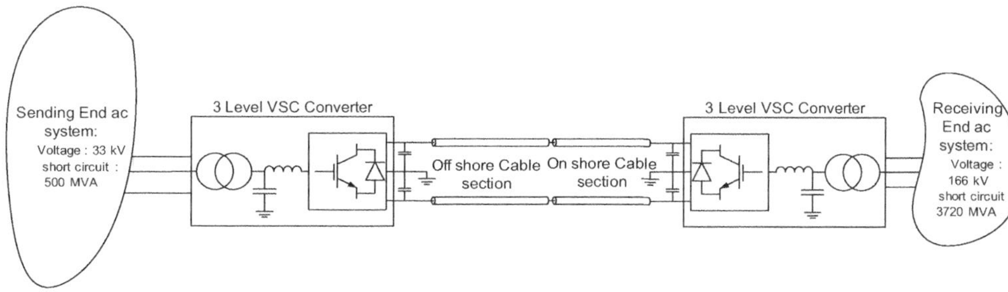

Figure 10.15 HVDC Light for an offshore wind farm

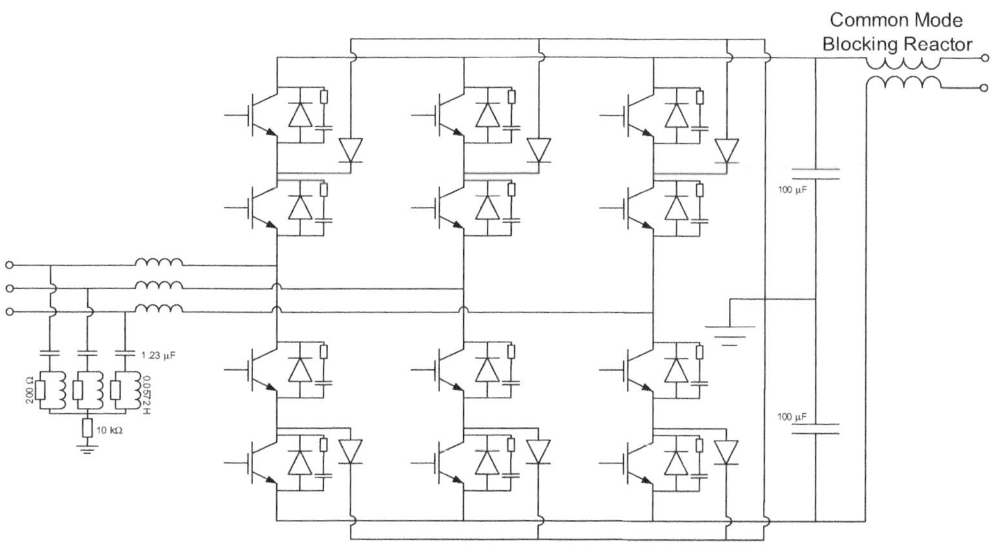

Figure 10.16 Sending end three-level VSC

($f = 650\,\text{Hz}$, $Q = 40$), a 3×30 MVAR tuned filter ($f = 1050\,\text{Hz}$, $Q = 30$) and a series impedance of $7.4\,\Omega$. The short-circuit capacities of the sending and receiving ends are 500 and 3720 MVA respectively.

Sending end controls

Figure 10.17(a) displays a block diagram of the voltage control loop. This loop estimates the voltage droop from the measured reactive power and AC terminal voltage. The voltage droop is added to the measured voltage, and the resulting signal filtered, before being compared with the reference voltage to calculate the voltage error. The voltage error is then passed through a PI controller and a limiter to set the modulation index. The real power controller is shown in Figure 10.17(b). The difference between the measured power and the specified power reference is passed through a PI controller to set the phase angle shift for the PWM control signal. The latter is combined with the modulation index derived in

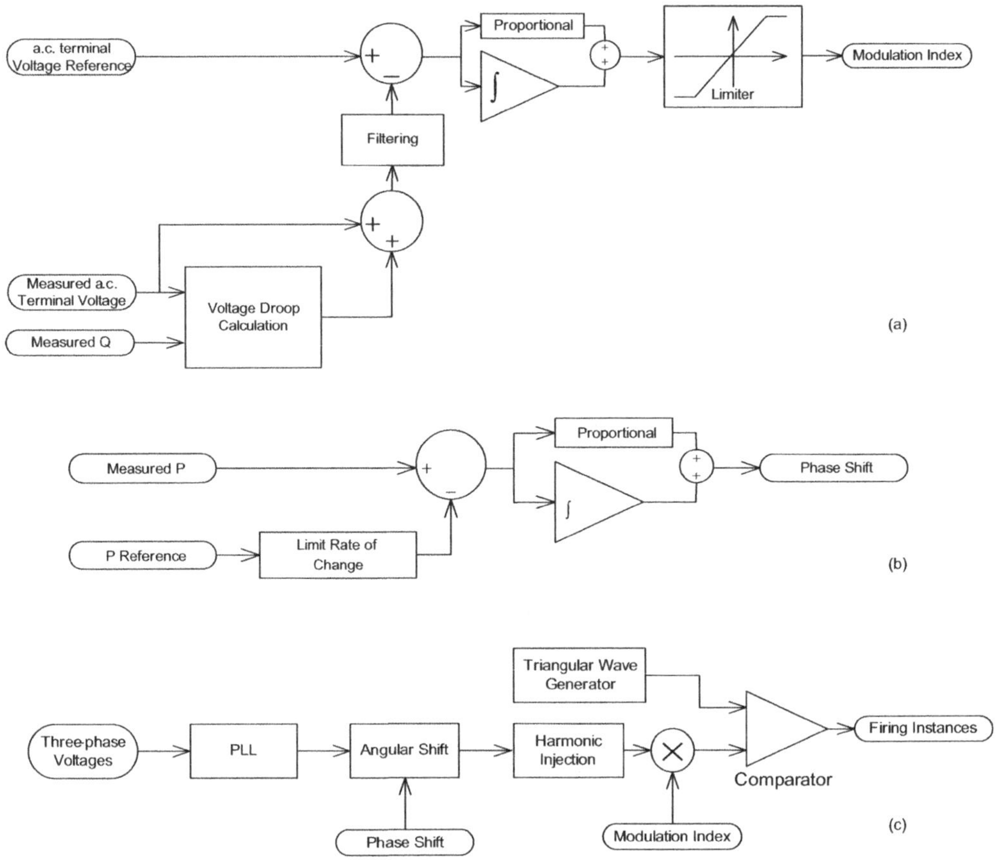

Figure 10.17 (a) Generation of modulation index based on AC terminal voltage error (sending end). (b) Generation of phase shift of control waveform (sending end). (c) Generation of firing instances (sending end)

Figure 10.17(a), as shown in Figure 10.17(c), to generate the firing instances. Instead of a purely sinusoidal control waveform, a harmonic injection PWM technique is used which adds third and ninth harmonics to produce a waveform with a flat top [11]; this is done to allow over modulation, while improving the resulting frequency spectra of the AC and DC waveforms. Moreover, the ratio of maximum peak value of the fundamental component to the amplitude of the unfiltered pulses is higher. The analytical expression of the control waveform is

$$y_{cntrol} = 1.15\sin(\omega t) + (0.271)(1.15)\sin(3\omega t) - (0.029)(1.15)\sin(9\omega t)$$

Receiving end controls

The AC voltage control loop, shown in Figure 10.18, is similar to that of the sending end except for the addition of a modulation index adjustment for valves 1, 3 and 5; this is

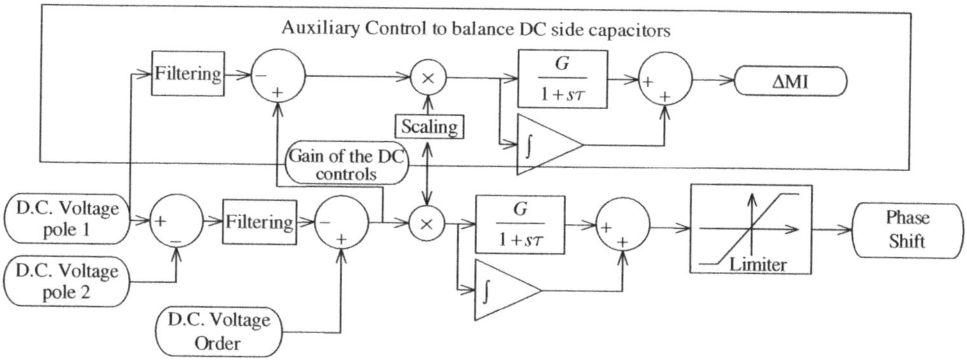

Figure 10.18 DC voltage control loop (receiving end)

done to maintain the DC side capacitor voltage balanced. It is achieved by controlling the magnitude of the positive side of the voltage, which involves adjusting the ON duration width of these three valves. The reference signal for this purpose (ΔMI) is calculated from an auxiliary control on the DC voltage loop; this control action is slower than that of the main DC voltage control loop.

The total DC capacitor voltage ($V_{dc1} + V_{dc2}$) is controlled by phase shift adjustment on the DC voltage control loop, to feed power into or out of the capacitors in order to maintain their total charge. Finally Figure 10.19 shows that two modulation indexes are added to the triangular waveform to produce the firing instances at the receiving end; the use of two modulation indexes makes the waveform comparison with the triangular waveforms slightly more complicated.

Test studies

The test system described above is used to demonstrate the performance of the wind farm using EMTDC simulation under normal and fault conditions. The simulation is first run for 3 seconds and the results in Figure 10.20 show the steady-state operating condition. The three levels are clearly evident in Figure 10.20(a) with reference to the waveform at the receiving

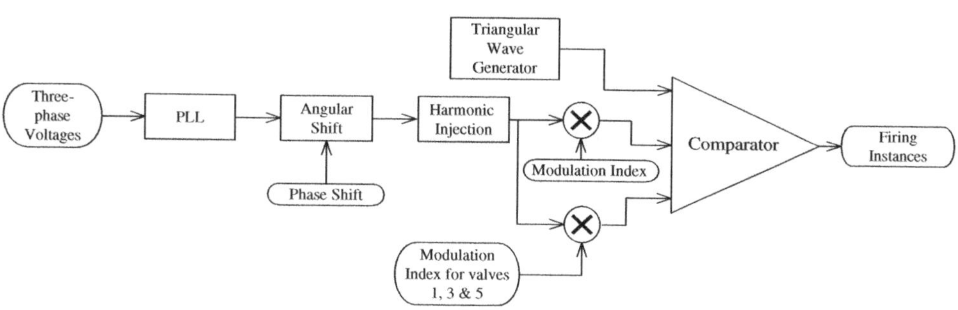

Figure 10.19 Generation of firing instances (receiving end)

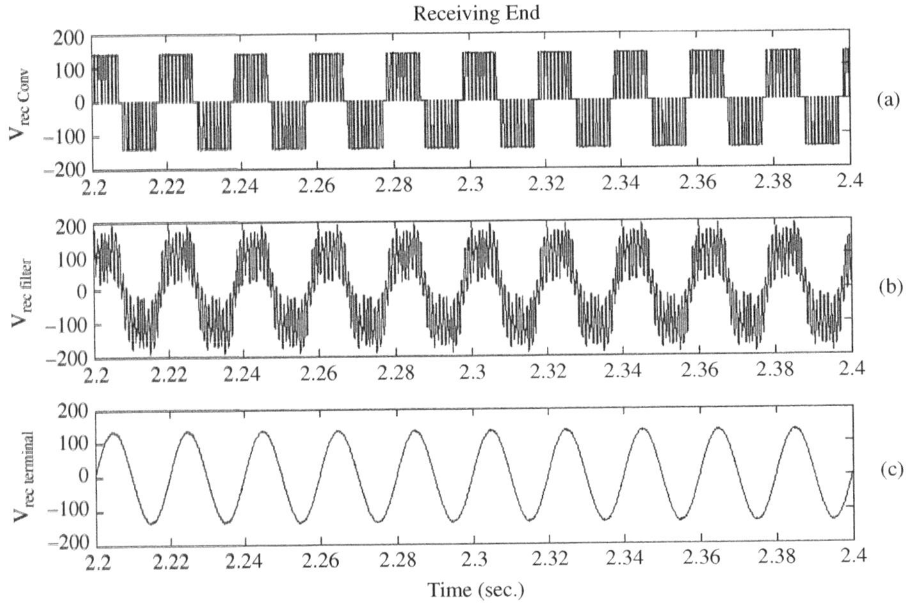

Figure 10.20 Receiving end voltages prior to fault application

end, measured between the converter and interface reactors. Figure 10.20(b) displays the AC voltage at the filter terminals and Figure 10.20(c) shows the terminal AC voltage waveform.

A single phase fault (phase a to ground) is then applied to the sending end system at the 3 seconds mark for a duration of 0.1 second, with a fault impedance of 1Ω. Figure 10.21 illustrates the three-phase rms voltage measurements and Figure 10.22 the real and reactive power during the disturbance. The controllers at both ends of the link increase their modulation index in an attempt to counter the drop in AC terminal voltage. The resulting drop in power at the sending end produces a negative power error signal which causes a more negative phase shift. The receiving end controller increases the phase shift (the voltage error is positive) so as to increase the DC voltage towards the specified reference.

The action of the controllers is to alter the modulation index and the phase shift of the PWM control waveform (as shown in Figure 10.23), to try and restore the voltages and power during the fault to their specified values. At fault removal (which occurs at 3.1 s) the voltage and power levels exceed the specified values, causing the controllers to reverse the direction they were driving the modulation index and phase shift of the PWM control waveform. These then overshoot the steady-state value. Hence the modulation index and phase shift follow a decaying sinusoid until they reach their steady-state value (over the next 2 s). The oscillatory nature is a function of the controller gains and time constants.

Finally, the behaviour of the DC voltage at the sending end and the DC current in pole 1 are illustrated in Figures 10.24(a) and (b) respectively

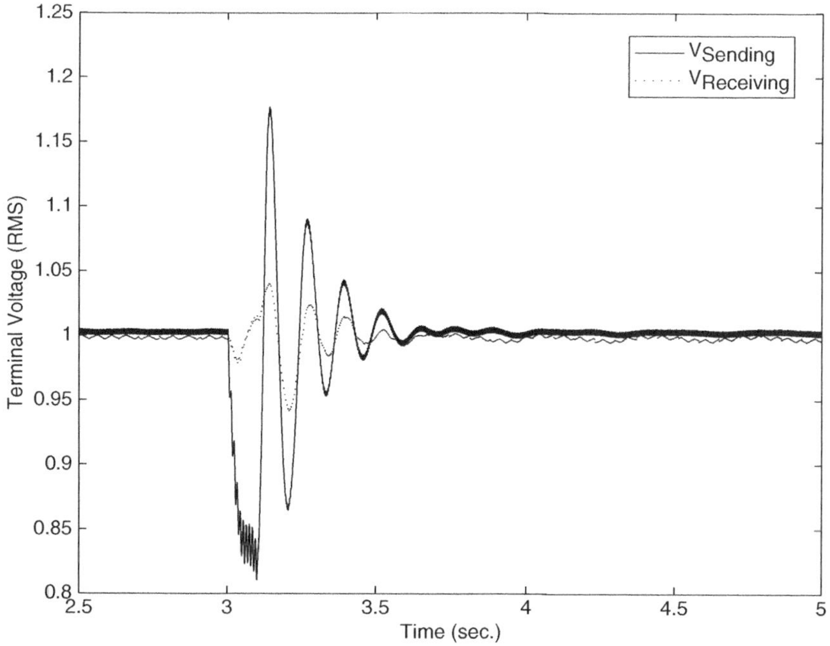

Figure 10.21 RMS three-phase voltage measurements following a phase-to-ground fault at the sending end

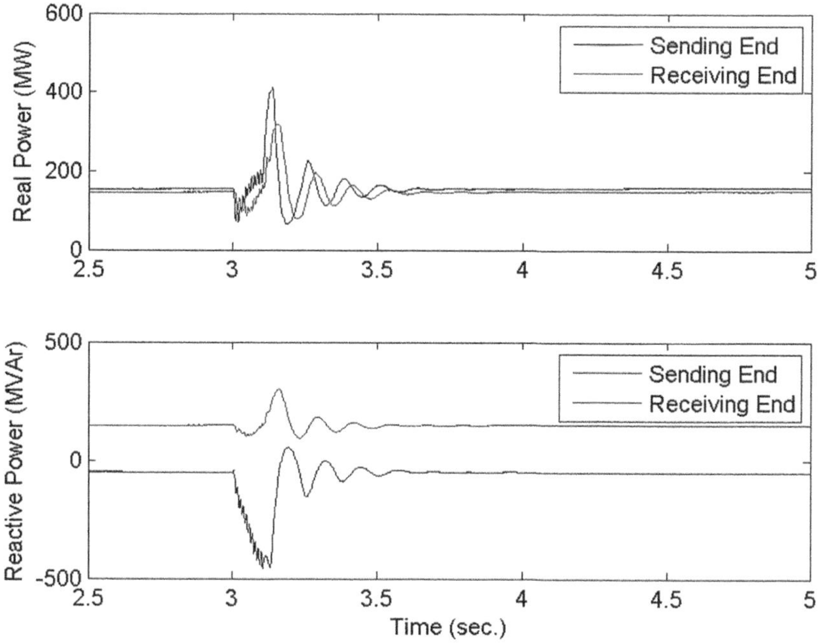

Figure 10.22 Real and reactive power during disturbance

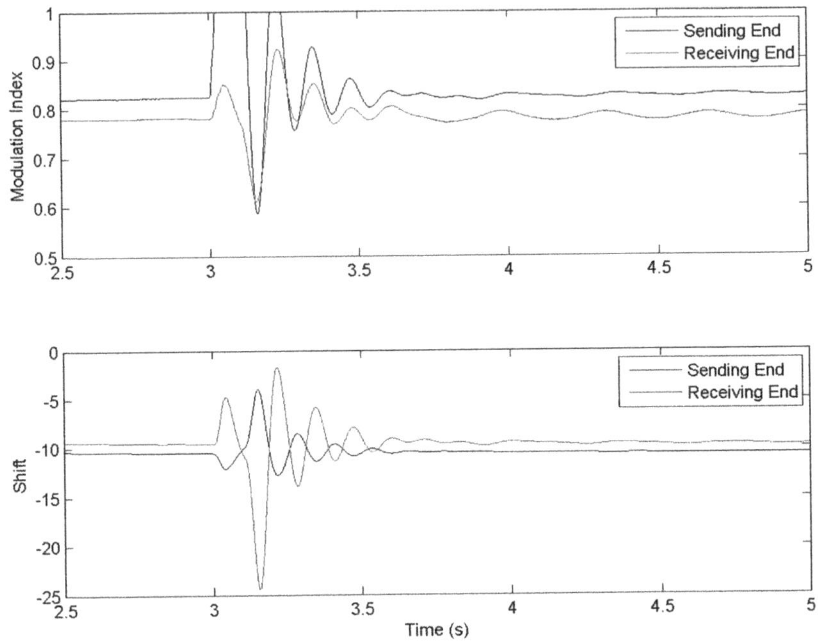

Figure 10.23 Modulation index and phase during disturbance

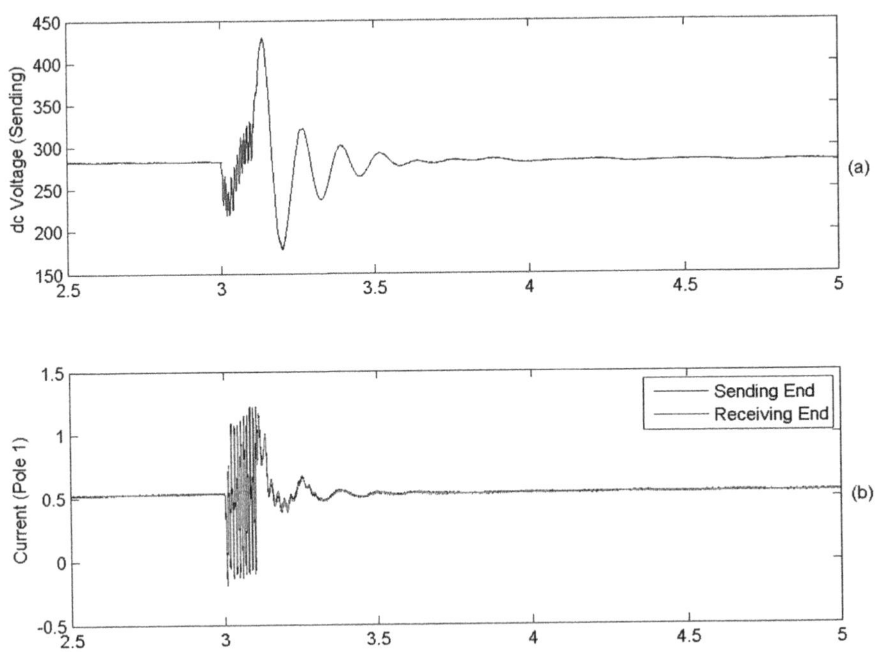

Figure 10.24 DC voltage (a) and currents (b) during the disturbance

10.6 Other VSC Projects

Japan's back-to-back VSC interconnection of different frequency systems

A 300 MW back-to-back (BTB) self-commutated VSC interconnector has been developed in Japan [12]. The circuit configuration is shown in Figure 10.25 and its main specifications are listed in Table 10.1. Each terminal consists of four self-commutated converters. The windings on the AC system side of the converter transformers are connected in series. On the DC side, the four converters are connected in parallel and a shunt capacitor is placed between the two terminals of the link. The converter uses a PWM scheme with a switching frequency of 450 Hz for the 50 Hz system and 540 Hz for the 60 Hz system. On each side of the link the active and reactive powers are controlled independently of each other.

The increased pulse number (nine) used reduces the harmonic content and improves the steady-state and dynamic performance of the converter.

Siemens switchgear factory and Flender-Werft

These two installations, using the VSC back-to-back configuration, feed power to the 60 Hz low-voltage network, one of them to a the test field of a Siemens switchgear factory in Frankfurt (Germany), and the other to a shipyard installation with Flender-Werft in Lübeck (Germany). In both cases the power is transferred from the 50 Hz low-voltage network. The schemes use single converters with a maximum power of 1.2 MW.

Stadtwerke Karlsruhe and Stadtwerke Ulm/Neu-Ulm

An HVDC VSC-based back-to-back connection is used to connect two stations in a 20 kV network in Karlsruhe (Germany). The use of a galvanic connection would have been difficult due to problems with residual currents in the neutral point compensation. In this case

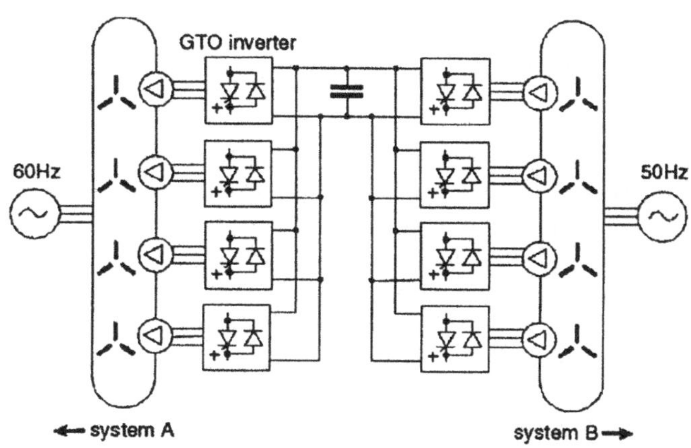

Figure 10.25 A back-to-back HVDC multi-pulse VSC interconnector

Table 10.1 Fundamental specifications of 300 MW self-commutated converter.

Item	Specification	
System	Rating capacity	300 MW–100 MVAR (316 MVA)
	Rating voltage	AC 275 kV
	Voltage distortion	Total < 1 % Each order < 0.5 %
Converter	Converter type	Voltage source type
	Power device	GTO
	Insulation	Air insulated
	Cooling	Pure-water cooling
	Control	9-pulse PWM control
	Configuration	4 stage (or 8 stage)
Transformer	Configuration	Direct step-up

the 2 MVA DC connection was found more economical than a conventional AC solution. Moreover, the DC alternative improves both the voltage quality and supply reliability.

A similar installation in operation connects the 10 kV networks of Ulm and Neu-Ulm, to transfer power between the separate networks in the municipalities of Ulm and Neu-Ulm at times of peak demand according to each network's possible spare capacity.

Industrial factory in Saudi Arabia

A multi-parallel VSC-based back-to-back HVDC link is used to connect a 33 kV, 60 Hz network and a 10 kV, 50 Hz network. The maximum power is 30 MVA.

Proposals for large offshore wind farms

Variable speed windmills of up to 5 MW are being considered. The active power is controlled by varying the angle of the blades. The connection via PWM-VSC also provides reactive power, typically at a 0.9 leading power factor, which gives a reactive power control range of up to 50 % of the active power.

There are proposals for offshore wind farms in the North Sea and the Baltic Sea with capacities between 200 and 1500 MW and at distances of up to 150 km.

As wind farms become larger and more distant from shore, there is more justification for using HVDC to transmit the power to the onshore network, particularly at power levels of 500 MW and beyond. At present the VSC technology cannot offer an economical solution for these power ratings; however, as this technology becomes more widely used for conventional applications and the ratings increase, its flexibility regarding reactive power control will make it more attractive for this application. It is also possible to combine the conventional and VSC-HVDC technologies, the latter for the offshore stations and the former for the onshore ones. Conventional HVDC can also be used in combination with an SVC or STATCOM to provide effective VAR and voltage control.

10.7 Potential for Multi-Terminal Sub-Transmission Systems

Due to continuous urbanisation, metropolitan power networks have to be continuously upgraded to meet the demand. However, the scarcity and increasing cost of land for new rights of way is making the expansion more difficult. Therefore, the present trend is to try and maximise the utilisation of existing assets, preferably deferring the expenditure on new capital investments. Also, in the deregulation environment, distribution services are being subjected to intense competition, especially for the connections to large consumers.

Conventional distribution system planning has no direct mechanism for the control of power flow along desired routes. Moreover, during equipment outages, the rerouting of the extra capacity available in other parts of the network cannot be effectively implemented.

These problems can be largely eliminated with the use of a multi-terminal DC sub-transmission system, either independent from or implanted within the existing AC network. The latter would add considerable flexibility to the operation of the AC system as well as provide voltage support at the points of interconnection of the AC and DC grids. The DC system would optimise the AC power flow and greatly improve its controllability and stability. Additionally, the converters could be used for active filtering of low-order harmonics present in the AC system. Finally, when economic energy storage becomes available, the converters should facilitate the exchange of power.

The VSC-based DC solution, with a constant DC transmission voltage always with the same polarity, is ideal for the implementation of the sub-transmission concept. A likely structure for an independent DC sub-transmission scheme is illustrated in Figure 10.26. The converters in the ring would operate as either rectifiers or inverters, thus acting as real/reactive power sources and sinks, extracting power from certain parts of the network and delivering it to the other zones. The converter operation could be rapidly changed from one mode to another as the situation demands. In the not too distant future the DC loop could also be designed as a high-temperature superconducting ring. If required, power sharing between the various rectifiers connected to the ring could be achieved by a droop control strategy, such that the power allocation among the rectifiers is proportional to their power rating.

Tapping can easily be accommodated in a multi-terminal VSC sub-transmission network. Also, as an alternative to a tap, the shield wire of a large bipolar scheme can be used as a smaller separate transmission link.

In conventional distribution systems all consumers are provided with approximately the same level of power quality and security, irrespective of their individual needs. Such an approach is now changing to provide an increased level of power quality to quality-sensitive consumers by using FACTS controllers. In this respect the DC sub-transmission scheme should provide an even greater flexibility.

A multi-terminal HVDC distribution configuration is perfectly suited to the connection of DC output types of power sources (such as photo voltaics) and of DC input loads (such as IT-related equipment). In both these cases the power exchange will be via DC–DC conversion, since the distribution system voltages are in kilovolts and the sources and loads in hundreds of volts. A general system [13], shown in Figure 10.27, will consist of distributed generation, a battery, a connection to the AC system and an internet data centre (iDC). Each of these components is connected, via a DC circuit breaker (DCCB), to the DC distribution system through an AC–DC converter or a DC–DC conditioner. The latter is required for level adjustment, since a typical distribution AC system voltage may be 6.6 kV, while the

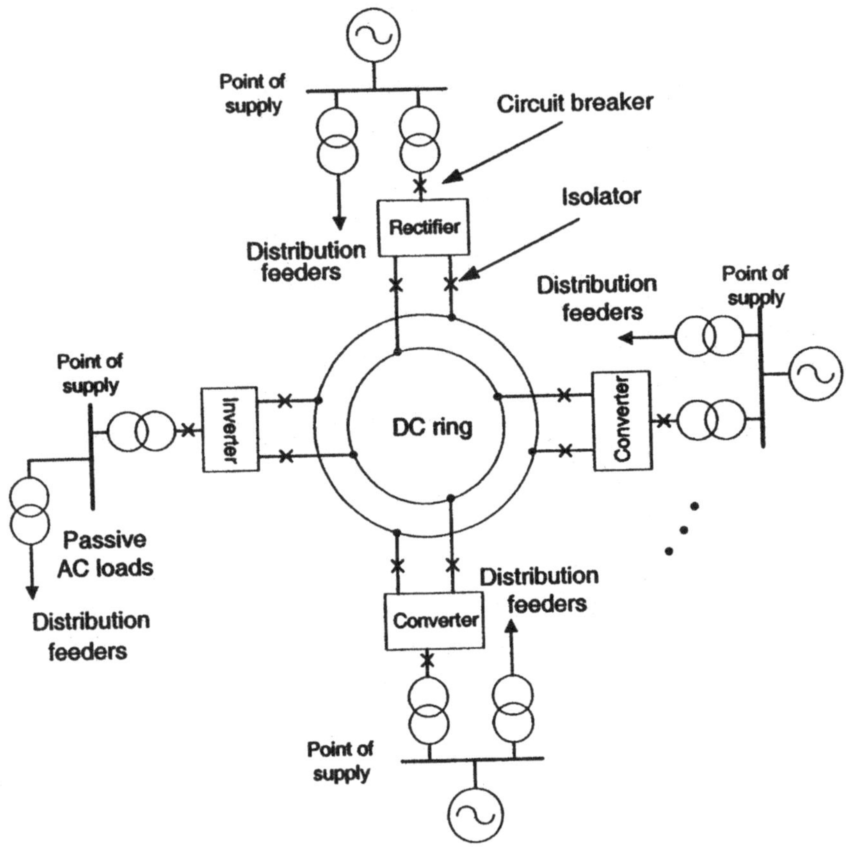

Figure 10.26 Low-power multi-terminal HVDC sub-transmission (Reproduced by permission of CIGRE.)

voltages of the distributed generation (such as a fuel cell) will be between 600 and 800 V and those of a the iDC load between 48 and 200 V.

10.8 Discussion

At present, VSC transmission is not competitive with LCC transmission for high voltage and power ratings. This is largely due to the cost of the power electronics components of the converter stations, which represents of the order of 50 % of the total cost of the terminal, as compared with the thyristor alternative of 25 %.

The provision of high-voltage (over 150 kV) and power transmission capacity (over 350 MW) relies at present on additional parallel VSC transmission schemes, with the attached extra cost of equipment and losses. To increase the VSC-PWM share of the market, further development is being carried out to produce switching components (in particular IGBTs) with higher voltage, peak turn-off currents and higher junction temperatures. The necessary investment has been encouraged by the operating reliability of the present schemes. The

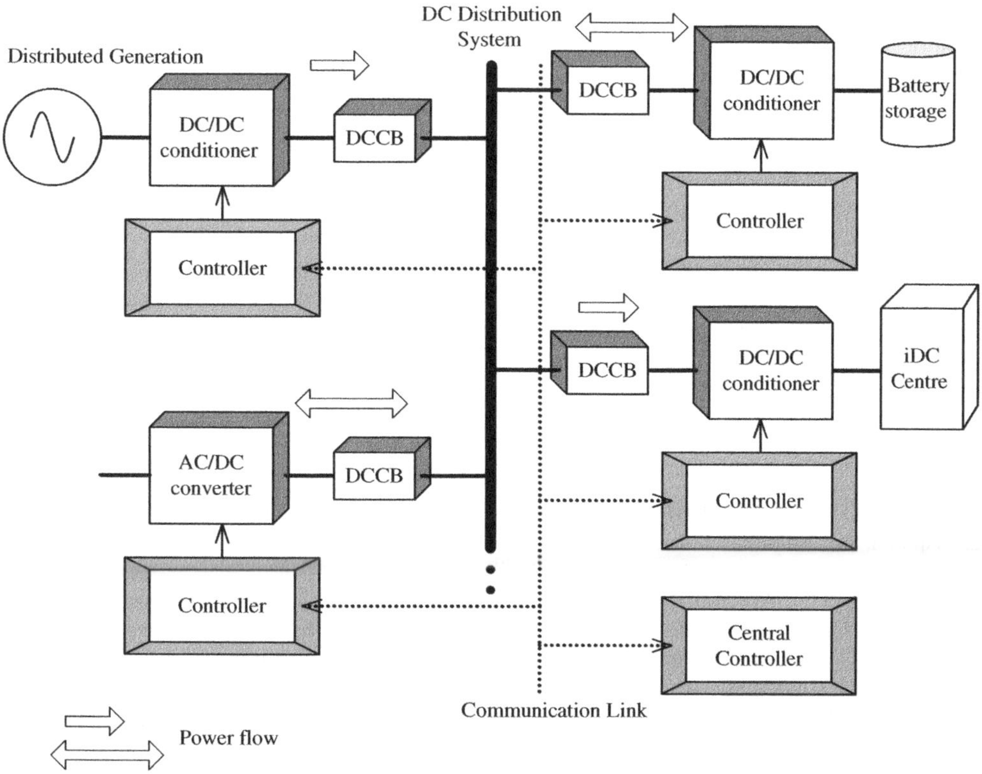

Figure 10.27 A multi-terminal DC distribution system

latest report [14] of the IGBT and DC cable technologies claims that ratings of ±300 kV, 1740 A and 1000 MW can already be achieved.

Of course, thyristor technology itself will not remain at a standstill and the expected catching up by IGBT technology may prove to be a moving target. This is exemplified by the projected ratings of the SiC solid-state devices shown in Chapter 2, which are greatly in favour of the SiC thyristor, with potential voltage ratings of up to 40 kV.

The power losses are higher in VSC transmission, mostly due to the use of higher frequency switching control. It is, therefore, important to assess the no-load and load losses as a function of power transmission and reactive power exchange with the AC systems over the life of the scheme. VSC transmission requires DC capacitors at both ends of the line to provide a stable voltage at the converter ends. It is not possible to run with a single capacitor in the middle.

Regarding the transmission medium, the VSC option is ideal for cable transmission as the voltage polarity does not change, and, therefore, very simple extruded cable technology can be used. On the other hand, overhead line transmission is the most economical option for very long distances. However, overhead VSC transmission is not being considered at present for the following reasons: (i) DC line short circuits need to be cleared by means of AC circuit breakers at both ends; (ii) the link ability to discharge from faults is slow and restarting can be delayed by up to 10 seconds; (iii) the high frequency generated from the

(PWM) converters causes high RFI. Underground (or submarine) cable does not experience DC short circuits other than catastrophic ones.

The PWM-VSC alternative offers complete independence of active and reactive power control, elimination of commutation failures, the possibility of connecting to a passive grid, absence of low-order harmonic filters and their related consequences (such as resonances and harmonic instabilities), and the provision of extra ancillary services (such as flicker and selected harmonics control).

Multi-level configurations offer some advantages in terms of efficiency, both in the converter (due to the lower switching frequencies involved) and in the transmission lines (due to the higher voltages used). These benefits have to be weighed against the greater number of switching components needed. Also the latter lack, at present, independent reactive power controllability at each end of the link.

References

1. CIGRE SC-B4 WG 37 (2004), 'VSC transmission', Final document, Paris.
2. Asplund, G., Eriksson, K. and Tollerz, O. (2000), 'Land and sea cable interconnections with HVDC Light', CEPSI 2000 Conference, Manila.
3. Johansson, S.G., Asplund, G., Jansson, E. and Rudervall, R. (2004), 'Power system stability benefits with VSC DC-transmission systems', CIGRE B4-204, Paris.
4. Asplund, G., Eriksson, K. and Svensson, K. (1997), 'DC transmission based on Voltage Source Converters', CIGRE SC-14 Colloquium, Johannesburg.
5. CIGRE SC-B4 WG 33 (2004), 'HVDC and FACTS for distribution systems', Interim report.
6. Iov, F., Sorensen, P., Hansen, A.D. and Blaabjerg, F. (2005), 'Modelling and control of VSC based DC connection for active stall wind farms to Grid', IPEC' 05, Niigata, Japan, 1078–1084.
7. Horiuchi, S. *et al.* (1996), 'Control system for high performance self-commutated power converter', CIGRE paper 14-304, Paris.
8. Arrillaga, J., Bowles, J., Campos Barros, J., Kanngiesser, J., and Ingram, L. (1993), 'Direct connection of generators to HVDC converters', *ELECTRA*, No. 149, 19–39.
9. Wyckman, M. (2003), 'Long distance cable laying Murray link case study', CIGRE AP-B4 Seminar, Colloquium on HVDC and FACTS, Melbourne.
10. Mattsson, I. *et al.* (2004), 'Murraylink, the longest underground HVDC cable in the world', CIGRE paper B4-103, Paris.
11. Railing, B.D. *et al.* (2004), 'Cross Sound cable project second generation VSC technology for HVDC', CIGRE paper B4-102, Paris.
12. Woodford, D., Electranix, Manitoba, Canada, Private communication.
13. Konishi, H., Kawazoe, H., Makino, Y. and Imaizumi, T. (2005), 'Study on multi-terminal dc transmission system applicable to distribution systems for specific area', IPEC' 05, Niigata, Japan, 1711–1715.
14. Jacobson, B., Jiang-Hafner, Y., Rey, P., Asplund, G., Jeroense, M., Gustafsson, A., and Bergkvist, M., (2006), 'HVDC with voltage source converters and extruded cables for up to ±300 kV and 1000 MW', CIGRE paper B4-105, Paris.

11

Multi-Level VSC and CSC Transmission

11.1 Introduction

The full range of AC voltage controllability provided by PWM-VSC, an important requirement in motor control, is not required in power transmission, where the AC system voltages are kept practically constant. Moreover, the use of higher switching frequencies and transmission voltage levels reduces the efficiency and increases the dv/dt rating of the PWM-VSC technology. Multi-level conversion is a possible alternative to PWM to overcome these problems [1]. However, the original multi-level diode clamping (MLDC-VSC) [2] and multi-level capacitor clamping (or flying capacitors) (MLCC-VSC) [3] configurations, described in Chapter 6, contain many extra switches and/or capacitors, as these are required independently for each of the three converter phases. The rating of these switches and capacitors is high because they form part of the main power circuit. They also suffer from capacitor balancing problems, which in the case of MLDC makes it practically impossible to exercise active power control for level numbers greater than three. Also the current rating of the clamping capacitors (in the MLCC case) is high, as the load current passes through them. Thus so far the multi-level application to HVDC transmission has been restricted to the three-level configuration and is complemented by PWM.

Chapter 7 has described simpler multi-level configurations based on the concept of multi-level reinjection (MLVR-VSC and MLCR-CSC), with potential application to VSC [4] and CSC [5] transmission. They share the multi-level properties of harmonic elimination, power loss reduction and use of the more robust thyristor-type self-commutating switching devices. As the reinjection circuit is operated at six times the fundamental frequency (for the double bridge converter configuration), it produces the multi-level output waveform with a substantially reduced number of switches.

Flexible Power Transmission: The HVDC Options J. Arrillaga, Y.H. Liu and N.R. Watson

Important additional properties of multi-level reinjection are:

- Adaptive balancing of the DC side capacitors in VSC transmission [6].
- Creation of a zero-voltage (ZVS) or a zero-current (ZCS) condition during the commutations, which greatly simplifies (or even eliminates) the snubber circuits. In particular, the ZCS condition adds self-commutating capability to the conventional thyristor converter and, therefore, opens up the possibility of thyristor-based self-commutating HVDC transmission technology.

However, unlike PWM-based transmission, the provision of reactive power control in multi-level conversion requires DC voltage variations. This is not a problem in applications using single converter stations, such as the STATCOM, but does not permit independent reactive power control at the two ends of a point-to-point HVDC interconnection.

This chapter discusses the characteristics of multi-level VSC and CSC transmission. Most of the content is generally applicable to the various configurations, but the test examples use as a basis the multi-level voltage and current reinjection schemes described in Chapter 7. Some new control concepts are also proposed to improve the reactive power controllability of multi-level HVDC transmission.

11.2 Multi-Level VSC Transmission [7]

11.2.1 Power Flow Considerations

The basic power transfer characteristics described in Section 10.2 apply equally to the VSC multi-level options. However, unlike the case of PWM-HVDC (where the two terminals of the link control the reactive power independently) this is not the case in multi-level HVDC schemes with single converter terminals. Each converter station only has one independent control variable, namely the phase displacement between the AC source and the converter output voltages, and this variable is used to control the active power flow between the AC source and the converter. Changes in the reactive power injection are made by varying the converter DC voltage, an action that affects simultaneously the two sides of the link; thus, with present controls the reactive power adjustments at the receiving and sending ends cannot be made independently of each other. The effect of this restriction on the voltage level and/or reactive power exchanges achievable at the terminals of the DC link are considered next to show that, other than in very weak systems, the multi-level option is adequate for HVDC transmission.

AC minimisation

The simplified equivalent of Figure 11.1 illustrates one terminal of a DC link based on the cascaded H-bridge multi-level voltage reinjection concept described in Chapter 7 (Section 7.5). However, the following analysis applies equally to the other multi-level configurations.

At the receiving end of the link (assumed the weaker side), if the AC source (V_{S2}) is permitted to reduce down to $(1-\varepsilon)V_{2rated}$, then the converter voltage (V_{C2}) must be adjusted

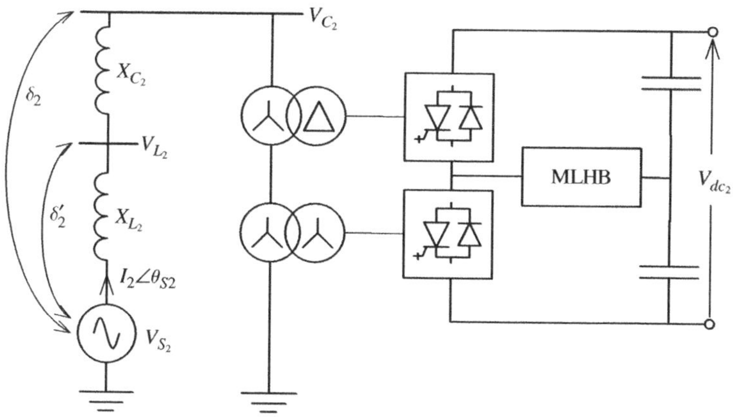

Figure 11.1 Simplified model of the receiving end of a multi-level VSC link

between $(1-\varepsilon)V_{2rated}$ and $(1+\varepsilon)V_{2rated}$ (depending on P_2) to achieve minimum AC at this end. Using the minimisation process described in Section 10.2.1 (Equations (10.13) to (10.16)), the required value of ε for this to happen can be derived from the expression

$$\frac{1+\varepsilon}{1-\varepsilon}=\sqrt{1+\frac{1}{(1-\varepsilon)^4(\mathrm{SCR}_2)^2}} \tag{11.1}$$

where SCR_2 is the short-circuit ratio at the receiving end of the link.

From Equation (11.1) the following values of ε are obtained for different system strengths:

SCR_2	1	2	2.5	3	4	5
ε	NC	0.0718	0.0436	0.0294	0.0161	0.0102

Similar to the current minimisation at the receiving end of the link, the minimum AC current (and therefore unity power factor operation) at the sending end (the side not controlling the DC voltage) occurs when the following relation is met:

$$\frac{V_{C1}}{V_{S1}}=\sqrt{1+\left[\frac{P_1}{P_{1rated}}\right]^2\left[\frac{V_{1rated}}{V_{S1}}\right]^4\left[\frac{X_{L1}+X_{C1}}{X_{1rated}}\right]^2}=\sqrt{1+\left[\frac{P_1}{P_{1rated}}\right]^2\left[\frac{V_{1rated}}{V_{S1}}\right]^4\left[\frac{1}{SCR_1}\right]^2} \tag{11.2}$$

When operating at the rated power, the receiving end maintains the relationship

$$\frac{V_{C2}}{V_{2rated}}=1+\varepsilon \tag{11.3}$$

and the sending end can only operate at unity power factor by fixing the source voltage to $V_{S1}=(1-\varepsilon')V_{1rated}$, where ε' is determined from the expression

$$\frac{1+\varepsilon}{1-\varepsilon'}=\sqrt{1+\left[\frac{V_{1rated}}{V_{S1}}\right]^4\left[\frac{1}{\mathrm{SCR}_1}\right]^2} \tag{11.4}$$

Some numerical results for the relationship between ε' and SCR_1 (for different receiving end system strengths) are listed below.

For $SCR_2 = 2$ and $\varepsilon = 0.0718$

SCR_1	2	2.5	3	4	5
ε'	0.0718	0.0177	−0.0105	−0.0377	−0.0501

For $SCR_2 = 2.5$ and $\varepsilon = 0.0436$

SCR_1	2.5	3	4	5
ε'	0.0436	0.0161	−0.0104	−0.0225

For $SCR_2 = 3$ and $\varepsilon = 0.0294$

SCR_1	3	4	5
ε'	0.0294	0.0033	−0.0086

Thus if the two AC voltage sources are kept within $V_{S1} = (1-\varepsilon')V_{1rated}$ and $V_{S2} = (1-\varepsilon)V_{2rated}$ respectively, the receiving end will operate at unity power factor, while the power factor of the sending end source will be

$$\cos[\theta_{S1}(p_1)] = \frac{\sin(\delta_1)}{\sqrt{1+\left[\frac{V_{C1}}{V_{S1}}\right]^2 - 2\frac{V_{C1}}{V_{S1}}\sqrt{1-\sin^2(\delta_1)}}} = \frac{\sin(\delta_1)}{\sqrt{1+\left[\frac{V_{C2}}{V_{2rated}}\right]^2 \frac{1}{(1-\varepsilon')^2} - 2\frac{V_{C2}}{V_{2rated}}\frac{1}{1-\varepsilon'}\sqrt{1-\sin^2(\delta_1)}}} \quad (11.5)$$

where

$$\sin(\delta_1) = \frac{P_1(X_{L1}+X_{C1})}{V_{C1}V_{S1}} = \frac{P_1/P_{1rated}}{\frac{V_{C1}}{V_{1rated}}\frac{V_{S1}}{V_{1rated}}(SCR_1)} = \frac{P_1/P_{1rated}}{\frac{V_{C2}}{V_{2rated}}(1-\varepsilon')(SCR_1)} \quad (11.6)$$

and

$$\frac{V_{C2}}{V_{2rated}} = (1-\varepsilon)\sqrt{1+\left[\frac{P_2}{P_{2rated}}\right]^2\left[\frac{1}{1-\varepsilon}\right]^4\left[\frac{1}{SCR_2}\right]^2} \quad (11.7)$$

By way of example, a link with SCRs of 2.5 and 3 at the receiving and sending ends, respectively, will be able to maintain unity power factor operation at the receiving end system for the full range of power transfer, whereas the sending end system power factor will vary from 0.998 to 1 for an active power range of between 0.5 and 1 pu. However, for very low-power transfers (below 0.1) the strong system will operate with a low power factor and provide a maximum reactive power level of 8.1 % when the power transfer is zero.

11.2.2 DC Link Control Characteristics

In line with VSC procedure, a multi-level converter uses the phase angle difference between the voltages across the interface transformer to control the active power flow, while the voltage magnitude difference controls the (generated or absorbed) reactive power. As the converter side AC voltage of the interface transformer is directly determined by the converter DC voltage and the DC voltages of an HVDC link are closely related, only one side has complete freedom to control the terminal voltage.

The converter terminal voltage (V_T) must be kept within a small margin to satisfy the local load. This is the voltage available for control purposes and is strongly related to the DC side voltage; if the transformer leakage reactance is small relative to the AC system reactance, controlling the link DC voltage and controlling the AC terminal voltage are very similar processes. In a well-designed AC system, adjusting the DC link voltage by a narrow margin should normally be sufficient to minimise the AC side and the DC side currents and thus reduce the transmission system losses.

In practice, the transformer terminal voltage V_{T2} at the receiving (the weaker) end of the link will be used as the control variable to minimise the ac currents.

Under normal operating conditions the real and imaginary components of the converter output currents directly determine the active and reactive powers and can thus be used as the control variables. Moreover, under abnormal conditions the direct control of current operates the system more safely than when power variables are used. The basic control structure of a single converter station based on the converter real and imaginary current components has been described in Section 7.3.5.

Figures 11.2(a) and (b) show the simplified control structures, without terminal voltage adjustment, for the converters at the sending and receiving ends of the link (for the purpose of verifying the effectiveness of the control strategy, in this section the AC system source voltages are adjusted to simplify the simulation).

The multi-pulse ramp signals of the converter firing logic are synchronised with the respective converter terminal voltages. Then the sending end real power P^*_{ref} and receiving end terminal voltage V^*_{ref} references are compared with their respective measured values. The errors are sent to PI controllers with saturation limits to derive the real and imaginary current components for the sending and receiving controllers respectively. Finally, the two PI controllers generate the phase angle displacement commands using the real and imaginary current errors of the two converters. The saturation property of the real and imaginary PI controllers ensures that the phase displacements are within safe operation ranges. The phase displacements are then sent to the converter logic to generate the gate firing signals.

The control structure required at the weaker end for the DC voltage is very much the same as that needed to control the terminal AC voltage, the only difference being that the monitored voltage signal required is V_{dc2} instead of V_{C2}.

The V_{C2} variation in the per unit system is given by

$$V^*_{ref} = (1-\varepsilon)\sqrt{1+\left[\frac{P_2}{P_{2rated}}\right]^2\left[\frac{1}{(1-\varepsilon)^4}\right]\left[\frac{1}{(\mathrm{SCR}_2)^2}\right]} \tag{11.8}$$

where only the per unit active power P_2/P_{2rated} varies, while ε and SCR_2 remain constant for a given system. Thus the reference voltage V^*_{ref} will be generated from the monitored active power P_2 in real time.

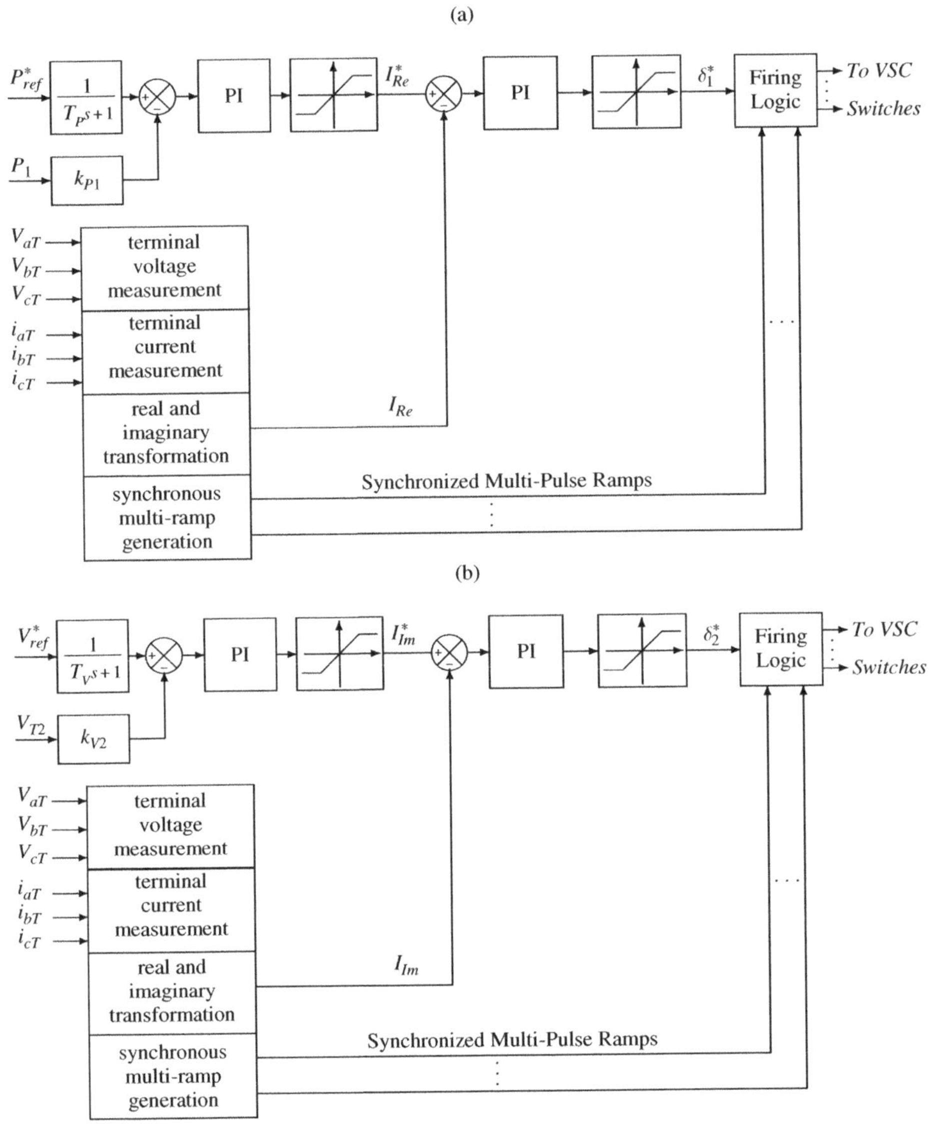

Figure 11.2 Control structures of the sending (a) and receiving (b) ends

11.2.3 Test System and Simulation Results

The circuit of Figure 11.3, used as a test system for the EMTDC simulation, has the following features and parameters:

- A total absence of filters.
- The DC line parameters are listed in Table 11.1.

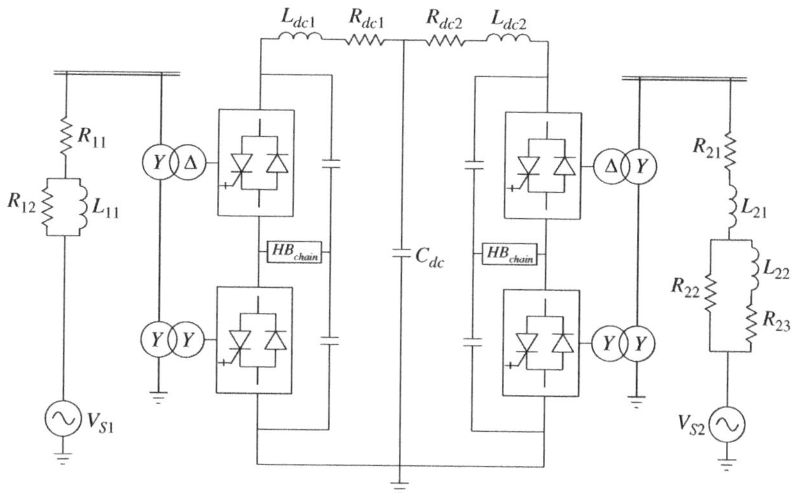

Figure 11.3 Multi-level VSC-HVDC configuration using cascaded H-bridge reinjection

Table 11.1 DC transmission line parameters

L_{dc1}	R_{dc1}	L_{dc2}	R_{dc2}	C_{dc}
0.5968 H	2.5 Ω	0.5968 H	2.5 Ω	26.0 μF

- The interface transformers are connected in series and their individual leakage reactances are 7 %.

- The AC system SCRs are chosen as 2.5 and 3 for the receiving and sending ends respectively; to achieve these SCRs the required AC system parameters are listed in Table 11.2.

- The rated voltages are 345 kV (at the sending end) and 230 kV (at the receiving end).

- As SCR_2 is set to 2.5, if the voltage deviation permitted at the receiving end is $V_{S2} = 230(1-\varepsilon) = 220\,\text{kV}$ (i.e. $\varepsilon = 0.0436$), then to compensate for the DC voltage difference caused by the DC line resistance, at the sending end ε' is set to zero and therefore $V_{S1} = 345\,\text{kV}$.

Table 11.2 Sending and receiving ends AC system parameters

R_{11}	R_{12}	L_{11}	R_{21}	R_{22}	R_{23}	L_{21}	L_{22}
2.13 Ω	3140 Ω	0.1 H	0.432 Ω	43.2 Ω	0.432 Ω	0.0275 H	0.0275 H

- All variables displayed in the graphs are in per unit. For convenience the reference parameters at the sending end are $1.673\sqrt{2}$ kA and 1000 MVA and those of the receiving end $2.673\sqrt{2}$ kA and 1000 MVA, while the DC side voltage and current references are 500 kV and 2 kA respectively.

Dynamic performance

The test system is controlled to follow the real power and receiving end terminal voltage orders; when the rated power is transferred from the 345 to the 230 kV AC system the two AC sources operate at optimal power factors.

Figure 11.4 shows EMTDC results for the active and reactive powers, as well as the voltage and current responses to a real power order change.

The simulation is carried out for 8 seconds of real-time operation. In this period the receiving end terminal voltage order is kept unchanged at 1 pu, the real power order is changed from 0 to 1 pu after 0.5 s, changed from 1 to −1 pu at 3.5 s and finally returned to zero at 6.5 s.

Figure 11.4(a) shows that the active powers of the sending P_1 and receiving P_2 ends follow very well the active power order P_{ref}. After about 0.5 s the active power reaches the power order without overshoot; when P_{ref} changes from 1 to −1 pu it reaches the new power level within 1 s with negligible overshoot and, finally, when the power order is returned to zero the response time is about 0.5 s.

Figure 11.4(b) shows the reactive powers Q_1 and Q_2, measured at the interface transformer terminals. In the absence of active power transfer, Q_2 is about 0.01 pu and Q_1 0.25 pu. When P_{ref} is set to 1 pu both converters provide appropriate reactive powers to their connected AC systems, thus minimising their respective AC. However, for the period when P_{ref} is −1, as the voltage control terminal has not been changed, the high power factor operation is lost.

Figure 11.4(c) shows the terminal voltages V_{T1}, V_{T2}. As the latter is under direct control, it is kept practically at 1 pu (apart from the small overshoots during the power order changes). The sending end voltage, in the absence of voltage control, experiences changes of up to about 5 % upon reaching the steady state and within 10 % in the dynamic region.

Figure 11.4(d) shows that the DC line current has the same shape as the active power.

Finally, Figures 11.4(e) and (f) display the DC voltages at both ends, as well as their respective reinjection H-bridge capacitor voltages. The DC voltages have the same shape as their terminal voltages. The reinjection capacitor voltages (apart from the ripple) follow the DC voltage and are kept balanced during the steady-state and dynamic conditions, i.e. there is no significant difference between the three capacitor voltages, except when the DC decays to zero, which shows a small difference. To emphasise these differences the normalising reference for these waveform is chosen as 500 kV/3.

The curves in Figure 11.5 are the terminal voltages and currents in steady state; graphs (a) and (b) correspond to 0 pu active power transfer, (c) and (d) to 1 pu and (e) and (f) to −1 pu. They illustrate the high quality of waveforms produced by the MLVR-VSC configuration.

Operation at high power factors

To verify the performance under high power factor operation, the step order of P_{ref} is sent to a non-linear integrator which changes its output at a constant increasing or decreasing

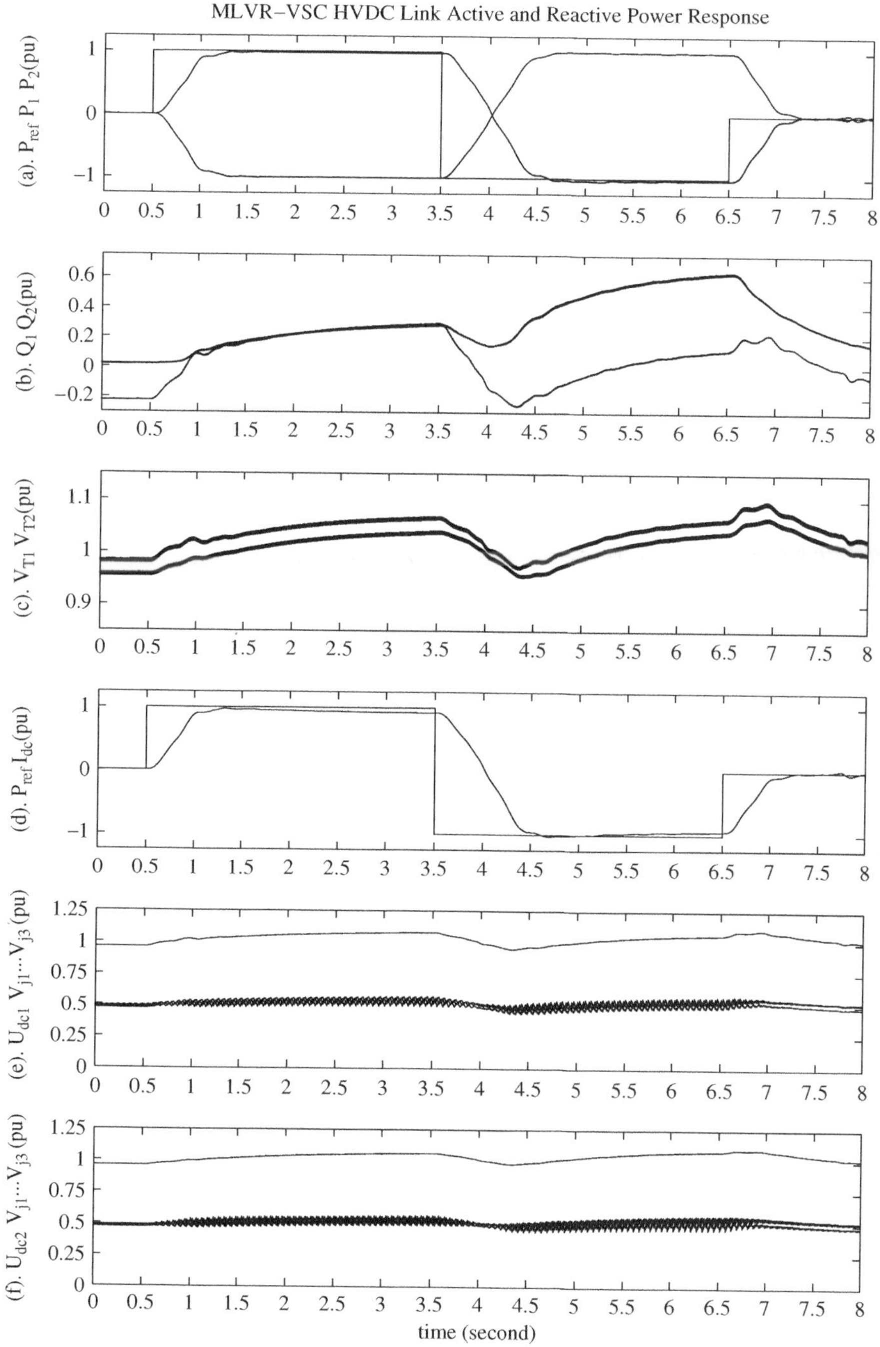

Figure 11.4 Dynamic response of the MLVR-VSC link to active power variations

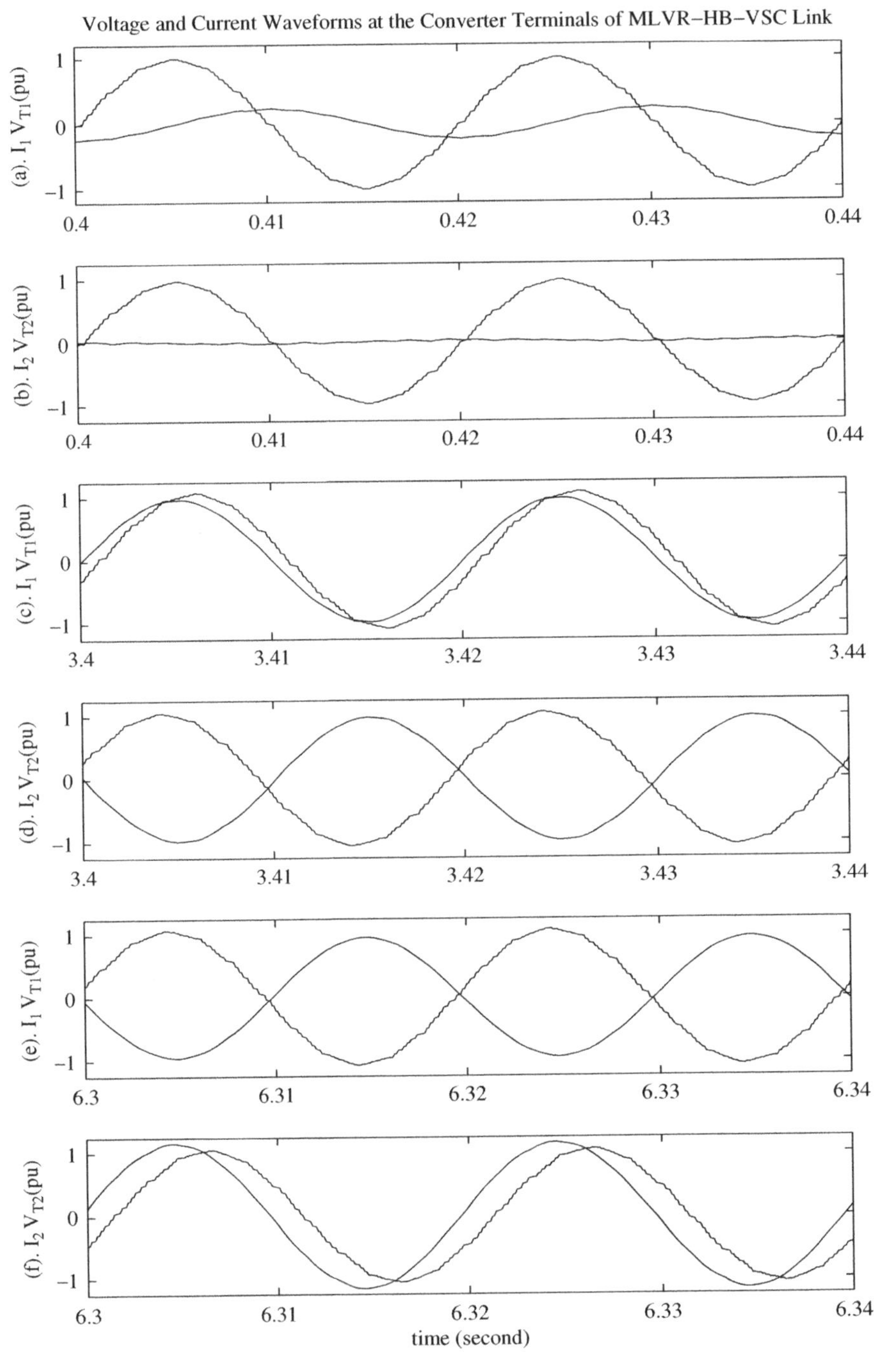

Figure 11.5 Steady-state voltage and current waveforms for different power factors

rate before its output reaches the input value and keeps the output unchanged after that. This integrator generates a slow growing active power order to force the link gradually to increase the active power in 10 seconds to its rated level. Therefore the full range of active power transfer performance is observed in detail. Figure 11.6 displays the following simulated results:

(a) $P_{C1}P_{C2}$, measured at the sending and receiving interface transformer terminals;

(b) $Q_{C1}Q_{C2}$, measured at the sending and receiving interface transformer terminals;

(c) $Q_{S1}Q_{S2}$, measured at the sending and receiving AC source terminals;

(d) $V_{T1}V_{T2}$, measured at the sending and receiving interface transformer terminals;

(e) $U_{dc1}U_{dc2}$, DC voltages measured at the sending and receiving converter terminals;

(f) $PF_{S1}PF_{S2}$, the power factors measured at the sending and receiving source terminals.

Figure 11.6(a) illustrates that P_{C1} and P_{C2} increase gradually following the active power order change from 0 to 1 pu. Figures 11.6(d) and (e) show that the DC voltages are always kept within $\pm 5\,\%$ of the rated value. Figure 11.6(f) shows that when the active power is over 10 %, PF_{S2} is at 0.998, whereas PF_{S1} is only 0.31; however, these values reach 0.999 and 0.9 respectively when the link transfer of active power is 0.4 pu and 0.999 and 0.980 when the power transfer is 0.6 pu. Above 0.8 pu power transfer PF_{S2} and PF_{S1} reach levels of 0.9999 and 0.999 respectively.

11.2.4 Provision of Independent Reactive Power Control

Due to the natural commutation restriction of thyristor-based technology, the converter terminals demand varying reactive power, which is dependent on the operating conditions of the DC output voltage and current.

In self-commutating conversion, each of the converter terminals can provide four-quadrant power controllability. In this case, appropriate control of the reactive power, as well as reducing transmission losses, can provide greater flexibility for the AC system, such as stabilising a particular bus voltage and operating at unity power factor to minimise the transmission current. A four-quadrant power-controlled terminal functions as a conventional HVDC terminal combined with a STATCOM.

The strength of the two AC systems connected by a DC link can be very different and, thus, ideally their reactive power requirements should be controlled independently. This condition is achieved in PWM-VSC transmission, but not in multi-level transmission, where the reactive power exchanges at the converter terminals are interdependent. While this may be considered a drawback to the prospective introduction of the technology, the main purpose of DC transmission is to interchange active power and the ratings of the converters will have little margin for the provision of reactive power.

The converter terminals of high-power HVDC transmission systems consist of two or more series- or parallel-connected groups. In conventional CSC schemes the two groups are

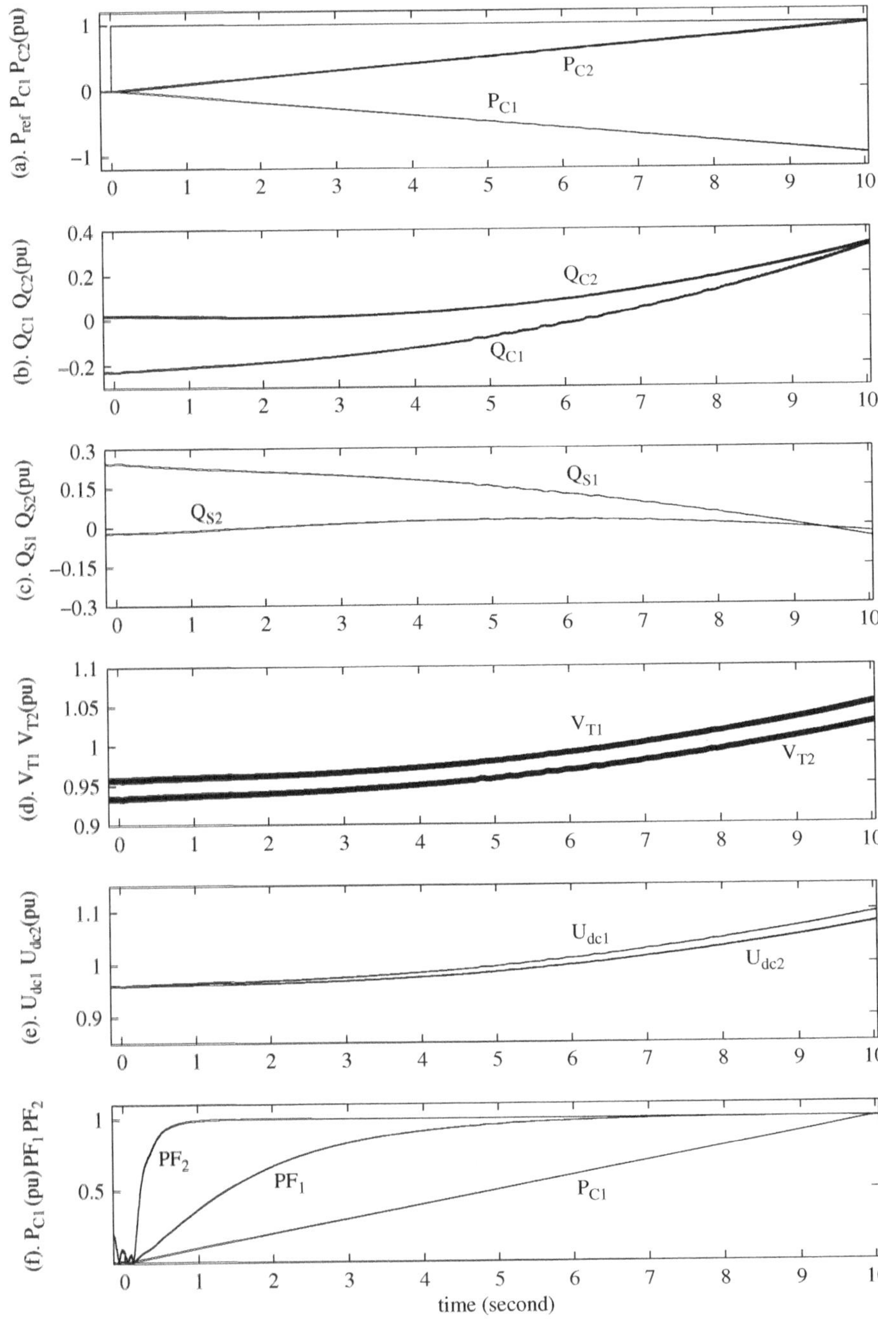

Figure 11.6 High power factor performance of the test link

under common firing angle control. If this option is used in fundamental-frequency-based self-commutating VSC transmission, independent reactive power control is not possible.

If the DC link is expected to provide independent reactive power control at the terminals, this could be achieved by introducing a chop in the converter output voltage waveform, which would alter the fundamental component; this type of control, however, would increase the harmonic content.

An alternative control option is available to double-group converter HVDC schemes. An example of such a configuration is shown in Figure 11.7, based on the cascaded H-bridge reinjection scheme discussed in Section 7.5.

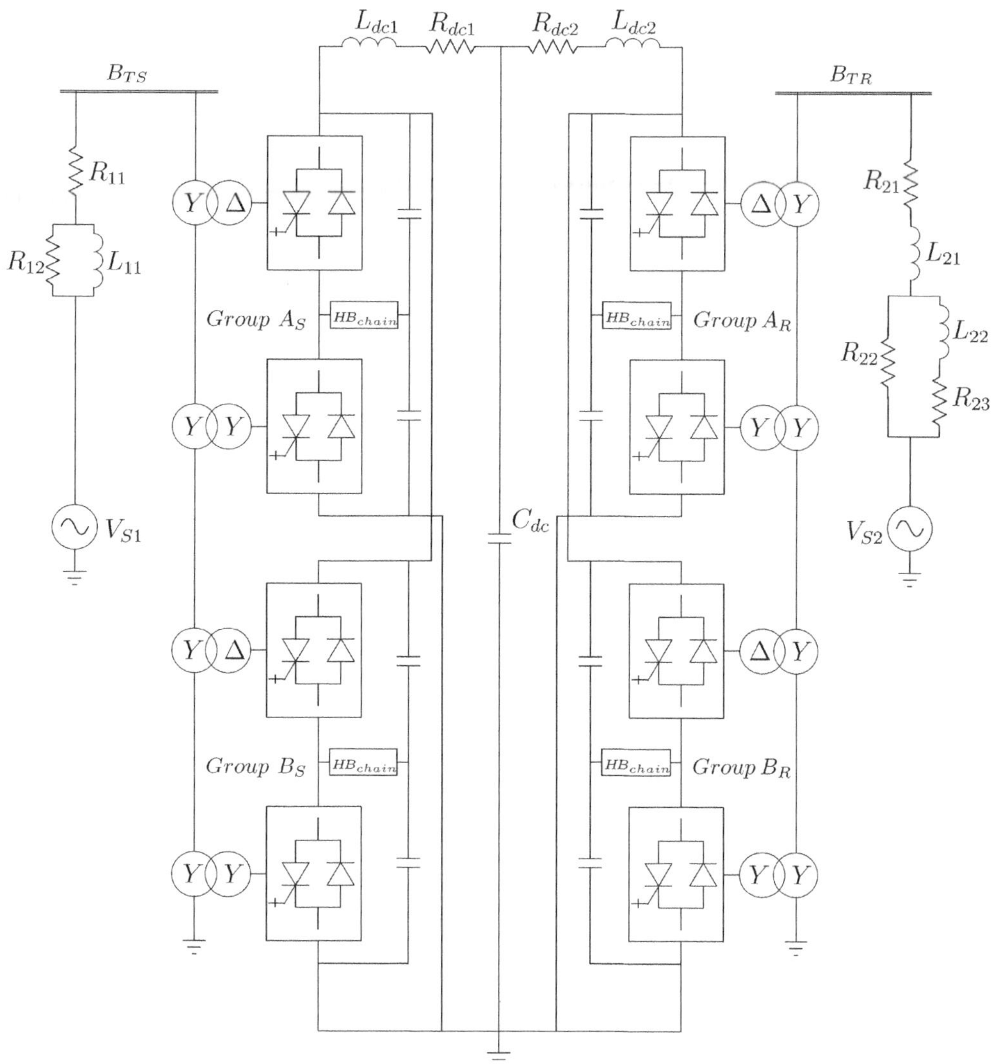

Figure 11.7 Test system used to verify the independence of reactive power control

In the proposed alternative there are two independent phase shift variables (one for each group), which make it possible to control the phase and amplitude of the combined AC voltage output and, thus, achieve more flexible control of the active and reactive powers.

By ignoring the resistance between the converter terminal and the source, Figure 11.8 shows the resulting voltage relationships. In this figure V_A, V_B are the AC voltages (per phase) of the two converter groups; these voltages are of the same magnitude (determined by the common DC voltage), but they are phase shifted with respect to each other. The vectorial addition of V_A and V_B produces a variable converter output voltage (V_C) that controls the reactive power independently of the DC voltage magnitude.

The following relationships apply to the phasor diagram of Figure 11.8:

$$|V_C| = 2|V_A|\cos\left(\frac{\delta_A - \delta_B}{2}\right) \tag{11.9}$$

$$\delta = \frac{\delta_A + \delta_B}{2} \tag{11.10}$$

$$P = \frac{V_C V_S \sin(\delta)}{X} = \frac{V_A V_S[\sin(\delta_A) + \sin(\delta_B)]}{X} \tag{11.11}$$

$$Q_C = \frac{V_C[V_C - V_S\cos(\delta)]}{X} = \frac{2V_A^2\{1+\cos[(\delta_A)-(\delta_B)]\} - V_A V_S[\cos(\delta_A)+\cos(\delta_B)]}{X} \tag{11.12}$$

where

V_S = the source voltage
P = the active power from V_S to V_C
Q_C = the reactive power from V_C to V_S.

It is clear that the amplitude of V_C can be controlled by $(\delta_A - \delta_B)$ and its phase angle by $(\delta_A + \delta_B)$. Therefore the double group multi-level converter at each end of the link provides

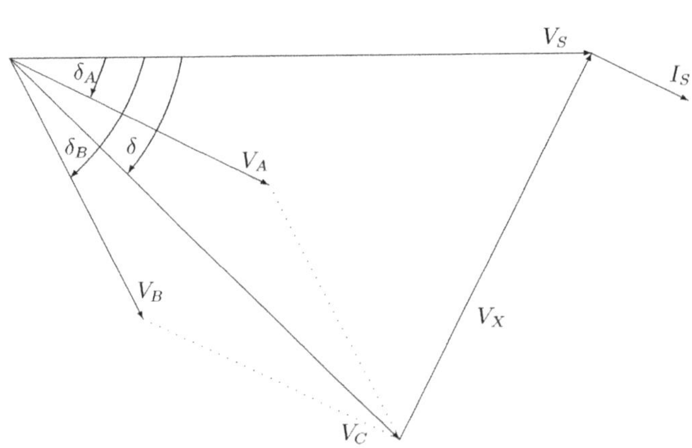

Figure 11.8 Phasor diagram with double group firing shift control

phase and amplitude voltage controllability without depending on its DC voltage (i.e. as in PWM-VSC transmission).

EMTDC verification

The test circuit, shown in Figure 11.7, is based on the CIGRE benchmark model modified to suit the MLVR-VSC features, i.e. the filters are removed and the interface transformers are connected in series on the AC side. The AC systems are relatively weak, with SCRs of 2.5 and 3 at the receiving and sending ends respectively. The AC and DC side parameters are the same as those of Section 11.2.3, which are shown in Tables 11.1 and 11.2, respectively, and the transformer leakage reactances are all 7 %.

A PSCAD model of the test system is used to derive the initial operating condition in the per unit system as specified in Section 11.2.3.

At each converter terminal the two phase shift variables are used to control the terminal voltage magnitude and phase angle between the converter terminal and source voltages. At the sending end, the control of the freely adjustable output voltages ensures that the active and reactive powers are controlled to follow the specified settings, while at the receiving end they are used to follow the required orders of DC voltage and reactive power.

The reactive power control flexibility achieved is shown in Figure 11.9. During the 6 second simulation period, the reactive power setting at the sending end (Q_{Sref}) (shown in graph (a)) is changed from 0 to 0.5 pu (after 1 s), from 0.5 to 0.3 (at 3 s) and from 0.3 to −0.3 (at 5 s), while the reactive power setting at the receiving end (Q_{Rref}) (shown in graph (b)) is changed from zero to −0.3 pu (at 0.1 s), from −0.2 to 0.2 pu (at 2.1 s) and from 0.2 to 0.5 pu (at 4 s). Graphs (a) and (b) also show that the reactive powers at the sending and receiving ends can be controlled independently; they follow their own orders well and the disturbances caused during the setting change dynamics are insignificant. The magnitudes of the converter terminal voltages, illustrated in graph (c), are as expected strongly related to the reactive powers generated (or absorbed) by the converter. Graph (d) contains the DC voltage variation at both ends of the link and graphs (e) and (f) the phase displacements between each of the groups and their respective transformer terminal voltages.

11.3 Multi-level CSC Transmission [8]

The pulse multiplication effect achieved by the DC reinjection concept described in Chapter 3 can eliminate the need for passive filters in line-commutated CSC, but at the expense of extra components and switches. Moreover, line-commutated converters require substantial reactive power compensation, much of which is provided by the filters, and this requirement has not encouraged the use of the original DC reinjection proposal.

Chapter 4 has already explained the applicability of the reinjection concept to self-commutated conversion and Chapter 7 has described its practical implementation in the form of MLCR conversion. As well as structural simplicity (when compared with other multi-level alternatives) MLCR provides the thyristor converter with self-commutating capability [9] and can, therefore, be an interesting flexible CSC alternative for long-distance bulk power HVDC transmission. Therefore the subject of multi-level CSC is discussed here with reference to the MLCR option.

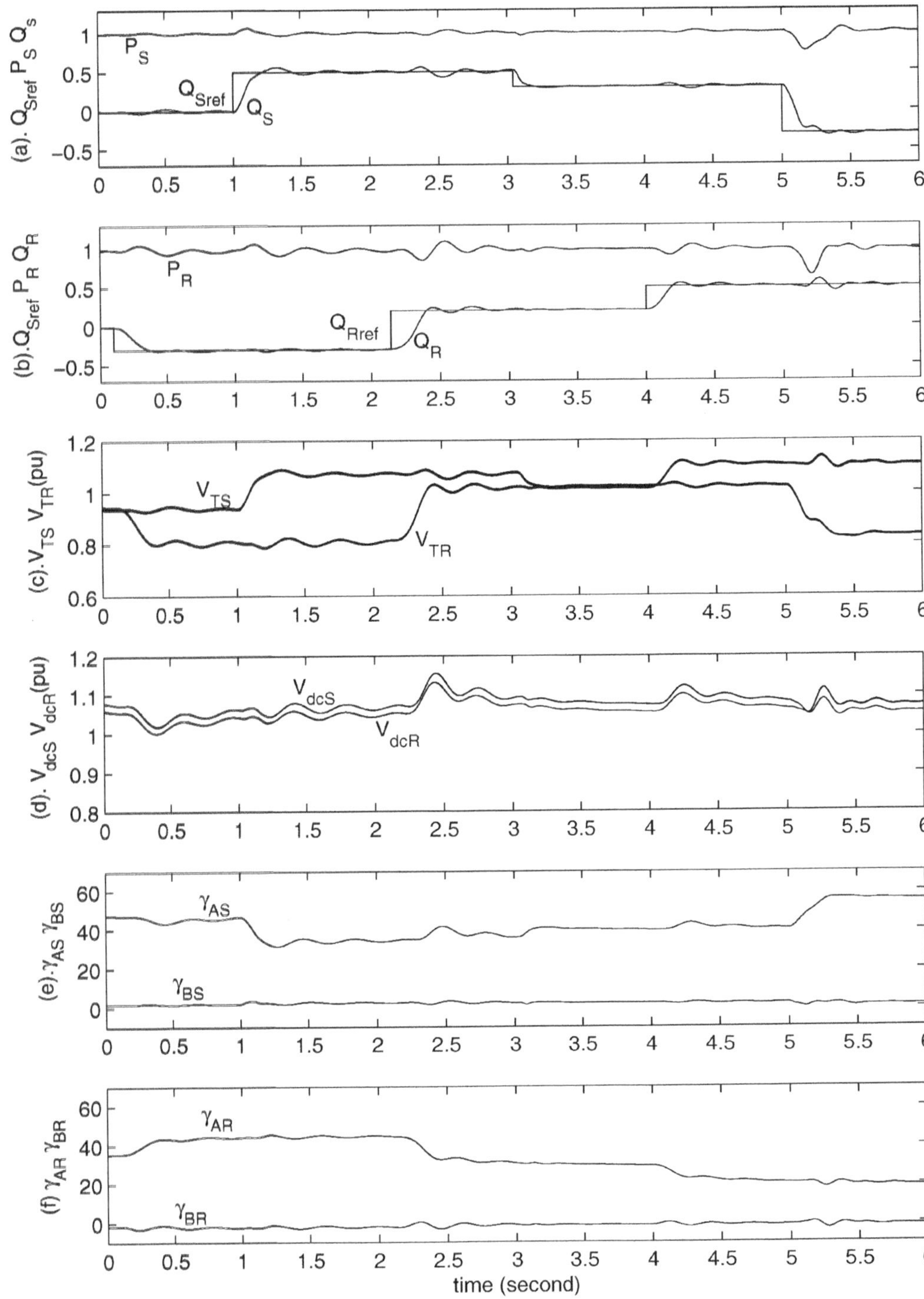

Figure 11.9 HVDC link response to changes in the reactive power settings under firing shift control

11.3.1 Dynamic Model

Figure 11.10 is a simplified diagram of an HVDC transmission system interconnecting two separate power systems represented by ideal sources (V_{S1}, V_{S2}), each in series with an impedance. The converter terminal voltages on the system side of the converter transformers are V_{T1} and V_{T2}. The AC output currents are specified by their real and imaginary components ($I_{\mathrm{Re}1}$, $I_{\mathrm{Im}1}$) and ($I_{\mathrm{Re}2}$, $I_{\mathrm{Im}2}$). The DC voltages at the converter terminals are V_{dc1} and V_{dc2} respectively and the common current I_{dc}.

The following formulation applies to the circuit of Figure 11.10:

$$P_1 = V_{dc1}I_{dc} = \frac{k_{v1}V_{T1}\cos(\theta_1)A(\theta_1,\theta_2)}{L_m s + R} \tag{11.13}$$

$$\begin{aligned} Q_1 &= \sqrt{3}V_{T1}I_{T1}\sin(\theta_1) = \sqrt{3}V_{T1}k_{i1}I_{dc}\sin(\theta_1) \\ &= \frac{\sqrt{3}V_{T1}\sin(\theta_1)k_{i1}A(\theta_1,\theta_2)}{L_m s + R} \end{aligned} \tag{11.14}$$

$$P_2 = V_{dc2}I_{dc} = \frac{k_{v2}V_{T2}\cos(\theta_2)A(\theta_1,\theta_2)}{L_m s + R} \tag{11.15}$$

$$\begin{aligned} Q_2 &= \sqrt{3}V_{T2}I_{T2}\sin(\theta_2) = \sqrt{3}V_{T2}k_{i2}I_{dc}\sin(\theta_2) \\ &= \frac{\sqrt{3}V_{T2}\sin(\theta_2)k_{i2}A(\theta_1,\theta_2)}{L_m s + R} \end{aligned} \tag{11.16}$$

$$I_{Re1} = \frac{k_{i1}\cos(\theta_1)A(\theta_1,\theta_2)}{L_m s + R} \tag{11.17}$$

$$I_{Im1} = \frac{k_{i1}\sin(\theta_1)A(\theta_1,\theta_2)}{L_m s + R} \tag{11.18}$$

$$I_{Re2} = \frac{k_{i2}\cos(\theta_2)A(\theta_1,\theta_2)}{L_m s + R} \tag{11.19}$$

$$I_{Im2} = \frac{k_{i2}\sin(\theta_2)A(\theta_1,\theta_2)}{L_m s + R} \tag{11.20}$$

where

V_T = the fundamental rms value (phase to phase) of the converter terminal voltage

θ = the phase angle difference between the AC current and terminal voltage (which, in the absence of commutation overlap, is also the firing angle and the power factor angle)

$$A(\theta_1,\theta_2) = |k_{v1}V_{T1}\cos\theta_1 - k_{v2}V_{T2}\cos\theta_2| \tag{11.21}$$

P = the real power transfer at the converter terminals and Q = the reactive power supplied to, or extracted from, the converter.

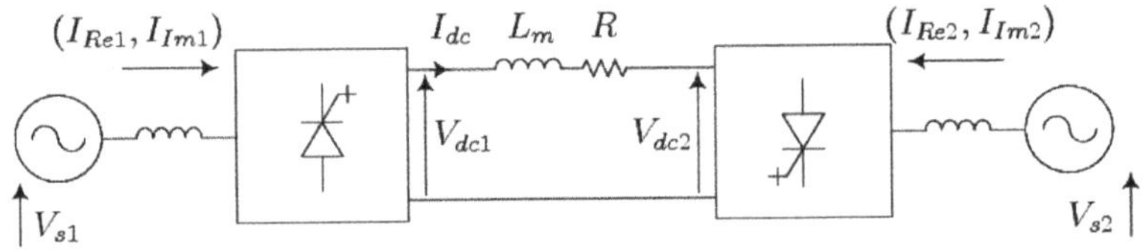

Figure 11.10 Simplified CSC-HVDC transmission link

11.3.2 Control Structure

Based on Equations (11.13) to (11.20), Figure 11.11 shows a block diagram of the control system in terms of active and reactive powers. This diagram is modified in Figure 11.12 to make it more directly applicable to the control of the active and reactive current components.

As shown in the figure, the real and reactive powers vary with the current I_{dc}, which in turn depends on the DC side voltages V_{dc1}, V_{dc2}. The DC voltage is a cosine function of θ, which varies in the range $-180° \leq \theta \leq 180°$. This makes the MLCR-CSC a very non-linear

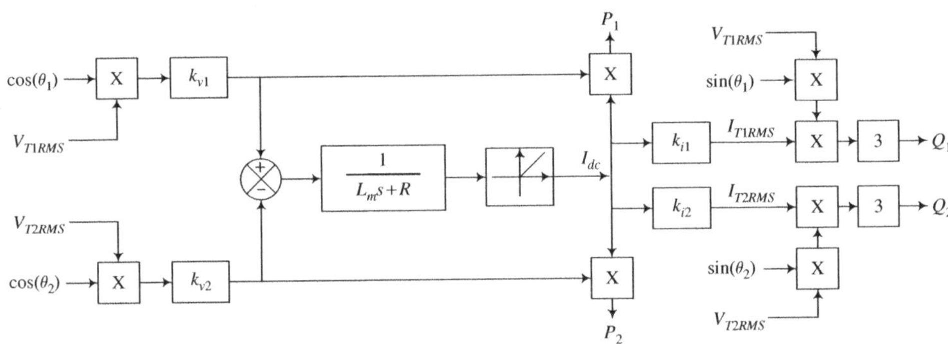

Figure 11.11 Block diagram for active and reactive power control

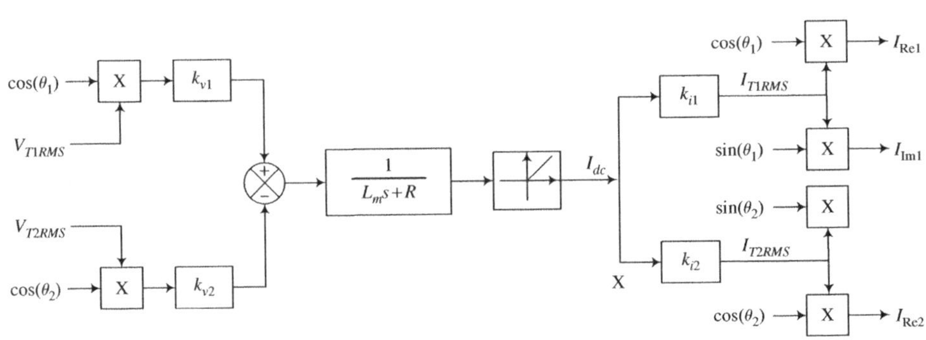

Figure 11.12 Block diagram for real and imaginary current control

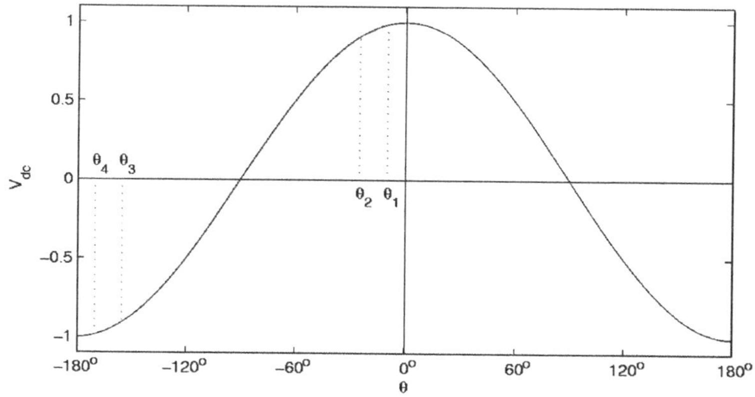

Figure 11.13 Operating regions of the converters at the sending and receiving ends

system. In practice, however, the DC voltages are kept within very narrow limits, such as shown in Figure 11.13. The sending end operates in a narrow band of $\theta_2 \leq \theta \leq \theta_1$ and the receiving end in the corresponding band of $\theta_4 \leq \theta \leq \theta_3$. Controlling the converter near unity power factor is difficult as there is hardly any change in the cosine function in this area, which produces practically no change in the DC.

Since the real and imaginary output current components at the sending and receiving ends of the link are interrelated, the real component of the sending end and the imaginary component of the receiving end can be set as the control parameters. Then the imaginary component of the former and the real component of the latter are dependent on the operating state.

The firing angles are placed on the negative side, enabling the two converters to supply some reactive power to their respective AC systems.

The control structure of Figure 11.14 shows that the measured output currents are transformed into their real and imaginary components using the measured source voltages as a reference. The latter are also used as a reference to synchronise the multi-pulse ramp signals sent to the converter valves' firing logic. The real P^* and reactive Q^* power references are divided by the source voltages to obtain the real and imaginary current orders. Finally, using the real and imaginary current errors, the PI controllers derive the $\Delta\theta_1, \Delta\theta_2$ signals to be added to the $-15°$ and $-165°$ settings to generate the firing instants to be sent to the CSC firing logic.

The DC control characteristic of the CSC schemes permits the link to operate safely during normal and abnormal conditions. Although, multi-level CSC has no independent phase current and voltage controllability, the AC side current can be controlled symmetrically under normal and AC fault conditions (even after losing one phase). Therefore multi-level CSC provides high reliability and fast recovery times.

11.3.3 Simulated Performance under Normal Operating Conditions

The test system used for the EMTDC simulation [10] consists of a two-terminal HVDC link, based on the simplified diagram of Figure 11.10, with each of the AC systems represented by a 500 kV (phase-to-phase) voltage source in series with a reactance calculated to give an SCR of 2.5 for a DC power rating of 1000 MW. The DC line is modelled by a smoothing reactance of 2 H in series with a 5 Ω resistance.

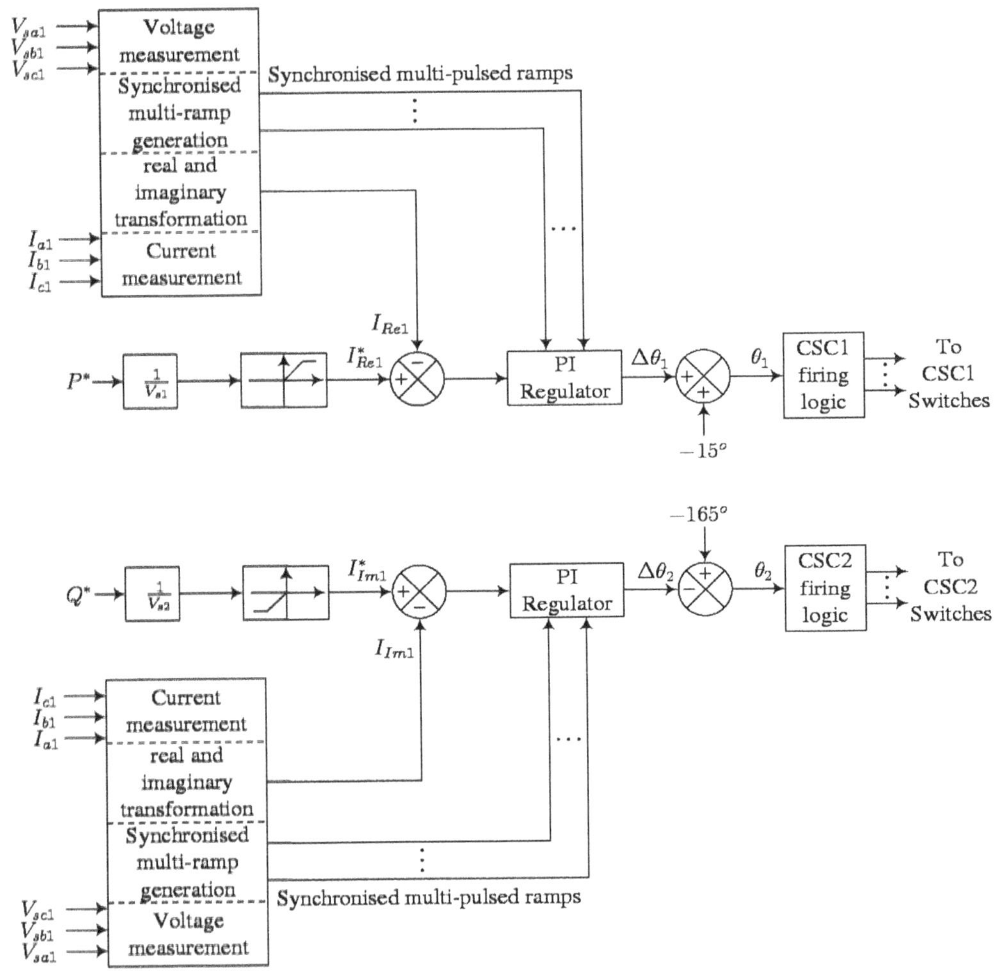

Figure 11.14 The MLCR-CSC HVDC system control structure

Response to active power changes

The dynamic response to step variations in the active power is demonstrated in Figure 11.15, which includes the following information:

P_{ref} = real power order
Q_{ref} = reactive power order
P_1, P_2 = active power at the sending and receiving ends
Q_1, Q_2 = reactive power at the sending and receiving ends.

Initially the link is operating under a real power order of 1000 MW at the sending station and a reactive power order of −800 MVAR at the receiving station (i.e. generating 800 MVAR).

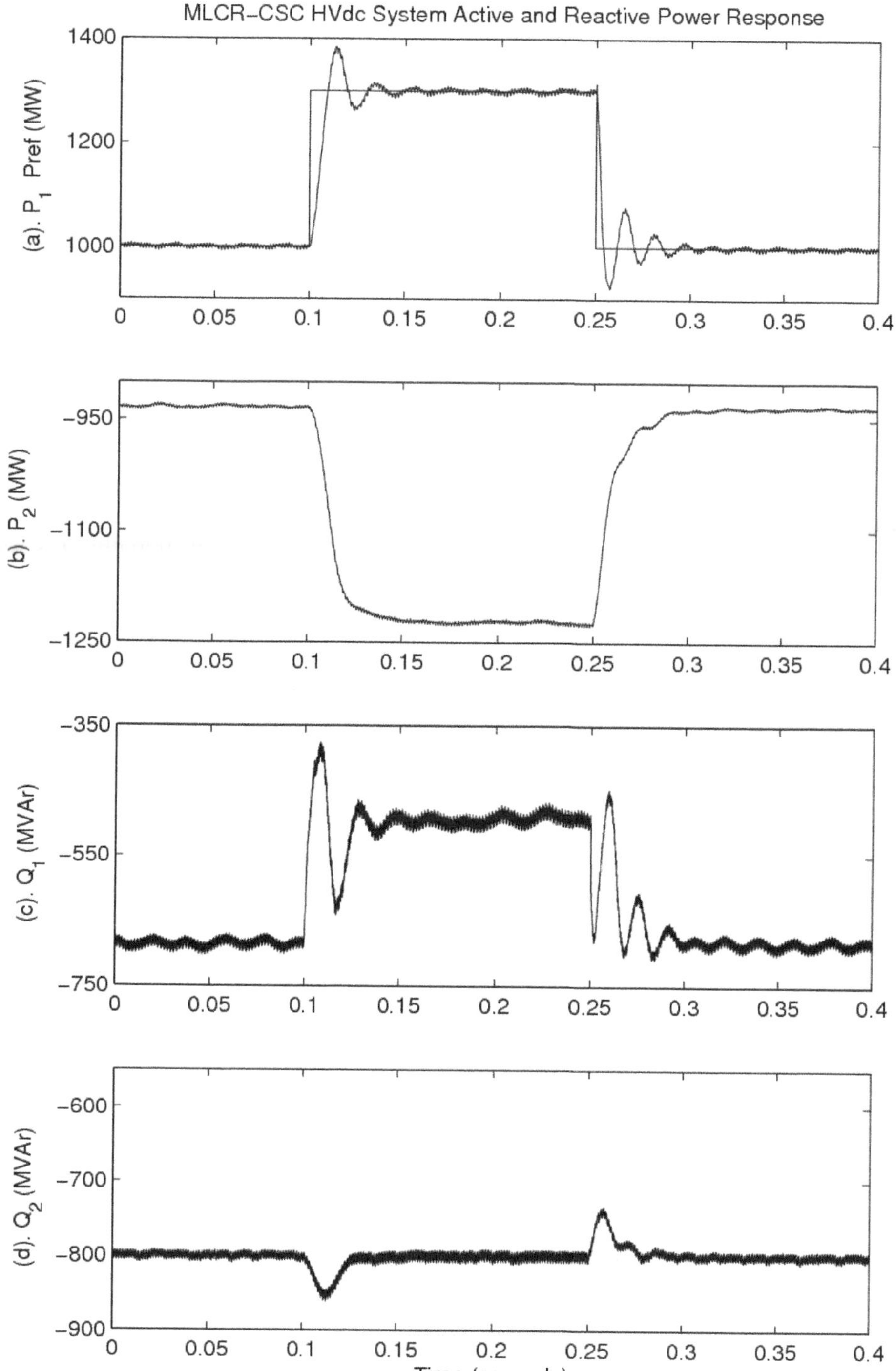

Figure 11.15 Real and reactive power response to a real power order

After 100 ms the real power order is changed to 1300 MW, and at 250 ms the power order is returned to the original setting. In each case the results (plots (a) and (b) of Figure 11.15) show that the system reaches a new steady-state condition after approximately 50 ms, with P_1 experiencing an overshoot of about 30 % of the step change, while P_2 shows no overshoot.

The effect of these changes on the reactive powers (plots (c) and (d) of Figure 11.15) shows a larger disturbance at the sending end (the power controlling station). Upon reaching the new steady-state condition (after 50 ms) the reactive power at the sending station settles at a lower level (−500 instead of the original −685 MVAR according to plot (c)). This drop in the supply of reactive power causes a corresponding reduction in the sending end system voltage (from 551.5 to 528.5 kV, i.e. 4 %). The voltage variation will, of course, depend on the magnitude of the active power change and the converter SCR. In general, therefore, the assistance of on-load tap-changing (OLTC) will be needed to keep the voltage within specified limits.

Response to reactive power changes

Figure 11.16 illustrates the response of the link to step changes in reactive power at the receiving station, while maintaining the active power setting constant. Initially Q is set at −800 MVAR, after 100 ms its value is reduced to −600 MVAR and at 250 ms is returned to its original level. Again plots (a) and (b) show that it takes approximately 50 ms for the system to reach a new steady sate, and that Q_2 experiences a 25 % overshoot whereas no overshoot is observed in Q_1. In this case the sending end station suffers a large reduction in the reactive power injection (from −685 down to −450 MVAR according to plot (b)), thereby causing the AC system voltage to drop by about 3.7 % (from 551.5 to 531 kV).

Figure 11.17 illustrates the result of incorporating OLTC control following a change in the power setting. To this effect, again the receiving system is subjected to a reduction in the reactive power setting from the initial −800 to −600 MVAR as before, but without returning it to its original value. Following the step change at 100 ms, plot (f) shows that the terminal voltage at the sending end (which controls the active power) drops down to 531 kV. However, in this case the transformer tap setting is changed from 1 to 0.925 after 250 ms (see plot (e)). It is clear from plot (f) that the terminal voltage is restored to its original value.

11.3.4 Simulated Performance Following Disturbances

Response to an AC system fault

Figure 11.18 illustrates the response of the MLCR-HVDC system to a three-phase-to-ground short circuit applied directly at the terminals of the receiving converter terminal, while the link is initially operating as specified in Section 11.3.3. In the simulation the fault is assumed to be detected at the instant 100 ms and cleared (by the AC circuit breakers) after another 100 ms. The collapse of the receiving end voltage will initially cause an increase in the DC current, which will soon reach the maximum setting and will, therefore, transfer the sending end control from constant power to constant current. No further corrective action is required, as the link continues to carry practically normal current during the fault and will be ready to resume normal operation soon after the fault is cleared.

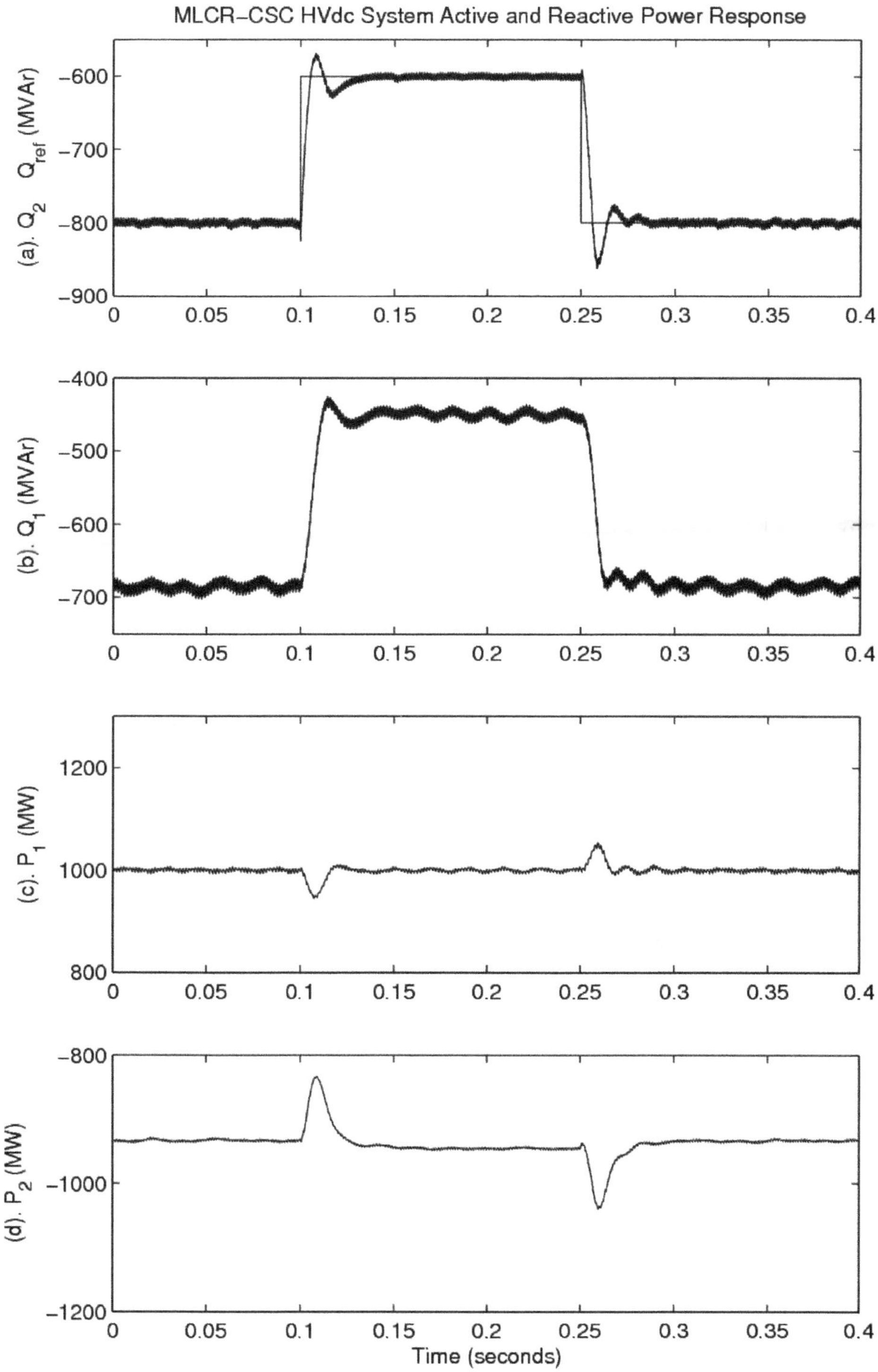

Figure 11.16 Real and reactive power response to a reactive power order

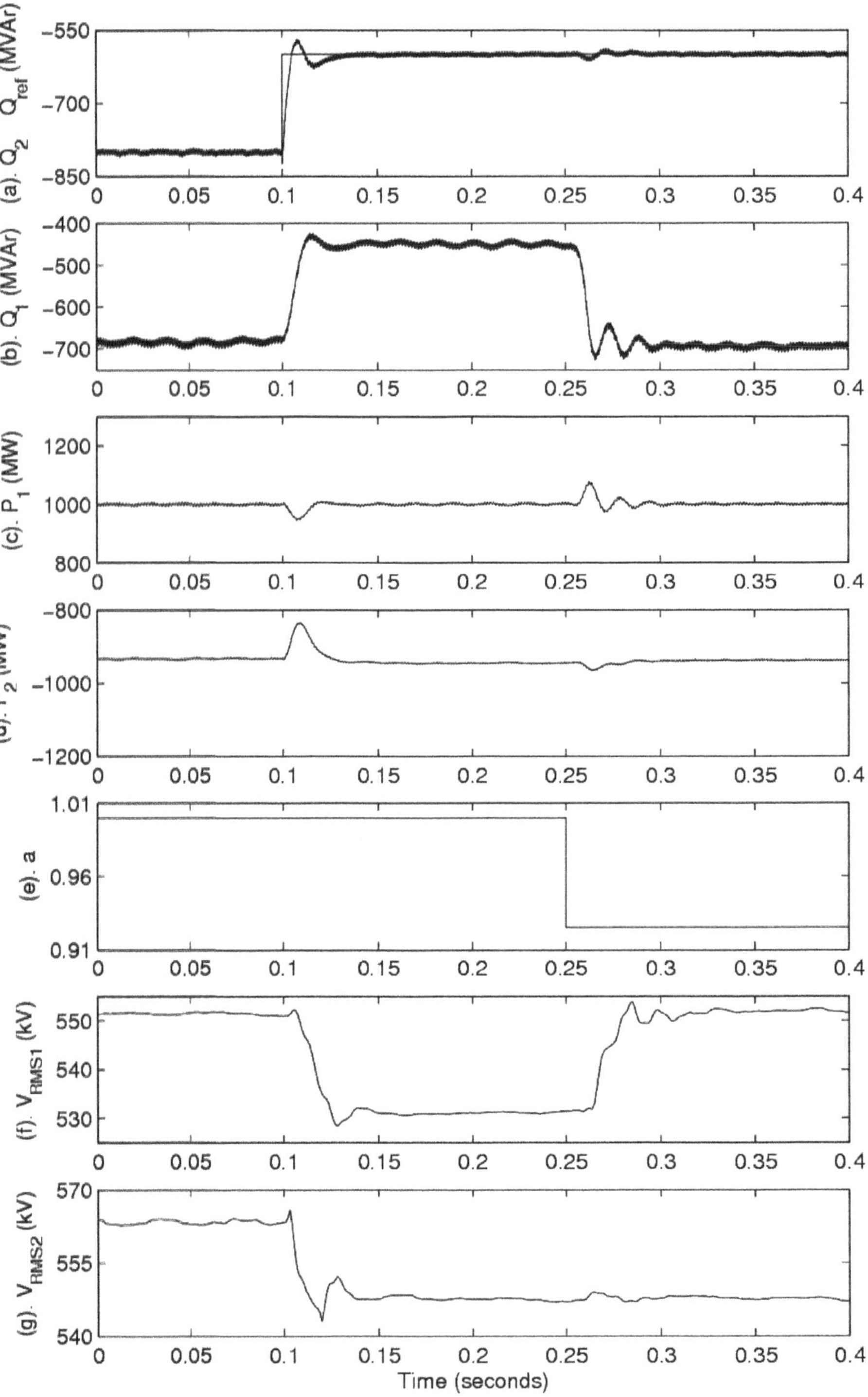

Figure 11.17 The effect of OLTC on the terminal voltage for a change in the reactive power order

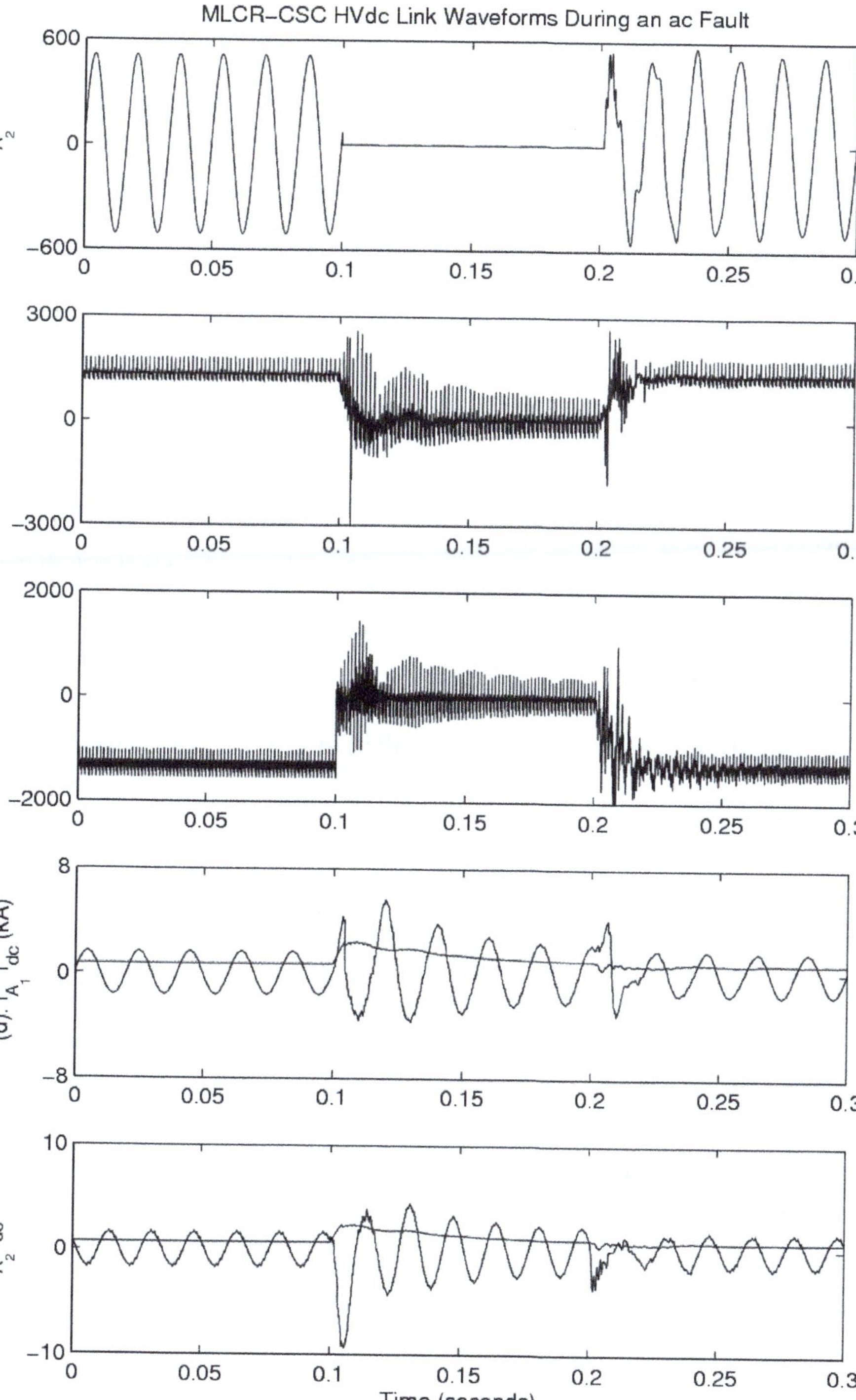

Figure 11.18 Response of MLCR-HVDC to an AC side fault

Plots (c) and (d) show that there is a 200 % overshoot in the DC current and corresponding increases in the AC current for about 20 ms. Following fault clearance, at 200 ms, the system takes approximately a further 50 ms to regain the pre-fault operating conditions.

Response to a DC system fault

A bipolar short circuit is placed in the midpoint of the DC line when the system is operating with a real power order of 1000 MW, controlled by the sending station, and a reactive power order of −800 MVAR at the receiving station.

The fault will initially increase the DC current to try and maintain the specified power flow. However, in the process, the maximum current limit will be reached and the converter will go into constant current control. Similarly the inverter end current will reduce and be kept at its minimum current setting. Therefore a small current (the difference between the rectifier and inverter current settings) will continue to flow at the fault point. Under those conditions the DC fault is not self-clearing.

Instead, upon detection of the fault assumed to occur at the 100 ms point, the sending end converter is temporarily ordered to go into inversion. This action clears the energy stored in the DC system faster and, thus, after a time allowed for the arc deionisation (arbitrarily set at 50 ms in the simulation), normal control operation resumes. The result of the simulation, shown in Figure 11.19, indicates that normal operating conditions are re-established after 100 ms from the instant of fault detection.

11.3.5 Reactive Power Control in Multi-Level CSC Transmission

The exchange of reactive power between a converter and the AC system is determined by the sine of the firing angle (θ). Altering θ has an immediate effect on the DC voltage level and, thus, to maintain the specified DC power transfer through the link, a corresponding change of firing angle must be made at the other end, which in turn affects its reactive power exchange with the AC system. Therefore, with conventional converter control, the reactive powers injected at the two ends of a multi-level CSC link are interdependent.

Most HVDC links used in large-power interconnections are of the bipolar type, each pole consisting of two 12-pulse groups, as shown in Figure 11.20. In these cases an interesting solution to the reactive power interdependence problem is the use of multi-group firing shift control, a subject discussed in the next section.

Multi-group firing shift control

The output current waveform produced by a converter consisting of two or more 12-pulse groups is not altered when a shift is introduced between the firings of the constituent groups of the converter station.

When a change of operating conditions at the receiving end demands more reactive power, and thus reduces the DC voltage, shifting the group firings of the sending end converter in opposite directions provides the required variation of the DC voltage, while maintaining the reactive power constant (due to the opposite polarity of the two firing angle corrections). A relatively small change of active power will be caused by the variation of the fundamental

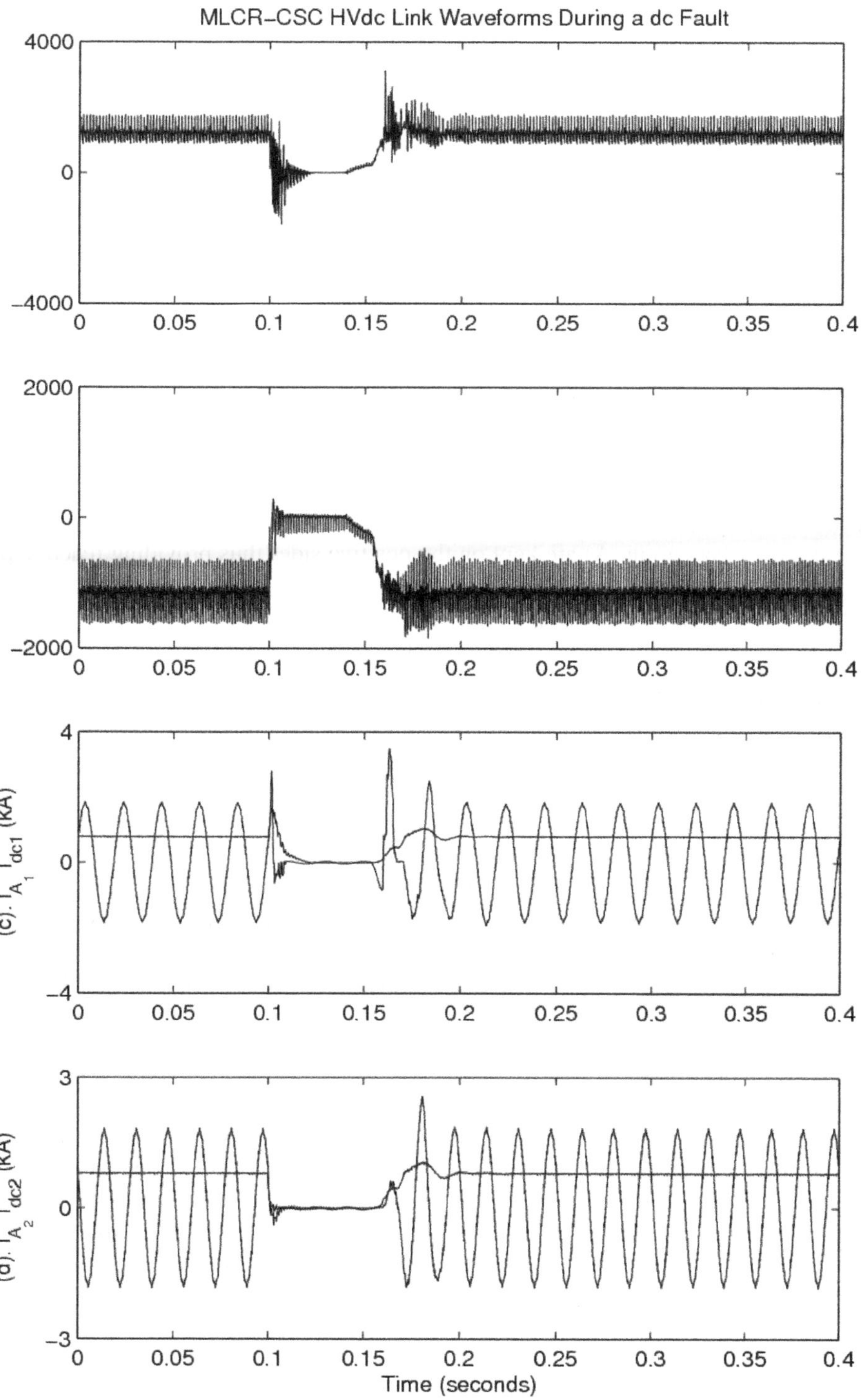

Figure 11.19 Response of MLCR-HVDC to a DC side fault

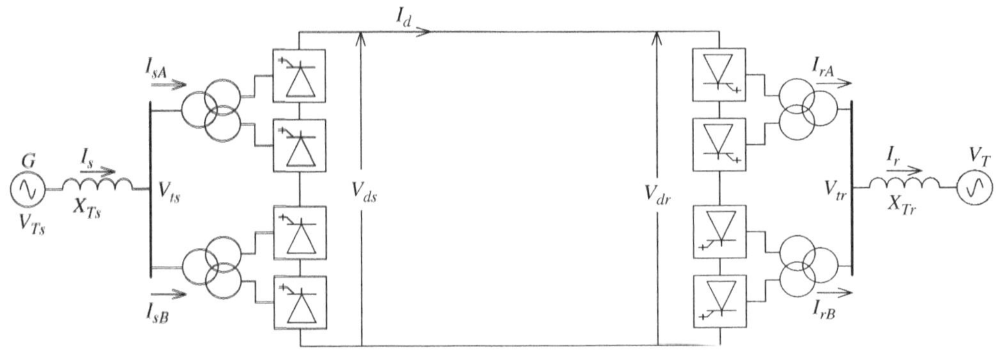

Figure 11.20 Simplified diagram of a CSC-based HVDC link with double group converters

current produced by the shift, but this change can be compensated for by a small extra correction of the two firing angles.

For the sending end converter to operate in the firing shift mode, the firing instants of one converter group (say group A) are kept on the positive side (thus providing reactive power), while the second group (say group B) may act as a source or sink of reactive power (i.e. the firing angle may be positive or negative).

EMTDC verification

The test circuit is a simplified HVDC link configuration with the two interconnected systems represented by a Thévenin circuit. Each terminal consists of two five-level MLCR converter groups, for which the circuit configuration and steady-state output current waveform are shown in Figures 7.25 and 7.26 of Chapter 7.

Using the rated power (1000 MW) and voltage (220 kV) levels as base values, the source voltages are set at 1.02 and 1.06 pu at the receiving and sending ends respectively; the series impedance on both sides is initially equal to 0.254 pu (plus the transformer leakage reactance which is equal to 0.1 pu) and the DC line is represented by a resistance of 0.1 pu in series with a 2 H smoothing inductor. The active power transfer is controlled at the sending end and the terminal AC voltage at the receiving end.

A PSCAD model of the test system under the above operating conditions is used to start the EMTDC simulation. Figure 11.21 illustrates the response of the HVDC link to variations in the receiving end series impedance. Graph (a) shows the set value of the series reactance at the receiving end, which is initially equal to 0.254 pu. Upon reaching the initial steady condition, the value of the receiving end series reactance is increased by 50 % (at 0.5 s) and by 100 % (at 1.5 s) and then returned to the original values at 2.5 and 3.5 s respectively. The dynamic response at the receiving end is shown in graphs (b) (active power), (c) (reactive power) and (d) (terminal voltage). The terminal voltage, active and reactive power responses at the sending end are shown in graphs (e), (f) and (g) respectively.

In each case, after the oscillation that follows the change of condition disappears, the sending end parameters are maintained constant on the steady state, which clearly indicates the independence of the reactive power control on both sides of the link provided by the proposed firing shift control.

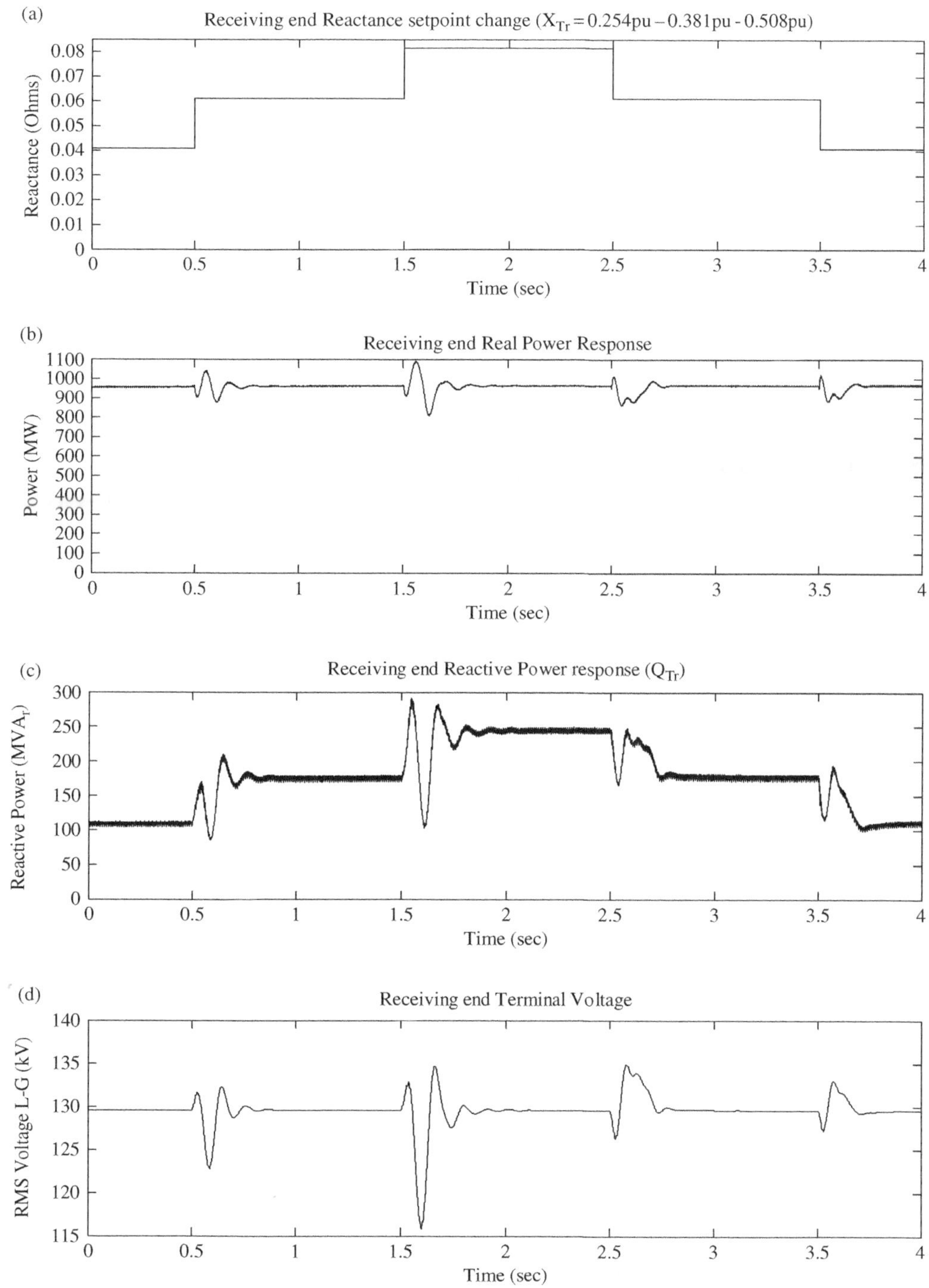

Figure 11.21 (a)–(g) Receiving and sending end responses to changes in the reactive power conditions at the receiving end

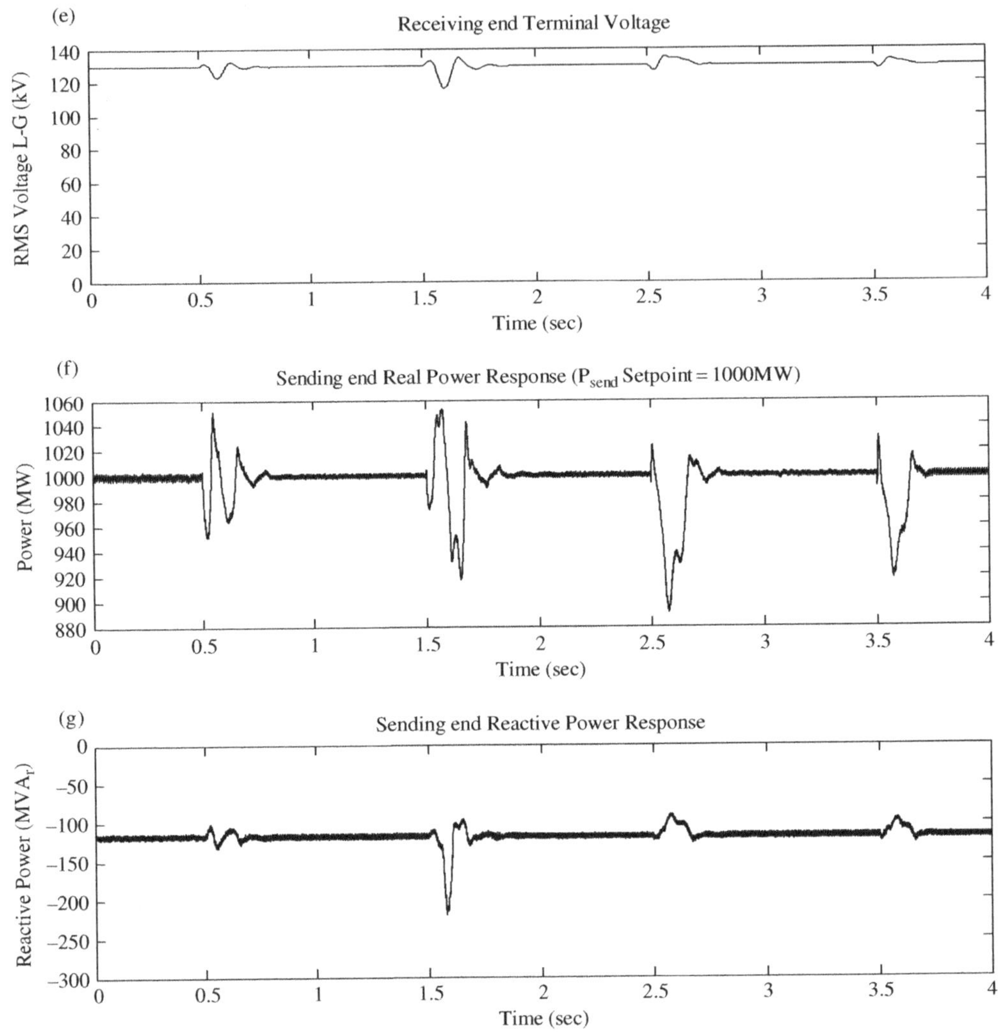

Figure 11.21 (Continued)

11.4 Summary

Due to their simpler switching structures and their ability to commutate at zero voltage, MLVR converters are better options for use in HVDC transmission than the earlier multi-level proposals discussed in Chapter 6. Of the MLVR alternatives, H-bridge reinjection has the simpler switching structure for very high-voltage applications and also permits active capacitor balancing.

MLCR-HVDC appears to be a good contender for large-power applications, owing to the zero-current switching condition that makes the conventional thyristor converter self-commutating, thus permitting the continued use of the robust thyristor technology for the main converter bridges. EMTDC simulation has shown that MLCR provides a fast response to step changes in the active and reactive power.

In common with other multi-level alternatives, the MLVR and MLCR converter topologies lack independent controllability of the reactive power at both ends of the link and, therefore, a large change of reactive power at one end of the link may cause unacceptable voltage variations at the other end. The use of OLTC control can alleviate this problem. In double group bipolar schemes a more flexible reactive power control can be achieved by the use of firing shift control between the two converter groups at the terminal stations.

The responses of the MLCR configuration to AC and DC short circuits are very much like those experienced by line-commutated HVDC links, and thus provide faster recoveries than PWM-HVDC and AC transmission schemes.

References

1. CIGRE SC-B4 WG 37 (2004), 'VSC transmission', Final document, Paris.
2. Jung, G. *et al.* (1991), 'A general circuit topology of multilevel inverter', IEEE Power Electronics Specialist Conference, MIT, USA, Vol. PESC 91, Record 96–103.
3. Meynard, T. and Foch, H. (1993), 'Imbricated cells multi-level voltage source inverters for high voltage applications', *European Power Electronics Journal*, **3**(2), 99–106.
4. Liu, Y.H., Arrillaga, J. and Watson, N.R. (2004), 'Multi-level voltage reinjection – a new concept in high power voltage source conversion', *Proceedings of the IEE, Generation, Transmission and Distribution*, **151**(3), 290–298.
5. Perera, L.B., Watson, N.R., Liu, Y.H. and Arrillaga, J. (2005), 'Multi-level current reinjection self-commutated HVDC converter', *Proceedings of the IEE, Generation, Transmission and Distribution*, **152**(5), 607–615.
6. Liu, Y.H., Arrillaga, J. and Watson, N.R. (2005), 'Capacitor voltage balancing in multi-level voltage reinjection (MLVR) converters', *IEEE Transactions on Power Delivery*, **20**(2), 1728–1737.
7. Liu, Y.H. Arrillaga, J., Watson, N.R. and Wood, A.R. (2006), 'Multi-level voltage reinjection VSC HVDC transmission', 8th IEE Conference on acdc, London, 130–134.
8. Liu, Y.H., Perera, L.B., Arrillaga, J. and Watson, N.R. (2007), 'A back to back HVDC link with multi-level current reinjection Converters', *IEEE Transactions on Power Delivery*, in press.
9. Arrillaga, J., Liu, Y.H., Perera, L.B. and Watson, N.R. (2006), 'A current reinjection scheme that adds self-commutation and pulse multiplication to the thyristor converter', *IEEE Transactions on Power Delivery*, **21**(3), 1593–1599.
10. Arrillaga, J. and Watson, N.R. (2001), *Computer modelling of electrical power systems*, Second Edition, John Wiley & Sons, Ltd, Chichester.

Index

Flexible Power Transmission: The HVDC Options J. Arrillaga, Y.H. Liu and N.R. Watson

Printed and bound in the UK by
CPI Antony Rowe, Eastbourne

Printed and bound by CPI Group (UK) Ltd, Croydon, CR0 4YY
16/09/2026
14963389-0001